INTEGRAL TRANSFORMS

for

ENGINEERS

and

APPLIED MATHEMATICIANS

Also by Larry E. Andrews

Special Functions for Engineers and Applied Mathematicians

Also in Optics:

Becherer: *Laser Radar**

Fantone: *Optical Fabrication Methods**

Fischer: *Optical System Design: A Practical Approach**

Gilmore: *Expert Vision Systems**

Hopkins/Smith/Weller: *Lens Design: A Resource Manual**

Jacobson: *Thin Films**

Johnson: *Infrared System Design and Analysis**

Ross: *Laser Communication and Ranging**

Vukobratovich: *Introduction to Opto-Mechanical Design**

*in the *Macmillan Series in Optical and Electro-Optical Engineering*
Robert E. Fischer and Warren J. Smith, Series Editors

INTEGRAL TRANSFORMS

for

ENGINEERS

and

APPLIED MATHEMATICIANS

L.C. ANDREWS
B.K. SHIVAMOGGI

University of Central Florida

MACMILLAN PUBLISHING COMPANY
A Division of Macmillan, Inc.
NEW YORK

Collier Macmillan Publishers
LONDON

ACKNOWLEDGMENT

The authors wish to express their appreciation for the use of some material from *Elementary Partial Differential Equations With Boundary Value Problems* by Larry C. Andrews, copyright © 1986 by Harcourt Brace Jovanovich, Inc. Reprinted by permission of the publisher.

Macmillan Publishing Company
866 Third Avenue, New York, New York 10022

Collier Macmillan Canada, Inc.
Collier Macmillan Publishers • London

Printed in the United States of America

Printing: 2 3 4 5 6 7 8 Year: 8 9 0 1 2 3 4

Library of Congress Cataloging-in-Publication Data

Andrews, Larry C.
Integral transforms for engineers and applied mathematicians.

Bibliography: p. 335
Includes index.
1. Integral transforms. I. Shivamoggi, Bhimsen K. II. Title.
QA432.A63 1988 515.7′23 87–20404
ISBN 0–02–949041–3

Contents

Preface

In recent years, integral transforms have become essential working tools of every engineer and applied scientist. The Laplace transform, which undoubtedly is the most familiar example, is basic to the solution of initial value problems. The Fourier transform, while being suited to solving boundary-value problems, is basic to the frequency spectrum analysis of time-varying waveforms. The purpose of this text is to introduce the use of integral transforms in obtaining solutions to problems governed by ordinary and partial differential equations and certain types of integral equations. Some other applications are also covered where appropriate.

The Laplace and Fourier transforms are by far the most widely used of all integral transforms. For this reason they have been given a more extensive treatment in this book than other integral transforms. However, there are several other integral transforms that also have been used successfully in the solution of certain boundary-value problems and in other applications. Included in this category are Mellin, Hankel, finite, and discrete transforms, which have also been given some discussion here.

The text is directed primarily toward senior and beginning graduate students in engineering sciences, physics, and mathematics who desire a deeper knowledge of transform methods than can be obtained in introductory courses in differential equations and other similar courses. It can also be used as a self-study text for practicing engineers and applied scientists who wish to learn more about the general theory and use of integral transforms. We assume the reader has a basic knowledge of differential equations and contour integration techniques from complex variables. However, most of the material involving complex variables occurs in separate sections so that much of the text can be accessible to those with a minimum background in complex variable methods. As an aid in this regard, we have included a brief appendix relevant to our use of the basic concepts and theory of complex variables in the text. Also, because of the close association of special functions and integral transforms, the first chapter is a short introduction to several of the special functions that arise quite frequently in applications. This is considered an optional chapter for those with some acquaintance with these functions, and thus it is possible to start the text with Chap. 2. Most chapters are independent of one another so that various arrangements of the material are possible.

Applications occur throughout the text and are drawn from the fields of mechanical vibration, heat conduction, potential theory, mechanics of solids and fluids, probability and statistics, and several other areas. A working knowledge in any of these areas is generally sufficient to work the examples and exercises.

In our treatment of integral transforms we have excised formal proofs in several places, but then usually make an appropriate reference for the more formal aspects of the theory. In the applications we often make the assumptions as to the commutability of certain limiting operations, and the derivation of a particular solution sometimes may not be rigorous. However, the approach adopted here is adequate in the usual applications in engineering and applied sciences. We have included a large number of worked examples and exercises to illustrate the versatility and adequacy of this approach in applications to physical problems.

We wish to thank Jack Repcheck, Senior Editor of Scientific and Technical Books department at Macmillan, for his assistance in getting this text published in a timely manner. We also wish to express our appreciation to the production staff of Macmillan for their fine efforts. Finally, we wish to acknowledge Martin Otte who corrected several errors during a final reading of the manuscript.

Introduction

The classical methods of solution of initial and boundary value problems in physics and engineering sciences have their roots in Fourier's pioneering work. An alternative approach through integral transform methods emerged primarily through Heaviside's efforts on operational techniques. In addition to being of great theoretical interest to mathematicians, integral transform methods have been found to provide easy and effective ways of solving a variety of problems arising in engineering and physical science. The use of an integral transform is somewhat analogous to that of logarithms. That is, a problem involving multiplication or division can be reduced to one involving the simpler processes of addition or subtraction by taking logarithms. After the solution has been obtained in the logarithm domain, the original solution can be recovered by finding an antilogarithm. In the same way, a problem involving derivatives can be reduced to a simpler problem involving only multiplication by polynomials in the transform variable by taking an integral transform, solving the problem in the transform domain, and then finding an inverse transform. Integral transforms arise in a natural way through the principle of linear superposition in constructing integral representations of solutions of linear differential equations.

By an *integral transform*, we mean a relation of the form*

$$\int_{-\infty}^{\infty} K(s,t)f(t)\,dt = F(s) \tag{0.1}$$

such that a given function $f(t)$ is transformed into another function $F(s)$ by means of an integral. The new function $F(s)$ is said to be the *transform* of $f(t)$, and $K(s,t)$ is called the *kernel* of the transformation. Both $K(s, t)$ and $f(t)$ must satisfy certain conditions to ensure existence of the integral and a unique transform function $F(s)$. Also, generally speaking, not more than one function $f(t)$ should yield the same transform $F(s)$. When both of the limits of integration in the defining integral are finite, we have what is called a *finite transform*.

Within the above guidelines there are a variety of kernels that may be used to define particular integral transforms for a wide class of functions $f(t)$. If the kernel is defined by

$$K(s,t) = \begin{cases} 0, & t < 0 \\ e^{-st}, & t \geq 0 \end{cases} \tag{0.2}$$

the resulting transform

$$\int_0^{\infty} e^{-st} f(t)\,dt = F(s) \tag{0.3}$$

is called the *Laplace transform*. When

$$K(s,t) = \frac{1}{\sqrt{2\pi}} e^{ist} \tag{0.4}$$

we generate the *Fourier transform*†

$$\frac{1}{\sqrt{2\pi}} \int_{-\infty}^{\infty} e^{ist} f(t)\,dt = F(s) \tag{0.5}$$

which, when t is restricted to the positive real line, leads to the *Fourier sine* and *Fourier cosine transforms*

$$\sqrt{\frac{2}{\pi}} \int_0^{\infty} f(t) \sin st\,dt = F(s) \tag{0.6}$$

---

* We will always interpret integrals like (0.1) as the *principle value* of the integral, defined in general by $PV \int_{-\infty}^{\infty} f(x)dx = \lim_{R\to\infty} \int_{-R}^{R} f(x)dx$.

† Other definitions of $K(s,t)$ for Fourier transforms involve the choices e^{ist}, e^{-ist}, $(1/2\pi)e^{ist}$, among others.

and

$$\sqrt{\frac{2}{\pi}} \int_0^\infty f(t)\cos st \, dt = F(s) \tag{0.7}$$

The Laplace and Fourier transforms are by far the most prominent in applications. Many other transforms have been developed, but most have limited applicability. In addition to the Laplace and Fourier transforms, the next most useful transforms are perhaps the *Hankel transform of order* ν

$$\int_0^\infty tJ_\nu(st)f(t) \, dt = F(s) \tag{0.8}$$

where $J_\nu(x)$ is the Bessel function of the first kind (see Sec. 1.4), and the *Mellin transform*

$$\int_0^\infty t^{s-1}f(t) \, dt = F(s) \tag{0.9}$$

The Hankel transform arises naturally in solving boundary value problems formulated in cylindrical coordinates while the Mellin transform is useful in the solution of certain potential problems formulated in wedge-shaped regions.

The integral transforms mentioned thus far are applicable to problems involving either semiinfinite or infinite domains. However, in applying the method of integral transforms to problems formulated on finite domains it is necessary to introduce finite intervals on the transform integral. Transforms of this nature are called *finite integral transforms.*

A basic problem in the use of integral transforms is to determine the function $f(t)$ when its transform $F(s)$ is known. We refer to this as the *inverse problem.* In many cases the solution of the inverse problem is another integral transform relation of the type

$$\int_D H(s,t)F(s) \, ds = f(t) \tag{0.10}$$

where $H(s,t)$ is another kernel and D is the domain of s. Such a result is called an *inversion formula* for the particular transform. For example, the inversion formula for the Fourier transform takes the form (see Sec. 2.4)

$$\frac{1}{\sqrt{2\pi}} \int_{-\infty}^\infty e^{-ist}F(s) \, ds = f(t) \tag{0.11}$$

which is very much like the transform itself in Eq. (0.5). This means that the problems of evaluating transforms or inverse transforms are essentially the same for Fourier transforms. This is not necessarily the case for other transforms like the Laplace transform, however, where the inversion formula is quite distinct from that of the transform integral. Also, in the case of finite transforms, the inverse transform is in the form of an infinite series.

The basic aim of the transform method is to transform a given problem into one that is easier to solve. In the case of an ordinary differential equation with constant coefficients, the transformed problem is algebraic. The effect of applying an integral transform to a partial differential equation is to reduce it to a partial differential equation in one less variable. The solution of the transformed problem in either case will be a function of the transformed variable and any remaining independent variables. Inversion of this solution produces the solution of the original problem.

The exponential Fourier transform does not incorporate any boundary conditions in transforming the derivatives. Thus, it is best suited for solving differential equations on infinite domains where the boundary conditions usually only require bounded solutions. On the other hand, the Fourier cosine and sine transforms are well suited for solving certain problems on semiinfinite domains where the governing differential equation involves only even-order derivatives. We will see that the Fourier transform lends itself nicely to solving boundary-value problems associated with the following partial differential equations:

(a) the *heat equation:*

$$\nabla^2 u = a^{-2}u_t - q(x,y,z,t) \tag{0.12}$$

(b) the *wave equation:*

$$\nabla^2 u = c^{-2}u_{tt} - q(x,y,z,t) \tag{0.13}$$

(c) the *potential equation:*

$$\nabla^2 u = 0 \tag{0.14}$$

In addition, it is useful in the solution of linear integral equations of the form

$$f(x) = u(x) - \lambda \int_{-\infty}^{\infty} k(x,t)u(t)\,dt \tag{0.15}$$

and certain ordinary differential equations. Interesting applications of these transform methods arise in hydrodynamics, heat conduction, potential theory, and elasticity theory, among other areas. The Fourier transform also lends itself to the theory of probability and statistics. For example, it turns out that the moments of a random variable X are merely the coefficients of $(it)^k/k!$ in the Maclaurin series expansion of the characteristic function $C(t)$ of the random variable X, and this function is related to the probability density function $p(x)$ by the Fourier transform relation

$$C(t) = \int_{-\infty}^{\infty} e^{itx}\, p(x)\, dx \tag{0.16}$$

While the Fourier transform is suited for boundary-value problems, the Laplace transform is suited for initial-value problems. However, there are other situations for which the Laplace transform can also be used, such as in the evaluation of certain integrals and in the solution of certain integral equations of convolution type like

$$\int_0^t u(\tau)k(t - \tau)\, d\tau = f(t), \qquad t > 0 \tag{0.17}$$

In addition to the transforms mentioned above, there are other less well known transforms like the *Hilbert transform* and the *Sturm–Liouville transform*, both of which are more limited in their usefulness than the Fourier and Laplace transforms. Also, discrete transforms like the *discrete Fourier transform* (which is the discrete analog of the Fourier transform) and the *Z transform* (which is the discrete analog of the Laplace transform) are becoming more prominent in various engineering applications where it is either impossible or inconvenient to use more conventional transforms.

Much of our initial discussion will evolve around the problem of calculating the transforms $F(s)$ of given functions $f(t)$, and also around the related problem of finding inverse transforms of various functions $F(s)$. Our primary objective is to introduce methods to use the integral transforms, rather than concerning ourselves too deeply with the general theory itself. Therefore, we do not attempt to present the basic theorems in their most general forms. However, the conditions put forth in the theorems are generally broad enough to embrace most of the functions that naturally arise in engineering and physical situations. Proofs of the theorems are provided when feasible, but are sometimes based on heuristic arguments instead of rigorous mathematical procedures. For example, often we have the need in our proofs for interchanging certain limit operations, like integration and summation, and in these situations we normally operate under the assumption that such interchanges are permissible.

1

Special Functions

1.1 *Introduction*

Most of the functions encountered in introductory analysis belong to the class of *elementary functions*. This class is composed of polynomials, rational functions, transcendental functions (trigonometric, exponential, logarithmic, and so on), and functions constructed by combining two or more of these functions through addition, subtraction, multiplication, division, or composition. Beyond these functions lies a class of *special functions* which are important in a variety of engineering and physics applications.

The use of integral transforms is heavily interlaced with special functions like the *gamma function*, *error function*, *Bessel functions*, and so forth. Also, functions such as the *Heaviside unit function* and the *impulse function*, which are employed in a variety of engineering applications, are briefly discussed. Hence, a brief review of (or introduction to) some of these special functions can be quite useful before discussing integral transforms themselves.*

---

* For a more thorough treatment of special functions, see L. C. Andrews, *Special Functions for Engineers and Applied Mathematicians*, New York: Macmillan, 1985.

1.2 *The Gamma Function*

One of the simplest but very important special functions is the *gamma function*. Although it has less direct application than some of the other special functions, knowledge of the properties of this function is a prerequisite for the study of Bessel functions and others which do have direct application.

Historically, the gamma function was discovered by L. Euler (1707–1783) in 1729 who was concerned with the problem of interpolating between the numbers

$$n! = \int_0^\infty e^{-t} t^n \, dt, \qquad n = 0,1,2,\ldots$$

with nonintegral values of n. His studies eventually led him to the gamma function relation

$$\Gamma(z) = \int_0^\infty e^{-t} t^{z-1} \, dt, \qquad \mathrm{Re}(z) > 0^* \tag{1.1}$$

later termed the *Eulerian integral of the second kind* by A. M. Legendre (1752–1833). Legendre is also responsible for the symbol Γ that is most often used for the gamma function. The variable z in Eq. (1.1) may be real or complex. Although the integral is improper, it has been shown that it converges uniformly for all values of z for which $\mathrm{Re}(z) > 0$. The function $\Gamma(z)$ is bounded and differentiable and, in fact, an analytic function throughout this domain.

By substituting $z + 1$ for z in Eq. (1.1) and performing an integration by parts, we obtain

$$\begin{aligned}\Gamma(z + 1) &= \int_0^\infty e^{-t} t^z \, dt \\ &= -e^{-t} t^z \Big|_0^\infty + z \int_0^\infty e^{-t} t^{z-1} \, dt\end{aligned}$$

from which we deduce the very simple but important recurrence formula

$$\Gamma(z + 1) = z\Gamma(z) \tag{1.2}$$

The value $z = 1$ in Eq. (1.1) greatly simplifies the integral and leads to the result

$$\Gamma(1) = \int_0^\infty e^{-t} \, dt = 1$$

---

* By Re(z), we mean the real part x of the complex variable $z = x + iy$. Similarly, Im(z) refers to the imaginary part y. See also Appendix A.

Hence, by repeated use of the recurrence formula (1.2), we see that $\Gamma(2) = 1$, $\Gamma(3) = 1 \cdot 2$, $\Gamma(4) = 1 \cdot 2 \cdot 3$, while in general

$$\Gamma(n + 1) = n!, \qquad n = 0,1,2,... \tag{1.3}$$

Thus, we see that the gamma function is Euler's extension of the factorial function to nonintegral values of n. In fact, the gamma function is an extension of the factorial function to all complex numbers with a positive real part. In the next section we will extend this domain even further.

1.2.1 *Analytic Continuation for Re(z) < 0*

An analytic continuation of $\Gamma(z)$ to the left of the imaginary axis can be accomplished through repeated use of the recurrence formula (1.2) expressed in the form

$$\Gamma(z) = \Gamma(z + 1)/z, \qquad z \neq 0 \tag{1.4}$$

The right-hand side of (1.4) is defined for all $\text{Re}(z) > -1$, $z \neq 0$, and thus this expression defines $\Gamma(z)$ in this domain. Replacing z by $z + 1$ in (1.4) yields

$$\Gamma(z + 1) = \Gamma(z + 2)/(z + 1), \qquad z \neq -1$$

and when substituted into (1.4), we obtain

$$\Gamma(z) = \Gamma(z + 2)/z(z + 1), \qquad z \neq 0, -1 \tag{1.5}$$

which now defines $\Gamma(z)$ for all $\text{Re}(z) > -2$, $z \neq 0, -1$. Continuing this process, we deduce that

$$\Gamma(z) = \frac{\Gamma(z + k)}{z(z + 1)(z + 2)\cdots(z + k - 1)}, \qquad z \neq 0, -1, ..., -k + 1 \tag{1.6}$$

where k is a positive integer.

Equation (1.6) can be used to define the gamma function at every z with a negative real part except at negative integers and zero. The values $z = 0, -1, -2, ...,$ are actually first-order *poles* of the function and thus

$$|\Gamma(-n)| = \infty, \qquad n = 0,1,2,... \tag{1.7}$$

The graph of the gamma function for $z = x$, a real variable, is sketched in Fig. 1.1.

1.2.2 *Additional Properties*

One of the most common uses of the gamma function is in the evaluation of certain integrals. That is, when it appears in applications it frequently has the form suggested by Eq. (1.1) or some variation of it. For example, if we set $t = u^2$ in (1.1), we find

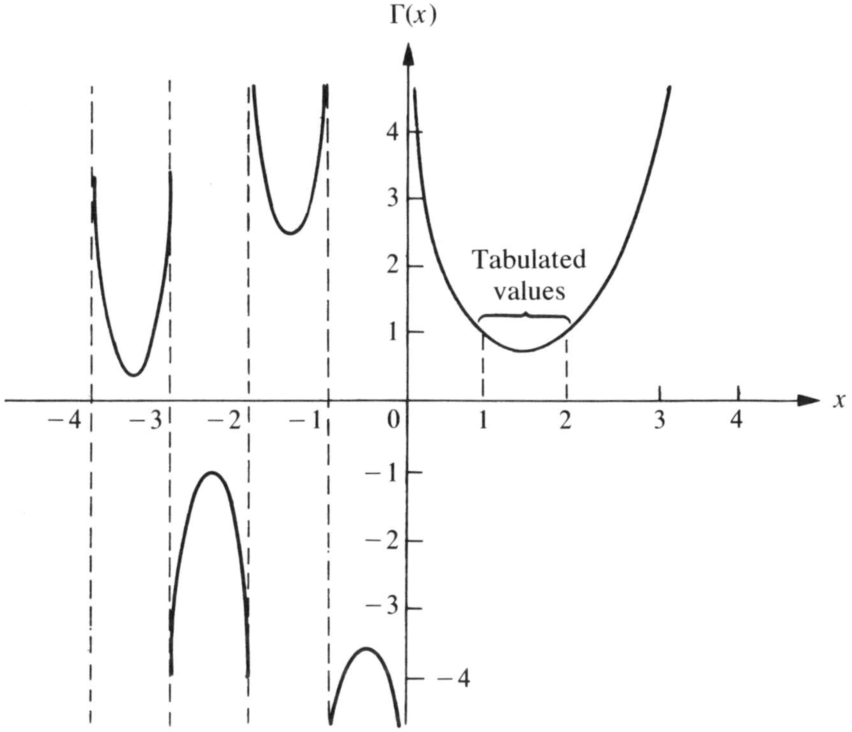

Figure 1.1 Graph of the gamma function

$$\Gamma(z) = 2\int_0^\infty e^{-u^2}u^{2z-1}\,du, \ \mathrm{Re}(z) > 0 \tag{1.8}$$

whereas the substitution $t = \log(1/u)$ in (1.1) yields*

$$\Gamma(z) = \int_0^1 \left(\log\frac{1}{u}\right)^{z-1} du, \ \mathrm{Re}(z) > 0 \tag{1.9}$$

Example 1.1: Evaluate the integral $I = \int_0^\infty e^{-x^3}x^4\,dx$ in terms of the gamma function.

Solution: By making the substitution $t = x^3$, we have

$$I = \frac{1}{3}\int_0^\infty e^{-t}t^{2/3}\,dt$$

* By log x, we mean the natural logarithm, often denoted by ln x.

and by comparison with Eq. (1.1), we deduce that

$$\int_0^\infty e^{-x^3}x^4\,dx = (1/3)\Gamma(5/3)$$

An interesting relation involving the product of two gamma functions can be derived by using the representation (1.8). We first write

$$\Gamma(x)\Gamma(y) = 2\int_0^\infty e^{-u^2}u^{2x-1}\,du \cdot 2\int_0^\infty e^{-v^2}v^{2y-1}\,dv$$

$$= 4\int_0^\infty\int_0^\infty e^{-(u^2+v^2)}u^{2x-1}v^{2y-1}\,du\,dv$$

where x and y are real variables. The change of variables

$$u = r\cos\theta, \qquad v = r\sin\theta$$

leads to

$$\Gamma(x)\Gamma(y) = 4\int_0^\infty\int_0^{\pi/2} e^{-r^2}r^{2x-1}\cos^{2x-1}\theta r^{2y-1}\sin^{2y-1}\theta\, rd\theta\, dr$$

$$= 4\int_0^\infty e^{-r^2}r^{2(x+y)-1}\,dr \cdot \int_0^{\pi/2} \cos^{2x-1}\theta\,\sin^{2y-1}\theta\,d\theta$$

which reduces to

$$\Gamma(x)\Gamma(y) = 2\Gamma(x+y)\int_0^{\pi/2}\cos^{2x-1}\theta\,\sin^{2y-1}\theta\,d\theta \tag{1.10}$$

Solving for the integral gives us the relation

$$\int_0^{\pi/2}\cos^{2x-1}\theta\,\sin^{2y-1}\theta\,d\theta = \frac{\Gamma(x)\Gamma(y)}{2\Gamma(x+y)}, \qquad x > 0,\ y > 0 \tag{1.11}$$

Example 1.2: Evaluate the integral $\int_0^{\pi/2}\sin^4\theta\,\cos^5\theta\,d\theta$.

Solution: Comparing this integral with that in Eq. (1.11), we see that $2x - 1 = 5$ and $2y - 1 = 4$; thus, $x = 3$ and $y = 5/2$. In terms of gamma functions, we have

$$\int_0^{\pi/2}\sin^4\theta\,\cos^5\theta\,d\theta = \frac{\Gamma(3)\Gamma(5/2)}{2\Gamma(11/2)}$$

But, $\Gamma(3) = 2$ and

$$\Gamma\left(\frac{11}{2}\right) = \Gamma\left(1+\frac{9}{2}\right) = \frac{9}{2}\cdot\frac{7}{2}\cdot\frac{5}{2}\Gamma\left(\frac{5}{2}\right)$$

which follows from repeated applications of the recurrence formula (1.2). Hence

$$\int_0^{\pi/2} \sin^4\theta \cos^5\theta \, d\theta = \frac{2}{9} \cdot \frac{2}{7} \cdot \frac{2}{5} = \frac{8}{315}$$

Finally, we wish to derive an important relation connecting the gamma function and the trigonometric functions. To do so, let us set $x = 1 - z$ and $y = z$ in Eq. (1.10), which leads to

$$\Gamma(z)\Gamma(1 - z) = 2 \int_0^{\pi/2} \tan^{2z-1}\theta \, d\theta \tag{1.12}$$

where z may be real or complex. The change of variable $u = \tan^2\theta$ converts this integral to the form

$$\Gamma(z)\Gamma(1 - z) = \int_0^{\infty} \frac{u^{z-1}}{1 + u} du, \qquad 0 < \mathrm{Re}(z) < 1 \tag{1.13}$$

where we (temporarily) restrict the real part of z as indicated in (1.13). The evaluation of this last integral can now be accomplished through contour integration in the complex plane.

Let us integrate the complex function

$$f(\zeta) = \frac{\zeta^{z-1}}{1 + \zeta}$$

which has a simple pole at $\zeta = -1$ and a branch point at $\zeta = 0$, around the contour shown in Fig. 1.2 and then let $\rho \to 0$ and $R \to \infty$. If we write $\zeta = u$ along the upper boundary of the cut along the positive real axis, then we must write $\zeta = ue^{2\pi i}$ along the lower boundary of this cut.

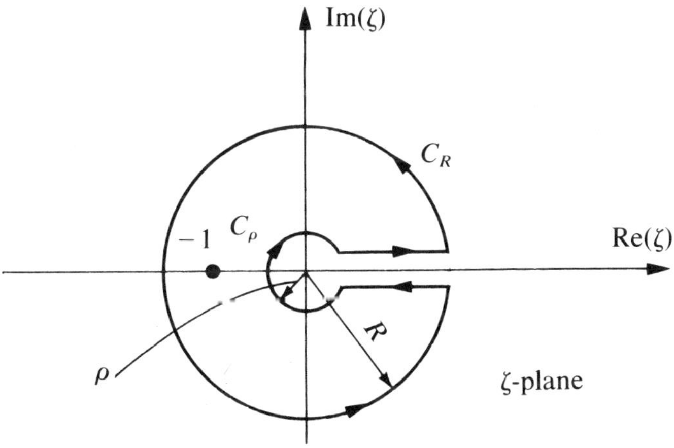

Figure 1.2 Contour for evaluating the integral in (1.13)

The integral of f over the entire closed path leads to

$$\oint_C f(\zeta)d\zeta = \int_{C_\rho} f(\zeta)d\zeta + \int_\rho^R f(u)du + \int_{C_R} f(\zeta)d\zeta + \int_R^\rho f(ue^{2\pi i})du \quad (1.14)$$

Based on Theorems A.3 and A.4 in Appendix A, we conclude that

$$\lim_{R\to\infty} \int_{C_R} f(\zeta)d\zeta = 0$$

$$\lim_{\rho\to 0} \int_{C_\rho} f(\zeta)d\zeta = 0$$

and since $f(\zeta)$ has a simple pole at $\zeta = -1$, we find in the limit as $R \to \infty$ and $\rho \to 0$ that (1.14) reduces to

$$\int_0^\infty \frac{u^{z-1}}{1+u}du + \int_\infty^0 \frac{(ue^{2\pi i})^{z-1}}{1+u}du = 2\pi i\text{Res}\{-1\}$$

or

$$(1 - e^{2\pi iz})\int_0^\infty \frac{u^{z-1}}{1+u}du = 2\pi i\text{Res}\{-1\} \quad (1.15)$$

Computing the residue, we obtain

$$\text{Res}\{-1\} = \lim_{\zeta\to -1} \zeta^{z-1} = (-1)^{z-1}$$

However, $(-1)^{z-1} = e^{\pi i(z-1)} = -e^{\pi iz}$, and hence (1.15) becomes

$$\int_0^\infty \frac{u^{z-1}}{1+u}du = \frac{-2\pi ie^{\pi iz}}{1 - e^{2\pi iz}} = \frac{-2\pi ie^{\pi iz}}{e^{\pi iz}(e^{-\pi iz} - e^{\pi iz})}$$

from which we deduce

$$\int_0^\infty \frac{u^{z-1}}{1+u}du = \frac{\pi}{\sin\pi z}, \qquad 0 < \text{Re}(z) < 1 \quad (1.16)$$

Combining the results of (1.13) and (1.16), we obtain

$$\Gamma(z)\Gamma(1 - z) = \pi/\sin\pi z \quad (1.17)$$

which is valid for all nonintegral values of z. By setting $z = 1/2$, we get

$$\Gamma(1/2)\Gamma(1/2) = \pi$$

which leads to the special value

$$\Gamma(1/2) = \sqrt{\pi} \quad (1.18)$$

Example 1.3: Evaluate the integral $\int_0^{\pi/2} \cot^{1/2}x\, dx$.

Solution: Making use of (1.11) and (1.17), we get

$$\int_0^{\pi/2} \cot^{1/2}x\, dx = \int_0^{\pi/2} \cos^{1/2}x \sin^{-1/2}x\, dx$$

$$= \frac{\Gamma(1/4)\Gamma(3/4)}{2\Gamma(1)}$$

$$= \frac{1}{2}\frac{\pi}{\sin(\pi/4)}$$

and hence

$$\int_0^{\pi/2} \cot^{1/2}x\, dx = \frac{\pi}{\sqrt{2}}$$

There are many other identities involving the gamma function which are too numerous to mention here. For reference purposes, a short list of basic identities follows.

Basic Identities for $\Gamma(z)$

(G1): $\Gamma(z) = \int_0^\infty e^{-t}t^{z-1}\, dt, \qquad \text{Re}(z) > 0$

(G2): $\Gamma(z) = 2\int_0^\infty e^{-t^2}t^{2z-1}\, dt, \qquad \text{Re}(z) > 0$

(G3): $\Gamma(z + 1) = z\Gamma(z)$

(G4): $\Gamma(n + 1) = n!, \qquad n = 0,1,2,\ldots$

(G5): $\Gamma(1/2) = \sqrt{\pi}$

(G6): $\Gamma(z) = \dfrac{\Gamma(z + k)}{z(z + 1)(z + 2)\cdots(z + k - 1)}, \qquad k = 1,2,3,\ldots$

(G7): $\sqrt{\pi}\,\Gamma(2z) = 2^{2z-1}\,\Gamma(z)\Gamma(z + \tfrac{1}{2})$

(G8): $\Gamma(n + 1/2) = \dfrac{(2n)!}{2^{2n}n!}\sqrt{\pi}, \qquad n = 0,1,2,\ldots$

(G9): $\Gamma(z)\Gamma(1 - z) = \dfrac{\pi}{\sin \pi z}, \qquad z$ nonintegral

(G10): $\Gamma(n + 1) \sim \sqrt{2\pi n}n^n e^{-n}$, $n \to \infty$* (Stirling's formula)

---

* The symbol $\sim$ means "behaves like" or "is asymptotic to."

EXERCISES 1.2

In Probs. 1–6, give numerical values for the expressions (use the result $\Gamma(1/2) = \sqrt{\pi}$ where necessary).

1. $\Gamma(6)$

2. $\Gamma(3/2)$

3. $\Gamma(7/2)$

4. $\Gamma(-1/2)$

5. $\Gamma(-9/2)$

6. $\Gamma(8/3)/\Gamma(2/3)$

In Probs. 7–9, verify the given identity.

7. $\Gamma(a + n) = a(a + 1)(a + 2)\cdots(a + n - 1)\Gamma(a), \qquad n = 1,2,3,\ldots$

8. $\Gamma(n - a)/\Gamma(-a) = (-1)^n a(a - 1)(a - 2)\cdots(a - n + 1), \qquad n = 1,2,3,\ldots$

9. $\Gamma(a)/\Gamma(a - n) = (a - 1)(a - 2)\cdots(a - n), \qquad n = 1,2,3,\ldots$

10. The *binomial coefficient* is defined by $(a \neq 0)$

$$\binom{a}{0} = 1, \binom{a}{k} = \frac{a(a-1)\cdots(a-k+1)}{k!}, \qquad k = 1,2,3,\ldots$$

Show that

(a) $\displaystyle\binom{n}{k} = \frac{n!}{k!(n-k)!}, \qquad n = 0,1,2,\ldots, \qquad k = 0,1,2,\ldots,n$

(b) $\displaystyle\binom{-1/2}{n} = \frac{(-1)^n(2n)!}{2^{2n}(n!)^2}, \qquad n = 0,1,2,\ldots$

(c) $\displaystyle\binom{a}{k} = \frac{\Gamma(a+1)}{k!\Gamma(a-k+1)}, \qquad k = 0,1,2,\ldots$

(d) $\displaystyle\binom{-a}{k} = (-1)^k\binom{a+k-1}{k}, \qquad k = 0,1,2,\ldots$

In Probs. 11–13, verify the given integral formula.

11. $\displaystyle\Gamma(x) = p^x \int_0^\infty e^{-pt}t^{x-1}\,dt, \qquad x > 0,\ p > 0$

12. $\displaystyle\Gamma(x) = \int_{-\infty}^\infty \exp(xt - e^t)dt, \qquad x > 0$

13. $\displaystyle\Gamma(x) = (\log b)^x \int_0^\infty t^{x-1}b^{-t}\,dt, \qquad x > 0, b > 1$

Hint: Let $u = t \log b$.

In Probs. 14 and 15, use properties of the gamma function to obtain the result.

14. $\int_a^\infty e^{2ax-x^2}\,dx = \frac{1}{2}\sqrt{\pi}e^{a^2}$ **15.** $\int_0^\infty \sqrt{x}e^{-x^3}\,dx = \frac{\sqrt{\pi}}{3}$

Hint: $2ax - x^2 = -(x-a)^2 + a^2$.

In Probs. 16–19, use Eq. (1.11) to evaluate the integral

16. $\int_0^{\pi/2} \sin^5 x\,dx$ **17.** $\int_0^{\pi/2} \sin x\cos^2 x\,dx$

18. $\int_0^{\pi/2} \sqrt{\sin 2x}\,dx$ **19.** $\int_0^\infty \frac{dx}{1+x^4}$

Hint: Let $x^2 = \tan\theta$.

20. Show that

$$\int_0^{\pi/2} \sin^{2n+1}\theta\,d\theta = \int_0^{\pi/2} \cos^{2n+1}\theta\,d\theta = \frac{2^{2n}(n!)^2}{(2n+1)!}, \qquad n = 0,1,2,\ldots$$

21. The *beta function* is defined by the integral

$$B(x,y) = \int_0^1 t^{x-1}(1-t)^{y-1}\,dt, \qquad x > 0,\ y > 0$$

Show that

(a) $B(x,y) = \Gamma(x)\Gamma(y)/\Gamma(x+y)$

Hint: Let $t = \cos^2\theta$.

(b) $B(x,y) = \int_0^\infty \frac{u^{x-1}}{(1+u)^{x+y}}\,du$

Hint: Let $t = u/(1+u)$.

22. Using the result of Prob. 21, evaluate the following:

(a) $B(2,3)$ (b) $B(1/2,1)$

(c) $B(2/3,1/3)$ (d) $B(3/4,1/4)$

23. By setting $y = x$ in Eq. (1.11),

(a) show that

$$\frac{\Gamma(x)\Gamma(x)}{2\Gamma(2x)} = 2^{1-2x}\int_0^{\pi/2} \sin^{2x-1}\phi\,d\phi$$

Hint: Use the identity $\sin x\cos x = \frac{1}{2}\sin 2x$.

(b) Evaluating the integral in (a), deduce the *Legendre duplication formula*

$$\sqrt{\pi}\,\Gamma(2x) = 2^{2x-1}\,\Gamma(x)\Gamma(x+1/2)$$

24. Using the result of Prob. 23, derive the relation

$$\Gamma(n + 1/2) = \frac{(2n)!}{2^{2n}n!}\sqrt{\pi}, \qquad n = 0,1,2,\ldots$$

25. Verify the gamma function relation ($n + \nu \neq -1,-2,-3,\ldots$)

$$\frac{\Gamma(2n + 2\nu + 1)}{\Gamma(n + \nu + 1)} = \frac{1}{\sqrt{\pi}} 2^{2n+2}\,\Gamma(n + \nu + 1/2)$$

1.3 *The Error Function and Related Functions*

The *error function* is defined by the integral

$$\mathrm{erf}(z) = \frac{2}{\sqrt{\pi}} \int_0^z e^{-t^2}\,dt \tag{1.19}$$

where the variable z may be real or complex.* This function is encountered in probability theory, the theory of errors, the theory of heat conduction, and various branches of mathematical physics.

By representing the exponential function in (1.19) in terms of its power series expansion, we have

$$\mathrm{erf}(z) = \frac{2}{\sqrt{\pi}} \int_0^z \sum_{n=0}^{\infty} \frac{(-1)^n}{n!} t^{2n}\,dt$$

from which we deduce (termwise integration of power series is permitted)

$$\mathrm{erf}(z) = \frac{2}{\sqrt{\pi}} \sum_{n=0}^{\infty} \frac{(-1)^n z^{2n+1}}{n!(2n + 1)} \tag{1.20}$$

This series converges everywhere in the finite complex plane; therefore, $\mathrm{erf}(z)$ is an *entire function*.

Examination of the series (1.20) reveals that the error function is an *odd function*, i.e.,

$$\mathrm{erf}(-z) = -\mathrm{erf}(z) \tag{1.21}$$

We also see that

$$\mathrm{erf}(\bar{z}) = \overline{\mathrm{erf}(z)} \tag{1.22}$$

where $\bar{z}$ denotes the complex conjugate.

When $z = 0$, it is clear that

$$\mathrm{erf}(0) = 0 \tag{1.23}$$

* The variable t can also be complex, although we generally assume it is real.

while for $z \to \infty$ [$|\arg(z)| < \pi/4$], it follows from properties of the gamma function that (in the limit)

$$\mathrm{erf}(\infty) = \frac{2}{\sqrt{\pi}} \int_0^\infty e^{-t^2}\, dt = \frac{\Gamma(1/2)}{\sqrt{\pi}} = 1 \tag{1.24}$$

The graph of erf(x), where x is real, is shown in Fig. 1.3.

1.3.1 *Complementary Error Function*

In some applications it is useful to introduce the *complementary error function*

$$\mathrm{erfc}(z) = \frac{2}{\sqrt{\pi}} \int_z^\infty e^{-t^2}\, dt \tag{1.25}$$

Using properties of integrals, it follows that

$$\mathrm{erfc}(z) = \frac{2}{\sqrt{\pi}} \int_0^\infty e^{-t^2}\, dt - \frac{2}{\sqrt{\pi}} \int_0^z e^{-t^2}\, dt,$$

from which we deduce

$$\mathrm{erfc}(z) = 1 - \mathrm{erf}(z). \tag{1.26}$$

Hence, all properties of erfc(z) are easily derived from those of erf(z).

1.3.2 *Fresnel Integrals*

Closely associated with the error function are the *Fresnel integrals*

$$C(x) = \int_0^x \cos(\pi t^2/2)\, dt \tag{1.27}$$

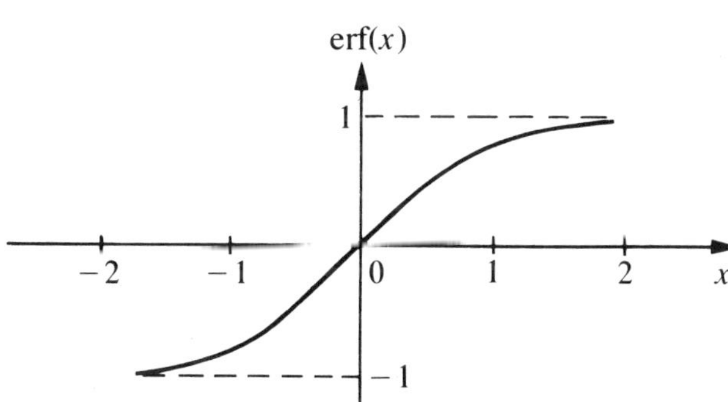

Figure 1.3 The error function

and

$$S(x) = \int_0^x \sin(\pi t^2/2)\, dt \tag{1.28}$$

These integrals come up in various branches of physics and engineering, such as in diffraction theory and the theory of vibrations, among others.

From definition, we have the immediate results

$$C(0) = S(0) = 0 \tag{1.29}$$

The derivatives of these functions are

$$C'(x) = \cos(\pi x^2/2), \qquad S'(x) = \sin(\pi x^2/2) \tag{1.30}$$

and thus we deduce that both $C(x)$ and $S(x)$ are oscillatory. Namely, $C(x)$ has extrema at the points where $x^2 = 2n + 1$ $(n = 0,1,2,\ldots)$, and $S(x)$ has extrema where $x^2 = 2n$ $(n = 1,2,3,\ldots)$. The largest maxima occur first and are found to be $C(1) = 0.77989\ldots$ and $S(\sqrt{2}) = 0.71397\ldots$. For $x \to \infty$, we can use the integral formulas (see Prob. 12 in Exer. 1.3)

$$\int_0^\infty \cos t^2 dt = \int_0^\infty \sin t^2 dt = \frac{1}{2}\sqrt{\frac{\pi}{2}}$$

to obtain the results

$$C(\infty) = S(\infty) = \tfrac{1}{2} \tag{1.31}$$

The graphs of $C(x)$ and $S(x)$ for positive real x are shown in Fig. 1.4.

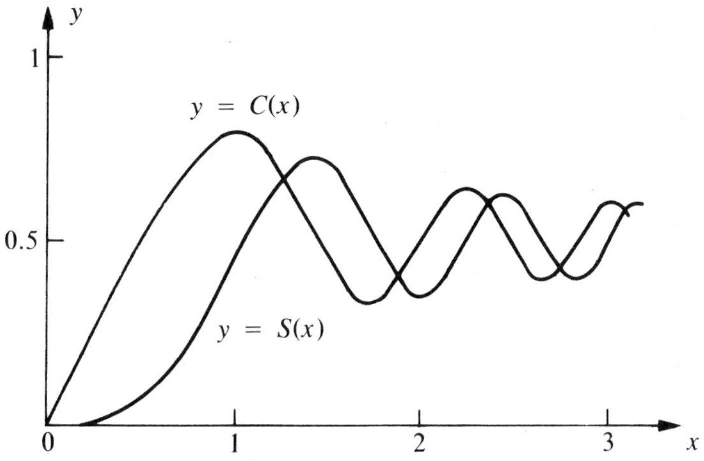

Figure 1.4 The Fresnel Integrals

EXERCISES 1.3

1. Show that

(a) $\displaystyle\int_{-a}^{a} e^{-t^2}\,dt = \sqrt{\pi}\,\mathrm{erf}(a)$

(b) $\displaystyle\int_{a}^{b} e^{-t^2}\,dt = \frac{\sqrt{\pi}}{2}[\mathrm{erf}(b) - \mathrm{erf}(a)] = \frac{\sqrt{\pi}}{2}[\mathrm{erfc}(a) - \mathrm{erfc}(b)]$

2. Show that

$$\frac{d}{dz}\,\mathrm{erf}(z) = \frac{2}{\sqrt{\pi}}e^{-z^2}$$

3. Use integration by parts to obtain

$$\int \mathrm{erf}(z)dz = z\,\mathrm{erf}(z) + \frac{1}{\sqrt{\pi}}e^{-z^2} + C$$

where C is a constant of integration.

4. Evaluate

(a) erfc(0)

(b) $\mathrm{erfc}(\infty)$

5. Show that ($p > 0$)

$$\int_0^\infty e^{-px}\,\mathrm{erf}(x)\,dx = \frac{1}{p}e^{p^2/4}\,\mathrm{erfc}(p/2)$$

Hint: Replace erf(x) by its integral representation and interchange the order of integration.

6. Show that ($p \geq 0$)

$$\int_0^\infty e^{-px-x^2/4}\,dx = \sqrt{\pi}\,e^{p^2}\,\mathrm{erfc}(p)$$

Hint: Write $x^2/4 + px = (x/2 + p)^2 - p^2$ and make the change of variable $u = x/2 + p$.

7. Using integration by parts,

(a) show that

$$\int_z^\infty e^{-t^2}\,dt = \frac{e^{-z^2}}{2z} - \frac{1}{2}\int_z^\infty \frac{e^{-t^2}}{t^2}\,dt$$

(b) By repeated integration by parts, derive the *asymptotic series*

$$\operatorname{erfc}(z) \sim \frac{e^{-z^2}}{\sqrt{\pi}\,z}\left[1 + \sum_{n=1}^{\infty} (-1)^n \frac{1 \times 3 \times \cdots \times (2n-1)}{(2z^2)^n}\right],$$

$$|z| \to \infty, \ |\arg(z)| < \pi/2$$

8. If X is a *normal* random variable, its probability density function is

$$p(x) = \frac{1}{\sqrt{2\pi}\sigma} e^{-(x-m)^2/2\sigma^2}$$

where m is the mean value of X and σ^2 the variance. The probability that $X \le y$ is defined by

$$P(X \le y) = \int_{-\infty}^{y} p(x)\, dx$$

(a) Show that

$$P(X \le y) = \frac{1}{2}\left[1 + \operatorname{erf}\left(\frac{y-m}{\sqrt{2}\sigma}\right)\right]$$

(b) What is the probability $P(X \le y)$ in the limit $y \to \infty$?

9. Considering the integral

$$I(a) = \int_{-\infty}^{\infty} \frac{e^{-a^2x^2}}{x^2 + b^2}\, dx, \qquad a \ge 0, b > 0$$

as a function of the parameter a:

(a) Show that I satisfies the first-order linear differential equation (DE)

$$\frac{dI}{da} - 2ab^2 I = -2\sqrt{\pi}.$$

(b) Evaluate $I(0)$ directly from the integral.

(c) Solve the DE in (a) subject to the initial condition in (b) to deduce that

$$I(a) = \frac{\pi}{b} e^{a^2b^2} \operatorname{erfc}(ab).$$

10. Show that the Fresnel integrals satisfy

(a) $C(-x) = -C(x)$ (b) $S(-x) = -S(x)$

11. Obtain the series representations

(a) $$C(x) = \sum_{n=0}^{\infty} \frac{(-1)^n(\pi/2)^{2n}}{(2n)!(4n+1)} x^{4n+1}$$

(b) $$S(x) = \sum_{n=0}^{\infty} \frac{(-1)^n(\pi/2)^{2n+1}}{(2n+1)!(4n+3)} x^{4n+3}$$

12. Establish the integral formula

$$\int_0^\infty e^{-a^2t^2}\,dt = \frac{\sqrt{\pi}}{2a}$$

Then, writing $a = (1 - i)/\sqrt{2}$ and separating into real and imaginary parts, deduce that

$$\int_0^\infty \cos t^2\,dt = \int_0^\infty \sin t^2\,dt = \frac{1}{2}\sqrt{\frac{\pi}{2}}$$

13. Using the definition of the error function (1.19), show that
(a) $\operatorname{erf}(\sqrt{i}x) = (1 + i)\,[C(x\sqrt{2/\pi}) - iS(x\sqrt{2/\pi})]$
(b) $\operatorname{erf}(\sqrt{-i}x) = (1 - i)\,[C(x\sqrt{2/\pi}) + iS(x\sqrt{2/\pi})]$
(c) $\operatorname{erf}(\sqrt{i}x) + \operatorname{erf}(\sqrt{-i}x) = 2\,[C(x\sqrt{2/\pi}) + S(x\sqrt{2/\pi})]$

1.4 *Bessel Functions*

Bessel functions are closely associated with problems possessing circular or cylindrical symmetry, such as the study of free vibrations of a circular membrane and finding the temperature distribution in a circular cylinder. These functions, of which there are several varieties, occur in so many additional areas of application in engineering and physical science that they are considered to be the most important functions beyond the elementary ones studied in calculus.

Bessel functions of the first kind are defined by the series

$$J_\nu(z) = \sum_{k=0}^{\infty} \frac{(-1)^k (z/2)^{2k+\nu}}{k!\,\Gamma(k + \nu + 1)} \tag{1.32}$$

where the parameter ν denotes the *order* of the given Bessel function. When $\nu = n$ $(n = 0,1,2,\ldots)$, Eq. (1.32) defines the Bessel function of integer order

$$J_n(z) = \sum_{k=0}^{\infty} \frac{(-1)^k (z/2)^{2k+n}}{k!\,(k + n)!}, \qquad n = 0,1,2,\ldots \tag{1.33}$$

the simplest representative of which is

$$J_0(z) = \sum_{k=0}^{\infty} \frac{(-1)^k (z/2)^{2k}}{(k!)^2}. \tag{1.34}$$

The graphs of $J_n(x)$, $n = 0,1,2$, are shown in Fig. 1.5, where x is real.

The parameter ν in (1.32) may also take on negative values. For example, when $\nu = -n$ $(n = 0,1,2,\ldots)$, we get

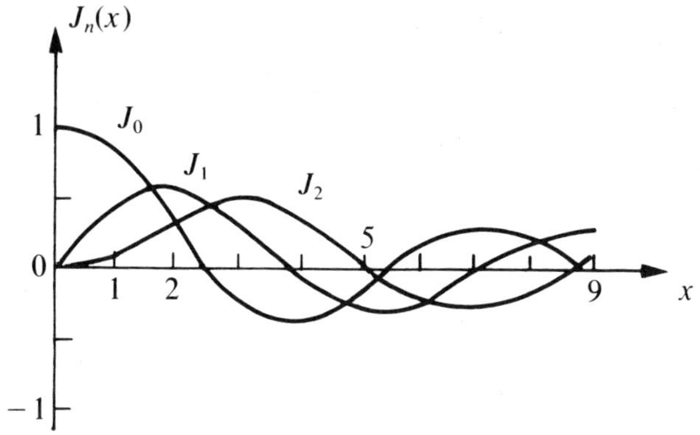

Figure 1.5 Graph of $J_n(x)$, $n = 0,1,2$

$$J_{-n}(z) = \sum_{k=0}^{\infty} \frac{(-1)^k (z/2)^{2k-n}}{k!\,(k-n)!}$$
$$= \sum_{k=n}^{\infty} \frac{(-1)^k (z/2)^{2k-n}}{k!(k-n)!}$$

where we have used the fact that $1/(k - n)! = 0$ $(k = 0,1,\ldots, n - 1)$ by virtue of Eq. (1.7). Finally, the change of index $k = m + n$ yields

$$J_{-n}(z) = \sum_{m=0}^{\infty} \frac{(-1)^{m+n} (z/2)^{2m+n}}{m!\,(m+n)!}$$

from which we deduce

$$J_{-n}(z) = (-1)^n J_n(z), \qquad n = 0,1,2,\ldots \tag{1.35}$$

However, this last relation applies only to integral-order Bessel functions.

Rewriting (1.32) in the form

$$J_\nu(z) = \left(\frac{z}{2}\right)^\nu \sum_{k=0}^{\infty} \frac{(-1)^k (z/2)^{2k}}{k!\Gamma(k+\nu+1)}$$

it can be shown that the series on the right converges in the whole z-plane. Therefore, the function $(2/z)^\nu J_\nu(z)$ is an *entire function* of z. However, this does not necessarily imply that $J_\nu(z)$ is entire. If $\nu < 0$ and nonintegral, then clearly $J_\nu(z)$ has an infinite discontinuity at $z = 0$, and hence, cannot represent an entire function. But, if $\nu = \pm n$, $n = 0,1,2,\ldots$, then it can be shown that $J_\nu(x)$ is entire — a result that depends upon the relation (1.35).

The Bessel functions are named in honor of F. W. Bessel (1784–1846) who in 1824 carried out the first systematic study of the properties of these functions and derived their governing differential equation (see Prob. 3 in Exer. 1.4). Nonetheless, Bessel functions were discovered years earlier by Euler and others who were concerned with various problems in mechanics, and the infinite series (1.34) was obtained by D. Bernoulli in 1703 — more than 120 years before Bessel's famous study — in connection with his investigation of the oscillatory behavior of a hanging chain.

1.4.1 *Basic Properties*

The Bessel functions satisfy a large number of basic identities such as

$$\frac{d}{dz}[z^{\nu}J_{\nu}(z)] = z^{\nu}J_{\nu-1}(z) \tag{1.36}$$

and

$$\frac{d}{dz}[z^{-\nu}J_{\nu}(z)] = -z^{-\nu}J_{\nu+1}(z) \tag{1.37}$$

both of which follow from termwise differentiation of the series for $z^{\nu}J_{\nu}(z)$ and $z^{-\nu}J_{\nu}(z)$ (see Prob. 1 in Exer. 1.4). If we carry out the differentiation in (1.36) and (1.37) and simplify the results, it follows that

$$J_{\nu}'(z) + \frac{\nu}{z}J_{\nu}(z) = J_{\nu-1}(z) \tag{1.38}$$

and

$$J_{\nu}'(z) - \frac{\nu}{z}J_{\nu}(z) = -J_{\nu+1}(z) \tag{1.39}$$

The substitution of $\nu = 0$ in (1.39) leads to the special result

$$J_0'(z) = -J_1(z) \tag{1.40}$$

Direct integration of (1.36) and (1.37) give us the integral relations

$$\int z^{\nu}J_{\nu-1}(z)\,dz = z^{\nu}J_{\nu}(z) + C \tag{1.41}$$

and

$$\int z^{-\nu}J_{\nu+1}(z)\,dz = -z^{-\nu}J_{\nu}(z) + C \tag{1.42}$$

where C denotes a constant of integration. As a general rule, any integral of the form

$$\int z^{m}J_{n}(z)\,dz, \qquad m + n > 0$$

where m and n are integers, can be evaluated with the use of (1.41) and (1.42), coupled with standard integration techniques such as integration by parts. When $m + n$ is odd, the integral can be evaluated in closed form, but will ultimately depend on the residual integral $\int J_0(z)dz$ when $m + n$ is even.

Example 1.4: Reduce $\int z^2 J_2(z)\,dz$ to an integral involving only $J_0(z)$.

Solution: To use (1.42), we first write

$$\int z^2 J_2(z)\,dz = \int z^3\,[z^{-1}J_2(z)]\,dz$$

and use integration by parts to get

$$\int z^2 J_2(z)\,dz = -z^2 J_1(z) + 3\int zJ_1(z)\,dz$$

A second integration by parts on the last integral yields

$$\int z^2 J_2(z)\,dz = -z^2 J_1(z) - 3zJ_0(z) + 3\int J_0(z)\,dz$$

The last integral involving $J_0(z)$ cannot be evaluated in closed form, and so our integration is complete.

For reference purposes, a short list of the basic identities of the Bessel function follows.

Basic Identities for $J_\nu(z)$

(J1): $J_\nu(z) = \sum_{k=0}^{\infty} \frac{(-1)^k\,(z/2)^{2k+\nu}}{k!\,\Gamma(k+\nu+1)}$

(J2): $J_0(0) = 1;\ J_\nu(0) = 0, \qquad \nu > 0$

(J3): $J_{-n}(z) = (-1)^n J_n(z), \qquad n = 0,1,2,\ldots$

(J4): $\frac{d}{dz}\,[z^\nu J_\nu(z)] = z^\nu J_{\nu-1}(z)$

(J5): $\frac{d}{dz}\,[z^{-\nu} J_\nu(z)] = -z^{-\nu} J_{\nu+1}(z)$

(J6): $J_\nu'(z) + \frac{\nu}{z} J_\nu(z) = J_{\nu-1}(z)$

(J7): $J_\nu'(z) - \frac{\nu}{z} J_\nu(z) = -J_{\nu+1}(z)$

(J8): $J_{\nu-1}(z) - J_{\nu+1}(z) = 2J_\nu'(z)$

(J9): $J_{\nu-1}(z) + J_{\nu+1}(z) = \frac{2}{\nu z} J_\nu(z)$

$$\text{(J10)}: \int z^{\nu} J_{\nu-1}(z)\,dz = z^{\nu} J_{\nu}(z) + C$$

$$\text{(J11)}: \int z^{-\nu} J_{\nu+1}(z)\,dz = -z^{-\nu} J_{\nu}(z) + C$$

$$\text{(J12)}: J_0(z) = \frac{1}{2\pi}\int_0^{2\pi} e^{iz\cos\theta}\,d\theta$$

$$\text{(J13)}: J_{\nu}(z) \sim \frac{(z/2)^{\nu}}{\Gamma(\nu+1)}, \qquad \nu \neq -1,-2,-3,\ldots,\ z \to 0$$

$$\text{(J14)}: J_{\nu}(z) \sim \sqrt{\frac{2}{\pi z}}\cos\left[z - (\nu + 1/2)\frac{\pi}{2}\right], \qquad |z| \to \infty,\ |\arg(z)| < \pi$$

Remark: In certain applications it is important to recognize another Bessel function $Y_{\nu}(x)$, called a *Bessel function of the second kind and order* ν. This function, defined by

$$Y_{\nu}(x) = \frac{J_{\nu}(x)\cos\nu\pi - J_{-\nu}(x)}{\sin\nu\pi}$$

is a linear combination of $J_{\nu}(x)$ and $J_{-\nu}(x)$, and therefore satisfies the same recurrence relations as $J_{\nu}(x)$.

1.4.2 *Modified Bessel Functions*

In certain applications the Bessel function $J_{\nu}(z)$ appears with a pure imaginary argument. By setting $z = iy$ in (1.32), we obtain

$$J_{\nu}(iy) = i^{\nu}\sum_{k=0}^{\infty}\frac{(y/2)^{2k+\nu}}{k!\Gamma(k+\nu+1)} \tag{1.43}$$

Except for the multiplicative factor i^{ν}, the right-hand side of (1.43) defines a real function, which is called the *modified Bessel function of the first kind* and denoted by the symbol $I_{\nu}(y)$. Thus,

$$I_{\nu}(y) = i^{-\nu}J_{\nu}(iy) \tag{1.44}$$

or by analytic continuation, we can generalize to complex arguments by writing

$$I_{\nu}(z) = \sum_{k=0}^{\infty}\frac{(z/2)^{2k+\nu}}{k!\Gamma(k+\nu+1)} \tag{1.45}$$

Comparing this series representation with Eq. (1.32) for $J_{\nu}(z)$, it would appear that $I_{\nu}(z)$ and $J_{\nu}(z)$ have many properties in common. Indeed, the modified Bessel functions satisfy relations analogous to all those for the standard Bessel functions. In particular, the modified Bessel functions satisfy properties similar to those for $J_{\nu}(z)$ given by (1.36)–(1.42) (see the exercises). The major distinction in these functions perhaps is exhibited by their graphs for real variable x. That is, the graph of $J_\nu(x)$ has an oscillatory behavior like that of a sine or cosine except for decreasing amplitude, while the graph of $I_\nu(x)$ shows no such oscillatory behavior (see Figs. 1.5 and 1.6). This is comparable to the distinction between circular functions and hyperbolic functions.

In certain applications, particularly in probability theory, we find the

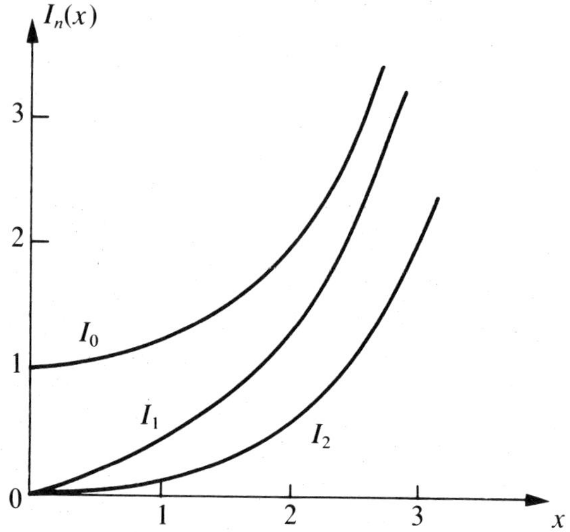

Figure 1.6 Graph of $I_n(x)$, $n = 0,1,2$

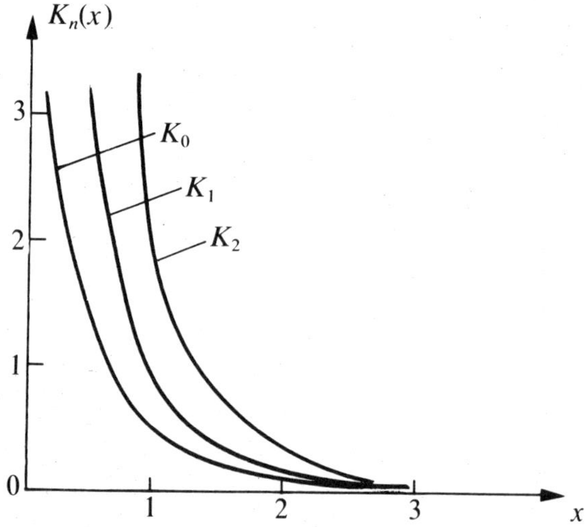

Figure 1.7 Graph of $K_n(x)$, $n = 0,1,2$

appearance of another modified Bessel function, which we define by

$$K_\nu(z) = \frac{\pi}{2}\frac{I_{-\nu}(z) - I_\nu(z)}{\sin \nu\pi} \tag{1.46}$$

This is called the *modified Bessel function of the second kind* and its graph for certain integer values of ν is given in Fig. 1.7. Some properties of this function are taken up in the exercises.

EXERCISES 1.4

1. Using the series representation (1.32), show that

(a) $\dfrac{d}{dz}[z^\nu J_\nu(z)] = z^\nu J_{\nu-1}(z)$

(b) $\dfrac{d}{dz}[z^{-\nu} J_\nu(z)] = -z^{-\nu} J_{\nu+1}(z)$

2. Based on (1.38) and (1.39), deduce that

(a) $2J_\nu'(z) = J_{\nu-1}(z) - J_{\nu+1}(z)$

(b) $\dfrac{2}{\nu z} J_\nu(z) = J_{\nu-1}(z) + J_{\nu+1}(z)$

3. Verify that the series (1.32) is a solution of *Bessel's differential equation*

$$z^2 J_\nu''(z) + zJ_\nu'(z) + (z^2 - \nu^2)J_\nu(z) = 0$$

4. By comparing series, deduce that

(a) $J_{1/2}(z) = \sqrt{\dfrac{2}{\pi z}}\sin z$ (b) $J_{-1/2}(z) = \sqrt{\dfrac{2}{\pi z}}\cos z$

5. Using the *Jacobi–Anger expansion*

$$e^{ix \sin\theta} = \sum_{n=-\infty}^{\infty} J_n(x)e^{in\theta}$$

show that

(a) $\cos(x \sin\theta) = J_0(x) + 2\sum_{n=1}^{\infty} J_{2n}(x)\cos(2n\theta)$

(b) $\sin(x \sin\theta) = 2\sum_{n=1}^{\infty} J_{2n-1}(x)\sin[(2n - 1)\theta]$

(c) $\cos x = J_0(x) + 2\sum_{n=1}^{\infty} (-1)^n J_{2n}(x)$

(d) $\sin x = 2\sum_{n=1}^{\infty} (-1)^n J_{2n-1}(x)$

6. By integrating both sides of the result of Prob. 5(a), deduce that

$$J_0(x) = \frac{1}{\pi}\int_0^{\pi} \cos(x \sin\theta)\, d\theta$$

7. By writing $\cos xt$ in an infinite series and using termwise integration, deduce that

$$J_0(x) = \frac{2}{\pi}\int_0^1 \frac{\cos xt}{\sqrt{1 - t^2}}\, dt$$

In Probs. 8–10, verify the given integral relation.

8. $\displaystyle\int x^2 J_0(x)\, dx = x^2 J_1(x) + xJ_0(x) - \int J_0(x)\, dx + C$

9. $\displaystyle\int x^3 J_0(x)\, dx = (x^3 - 4x)J_1(x) + 2x^2 J_0(x) + C$

10. $\displaystyle\int x^{-2} J_2(x)\, dx = -\frac{2}{3x^2} J_1(x) - \frac{1}{3}J_1(x) + \frac{1}{3x} J_0(x) + \frac{1}{3}\int J_0(x)\, dx + C$

11. Using the series representation (1.34), show that

$$\int_0^{\infty} e^{-x^2} xJ_0(2x)\, dx = 1/2e$$

12. By expressing $J_0(bx)$ in its series representation, use termwise integration to show that

$$\int_0^{\infty} e^{-ax} J_0(bx)\, dx = 1/\sqrt{a^2 + b^2}, \qquad a > 0,\ b > 0$$

13. By formally setting $a = ic$ in the result of Prob. 12, deduce that

(a) $\displaystyle\int_0^{\infty} \cos(cx)J_0(bx)\, dx = \begin{cases} 1/\sqrt{b^2 - c^2} & b > c \\ 0, & b < c. \end{cases}$

(b) $\displaystyle\int_0^{\infty} \sin(cx)J_0(bx)\, dx = \begin{cases} 0, & b > c \\ 1/\sqrt{b^2 - c^2}, & b < c. \end{cases}$

14. Integrate both sides of Prob. 13(a) with respect to c to deduce that

$$\int_0^{\infty} \frac{\sin x}{x} J_0(bx)\, dx = \begin{cases} \pi/2, & 0 < b < 1 \\ \sin^{-1}(1/b), & b > 1 \end{cases}$$

15. Using the series representation (1.45), show that

(a) $\dfrac{d}{dz}[z^{\nu} I_{\nu}(z)] = z^{\nu} I_{\nu-1}(z)$

(b) $\dfrac{d}{dz}[z^{-\nu} I_{\nu}(z)] = z^{-\nu} I_{\nu+1}(z)$

16. Show that $I_{-n}(z) = I_n(z), \quad n = 0,1,2,\ldots$

17. Using the results of Prob. 15, show that

(a) $I'_\nu(z) = I_{\nu-1}(z) - \dfrac{\nu}{z} I_\nu(z)$

(b) $I'_\nu(z) = I_{\nu+1}(z) + \dfrac{\nu}{z} I_\nu(z)$

(c) $I'_\nu(z) = \frac{1}{2}\,[I_{\nu-1}(z) + I_{\nu+1}(z)]$

(d) $I_{\nu-1}(z) - I_{\nu+1}(z) = \dfrac{2}{\nu z} I_\nu(z)$

18. By comparing infinite series, show that

(a) $I_{1/2}(z) = \sqrt{\dfrac{2}{\pi z}}\,\sinh z$ (b) $I_{-1/2}(z) = \sqrt{\dfrac{2}{\pi z}}\,\cosh z$

19. Show that

(a) $K_{-\nu}(z) = K_\nu(z)$ (b) $K_{1/2}(z) = \sqrt{\dfrac{\pi}{2z}}\,e^{-z}$

20. Show that

(a) $\dfrac{d}{dz}\,[z^\nu K_\nu(z)] = -z^\nu K_{\nu-1}(z)$

(b) $\dfrac{d}{dz}\,[z^{-\nu} K_\nu(z)] = -z^{-\nu} K_{\nu+1}(z)$

1.5 *Useful Engineering Functions*

In the solution of various engineering and scientific problems it is helpful to employ the use of special notation to identify functions that must be prescribed in a piecewise fashion. There are a variety of functions for which special notation has become standard, such as the step function, rectangle function, signum function, ramp function, sinc function, impulse function, and so on. In our brief introduction to such functions we primarily will discuss the step function (and related functions) and the impulse function. Because of their frequent occurrence in applications involving the time domain, we will designate the independent variable by t for discussion purposes only.

1.5.1 *Heaviside Unit Function*

Discontinuous functions occur quite naturally in circuit analysis problems as well as in some problems involving mechanical systems. In order to deal effectively with functions having finite jump discontinuities, it is

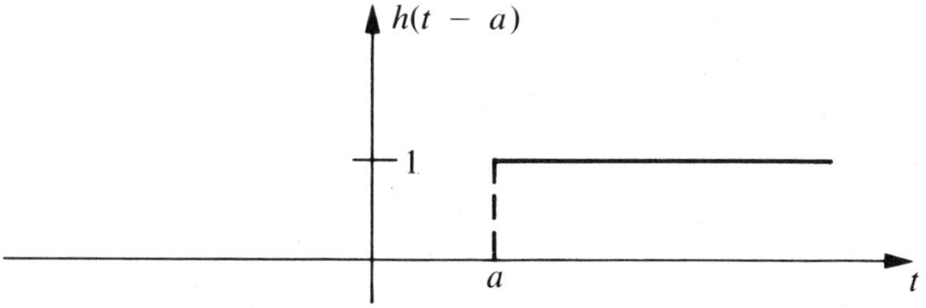

Figure 1.8 Heaviside unit function

helpful to introduce the *unit step function*, also widely known as the *Heaviside unit function* in honor of its discoverer.* We denote this function by the symbol $h(t - a)$ and define it by (see Fig. 1.8)

$$h(t - a) = \begin{cases} 0, & t < a \\ 1, & t > a \end{cases} \tag{1.47}$$

This function has a jump discontinuity at $t = a$ of unit magnitude. The main utility of the Heaviside unit function is that it acts like a "switch" to turn another function on or off at some time. For instance, the function

$$f(t) = h(t - 1)\cos 2\pi t$$

is clearly zero for $t < 1$ and assumes the graph of the cosine function for $t > 1$ as shown in Fig. 1.9.

A related function is the *rectangle function* defined by

$$f(t) = \begin{cases} 1, & a < t < b \\ 0, & \text{otherwise} \end{cases} \tag{1.48}$$

Although it is customary to do so, we will not introduce any special

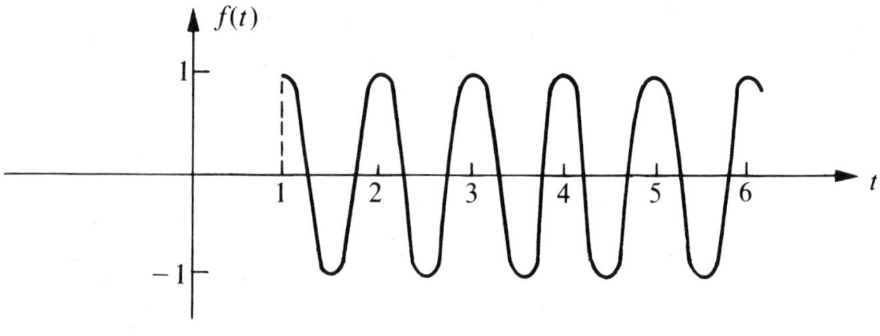

Figure 1.9 Graph of $f(t) = h(t - 1)\cos 2\pi t$

* See the footnote about Oliver Heaviside on p. 162.

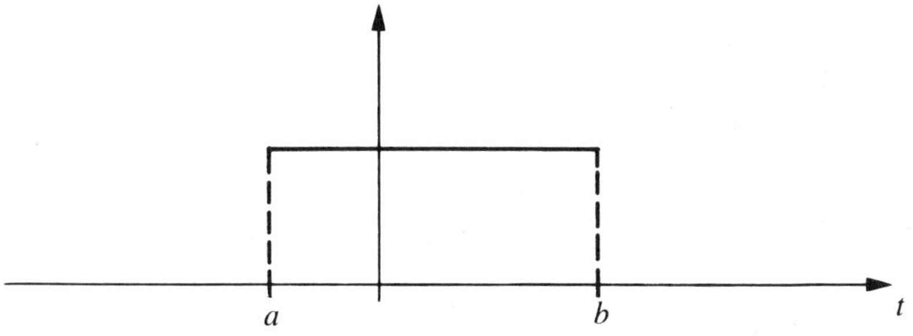

Figure 1.10 Rectangle function

notation for this function since it can be easily expressed in terms of the Heaviside unit function as (see Fig. 1.10)

$$f(t) = h(t - a) - h(t - b). \tag{1.49}$$

The rectangle function (1.49) is useful in describing other functions which are defined piecewise. For example, the function

$$f(t) = \begin{cases} f_1(t), & t < a \\ f_2(t), & a < t < b \\ f_3(t), & t > b \end{cases} \tag{1.50}$$

has the representation

$$\begin{aligned} f(t) &= f_1(t)[1 - h(t - a)] + f_2(t)[h(t - a) - h(t - b)] + f_3(t)h(t - b) \\ &= f_1(t) + [f_2(t) - f_1(t)]h(t - a) + [f_3(t) - f_2(t)]h(t - b) \end{aligned} \tag{1.51}$$

1.5.2 *Impulse Function*

In certain applications it is convenient to introduce the concept of an impulse function which is the result of a sudden excitation administered to a system, such as a sharp blow or voltage surge. Let us imagine that the sudden excitation, which we will denote by $d_a(t)$, has a nonzero value over the short interval of time $a - \varepsilon < t < a + \varepsilon$, but is otherwise zero. The total *impulse* (force times duration) imparted to the system is thus defined by

$$I = \int_{-\infty}^{\infty} d_a(t)\,dt = \int_{a-\varepsilon}^{a+\varepsilon} d_a(t)\,dt \qquad (\varepsilon > 0) \tag{1.52}$$

The value of I is a measure of the strength of the sudden excitation.

In order to provide a mathematical model of the function $d_a(t)$, it is convenient to think of it as having a constant value over the interval $a - \varepsilon \le t \le a + \varepsilon$ (see Fig. 1.11). Furthermore, we wish to choose

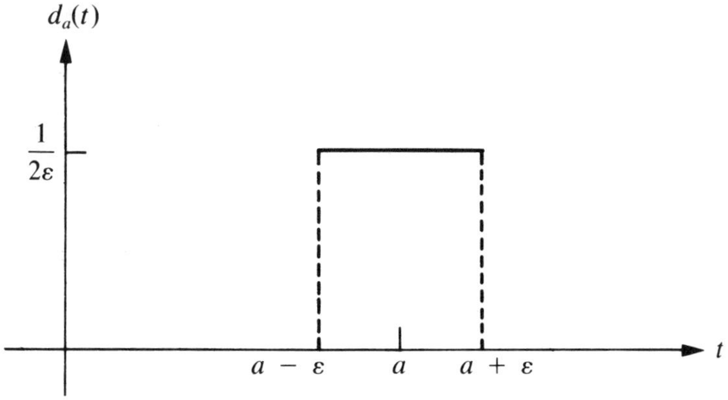

Figure 1.11 Impulse function

this constant value in such a way that the total impulse given by (1.52) is unity. Hence, we write

$$d_a(t) = (1/2\varepsilon)[h(t - a + \varepsilon) - h(t - a - \varepsilon)] \tag{1.53}$$

Now let us idealize the function $d_a(t)$ by requiring it to act over shorter and shorter intervals of time by allowing $\varepsilon \to 0$. Although the interval about $t = a$ is shrinking to zero, we still want $I = 1$; i.e.,

$$\lim_{\varepsilon\to 0} I = \lim_{\varepsilon\to 0} \int_{-\infty}^{\infty} d_a(t)\, dt = 1 \tag{1.54}$$

We can use the results of this limit process to define an "idealized" *unit impulse function*, $\delta(t - a)$, which has the property of imparting a unit impulse to the system at time $t = a$ but being zero for all other values of t. The defining properties of this function are therefore

$$\delta(t - a) = 0, \qquad t \neq a \tag{1.55}$$
$$\int_{-\infty}^{\infty} \delta(t - a)\, dt = 1$$

By a similar kind of limit process, it is possible to define the integral of a product of the unit impulse function and any continuous and bounded function f; i.e.,

$$\int_{-\infty}^{\infty} \delta(t - a) f(t)\, dt = \lim_{\varepsilon\to 0} \int_{-\infty}^{\infty} d_a(t) f(t)\, dt$$
$$= \lim_{\varepsilon\to 0} \frac{1}{2\varepsilon} \int_{a-\varepsilon}^{a+\varepsilon} f(t)\, dt$$

Recalling that

$$\int_a^b f(t)\, dt = f(\xi)(b - a), \qquad a < \xi < b \tag{1.56}$$

which is the *mean value theorem* of the integral calculus, we find that

$$\int_{-\infty}^{\infty} \delta(t - a)f(t)\,dt = \lim_{\varepsilon\to 0} \frac{1}{2\varepsilon} \cdot f(\xi) \cdot 2\varepsilon$$

for some ξ in the interval $a - \varepsilon < \xi < a + \varepsilon$. Consequently, in the limit we see that $\xi \to a$, and deduce the *sifting property*

$$\int_{-\infty}^{\infty} \delta(t - a)f(t)\,dt = f(a). \tag{1.57}$$

Obviously the "function" $\delta(t - a)$, also known as the *Dirac delta function*,* is not a function in the usual sense of the word. It has significance only as part of an integrand. In dealing with this function, therefore, it is best to avoid the idea of assigning "functional values" and instead refer to its integral property (1.57), even though it has no meaning as an ordinary integral. Following more rigorous lines, the impulse function can be defined as a limit of an infinite sequence of well-behaved functions (see Probs. 11 and 12 in Exer. 1.5).

There are certain operational properties of the impulse function that prove useful in practice. Our derivations of such properties, however, will be based strictly upon formal manipulations of the symbols, i.e., they will not be rigorous. To begin we make the observation

$$\int_{-\infty}^{\infty} \delta(t - a)f(t)\,dt = \int_{-\infty}^{\infty} \delta(t)f(t + a)\,dt \tag{1.58}$$

which follows from a simple shift in variable. Next, we write

$$\begin{aligned}\int_{-\infty}^{\infty} [f(t)\delta(t - a)]g(t)\,dt &= \int_{-\infty}^{\infty} \delta(t - a)[f(t)g(t)]\,dt \\ &= f(a)g(a)\end{aligned}$$

Since

$$g(a) = \int_{-\infty}^{\infty} \delta(t - a)g(t)\,dt$$

we see that

$$\int_{-\infty}^{\infty} [f(t)\delta(t - a)]g(t)\,dt = \int_{-\infty}^{\infty} [f(a)\delta(t - a)]g(t)\,dt$$

from which we formally deduce

$$f(t)\delta(t - a) = f(a)\delta(t - a) \tag{1.59}$$

* Named after Paul A. M. Dirac (1902–1984), who was awarded the Nobel prize (with E. Schrödinger) in 1933 for his work in quantum mechanics.

Finally, we wish to develop a formal relationship between the impulse function and the Heaviside unit function. To do so, let us use integration by parts on the expression

$$\int_{-\infty}^{\infty} h'(t-a)f(t)\,dt = h(t-a)f(t)\Big|_{-\infty}^{\infty} - \int_{-\infty}^{\infty} h(t-a)f'(t)\,dt$$

$$= f(\infty) - \int_{a}^{\infty} f'(t)\,dt$$

$$= f(a)$$

where we are assuming that $f(t)$ is both continuous and bounded. By comparison of this result with Eq. (1.57), we deduce that

$$\int_{-\infty}^{\infty} h'(t-a)f(t)\,dt = \int_{-\infty}^{\infty} \delta(t-a)f(t)\,dt \tag{1.60}$$

which suggests the formal relation

$$h'(t-a) = \delta(t-a) \tag{1.61}$$

Thus, in a purely formal sense we have extended the definition of derivative to include discontinuous functions. That is, for $t \neq a$, the derivative of $h(t-a)$ is clearly zero, and at $t = a$ the derivative is not defined in the usual sense. We now say that the derivative of a function with a finite jump discontinuity at some point will result in the presence of an impulse function at that point. This generalized form of the derivative can be very useful in practice in that it enables us to treat discontinuous functions as if they were continuous functions; i.e., we do not have to treat them piecewise (see Prob. 10 in Exer. 1.5).

EXERCISES 1.5

1. Show that

$$h(t) = \tfrac{1}{2}[1 + \operatorname{sgn}(t)]$$

where the *signum function* is defined by

$$\operatorname{sgn}(t) = \begin{cases} -1, & t < 0 \\ 1, & t > 0 \end{cases}$$

2. Show that

$$\int_a^b \delta(t-t_0)f(t)\,dt = \begin{cases} f(t_0), & a < t_0 < b \\ 0, & \text{otherwise} \end{cases}$$

3. Show that

$$\int_{-\infty}^{\infty} \delta(at)f(t)\,dt = \frac{1}{|a|}f(0)$$

4. Based on Eq. (1.59), deduce that
(a) $t\delta(t) = 0$.
(b) $\delta(at) = (1/|a|)\,\delta(t)$.
(c) $\delta(t - a) = \delta(a - t)$.

5. Show that if $g(t)$ is a monotonic function for which $g(a) = 0$, then
(a) $\delta[g(t)] = \dfrac{1}{|g'(a)|}\,\delta(t - a)$.
(b) From (a), deduce that $\delta(at - b) = (1/|a|)\delta(t - b/a)$.

6. Show that the signum function defined in Prob. 1 satisfies

$$\frac{d}{dt}\,\mathrm{sgn}(t) = 2\delta(t)$$

7. Using integration by parts, show formally that
(a) $\displaystyle\int_{-\infty}^{\infty} \delta'(t)f(t)\,dt = -f'(0)$
(b) $\displaystyle\int_{-\infty}^{\infty} \delta^{(n)}(t)f(t)\,dt = (-1)^n f^{(n)}(0), \qquad n = 1,2,3,\ldots$

8. If $f(t)$ is a continuous differentiable function, show that it satisfies the product rule

$$\frac{d}{dt}\,[\delta(t)f(t)] = \delta(t)f'(t) + \delta'(t)f(t)$$

9. Show that

$$\delta'(t)f(t) = f(0)\delta'(t) - f'(0)\delta(t)$$

Hint: Use (1.59) and Prob. 8.

10. Let

$$g(t) = f(t) - \sum_{k=1}^{n} a_k\,h(t - t_k)$$

where $f(t)$ is a piecewise continuous function having jump discontinuities of magnitude $a_1, a_2, \ldots, a_n$ at the points $t_1, t_2, \ldots, t_n$. Assuming that $f'(t)$ is defined everywhere except at these discontinuities,
(a) show that $g(t)$ is everywhere continuous and that $g'(t) = f'(t)$ except for a finite number of points.
(b) Deduce that the generalized derivative of the piecewise differentiable function $f(t)$ with finite jumps is the ordinary derivative, where it exists, plus the sum of impulse functions at the discontinuities multiplied by the magnitude of the jumps.

11. Consider the sequence of rectangle functions defined by ($n = 1,2,3,\ldots$)

$$\psi_n(t) = \begin{cases} n/2, & |t| < 1/n \\ 0, & |t| > 1/n \end{cases}$$

(a) Show that for each n the area enclosed by the rectangle is unity, and deduce that

$$\lim_{n \to \infty} \int_{-\infty}^{\infty} \psi_n(t)\, dt = 1$$

(b) More generally, if f is any function continuous at $t = 0$ and everywhere bounded, show that

$$\lim_{n \to \infty} \int_{-\infty}^{\infty} \psi_n(t) f(t)\, dt = f(0)$$

12. Any sequence of continuous and differentiable functions $\psi_1(t)$, $\psi_2(t)$, ..., $\psi_n(t)$, ... satisfying the conditions of Prob. 11(a,b), is called a *delta sequence*. Show that the following sequences are delta sequences:

(a) $\psi_n(t) = n/\pi(1 + n^2t^2), \quad n = 1,2,3,\ldots$

(b) $\psi_n(t) = (n/\sqrt{\pi})e^{-n^2t^2}, \quad n = 1,2,3,\ldots$

2

Fourier Integrals and Fourier Transforms

2.1 *Introduction*

The concept of an infinite series dates back as far as the ancient Greeks such as Archimedes (287–212 B.C.), who summed a geometric series in order to compute the area under a parabolic arc. In the eighteenth century, power series expansions for functions like e^x, $\sin x$, and $\arctan x$ were first published by the Scottish mathematician C. Maclaurin (1698–1746), and British mathematician B. Taylor (1685–1731) generalized this work by providing power series expansions about some point other than $x = 0$.

By the middle of the eighteenth century it became important to study the possibility of representing a given function by infinite series other than power series. D. Bernoulli (1700–1783) showed that the mathematical conditions imposed by physical considerations in solving the vibrating-string problem were formally satisfied by functions represented as infinite series involving sinusoidal functions. In the early 1800s, the French physicist J. Fourier* came across similar representations and announced

* Jean Baptiste Joseph Fourier (1768–1830) is known mainly for his work on the representation of functions by trigonometric series in his studies on the theory of heat conduction. His basic papers, presented to the Academy of Sciences in Paris in 1807 and 1811, were criticized by the referees for a lack of rigor and consequently were not published then. However, when publishing the classic *Théorie analytique de la Chaleur* in 1822, he also incorporated his earlier work almost without change.

in his work on heat conduction that an "arbitrary function" could be expanded in a series of sinusoidal functions. Some of Fourier's work lacked rigor, but nevertheless he provided the first real impetus to the subject now bearing his name. The *Fourier integral* was also first introduced by Fourier as an attempt to generalize his results from finite intervals to infinite intervals. The *Fourier transform*, while appearing in some early writings of A. L. Cauchy (1789–1857) and P. S. de Laplace (1749–1827), also appears in the work of Fourier.

In this chapter we will discuss Fourier integral representations and Fourier transforms, followed by a chapter on applications involving the Fourier transform.

2.2 *Fourier Integral Representations*

An important problem in mathematical analysis is the determination of various representations of a given function f. For example, a particular representation may reveal information about the function that is not as obvious by another representation. In the calculus we are taught that certain functions have *power series representations* of the form

$$f(x) = \sum_{n=0}^{\infty} c_n x^n \tag{2.1}$$

where

$$c_n = f^{(n)}(0)/n!, \qquad n = 0,1,2,\ldots$$

Power series such as this are useful for numerical calculations in addition to various other uses. If the function f is periodic with period $2p$, it may have a *Fourier series representation**

$$f(x) = \frac{1}{2}a_0 + \sum_{n=1}^{\infty}\left(a_n \cos\frac{n\pi x}{p} + b_n \sin\frac{n\pi x}{p}\right) \tag{2.2}$$

where

$$a_n = \frac{1}{p}\int_{-p}^{p} f(t)\cos\frac{n\pi t}{p}\,dt, \qquad n = 0,1,2,\ldots \tag{2.3}$$

and

$$b_n = \frac{1}{p}\int_{-p}^{p} f(t)\sin\frac{n\pi t}{p}\,dt, \qquad n = 1,2,3,\ldots \tag{2.4}$$

---

* For a general discussion of Fourier series, see L. C. Andrews, *Elementary Partial Differential Equations with Boundary Value Problems*, Orlando: Academic Press, 1986.

The theory of Fourier series shows that a periodic function satisfying certain minimal requirements can be represented by the infinite sum of sinusoidal functions given in (2.2). The formal* limit of this representation as the period tends to infinity can be used to introduce the notion of a *Fourier integral representation*. In other words, while periodic functions defined on the entire real axis have Fourier series representations, aperiodic functions similarly defined have Fourier integral representations.

If f and f' are piecewise continuous functions on some interval $[-p, p]$, we say that f is *piecewise smooth*. If f has this property and is periodic with period $2p$, it has the Fourier series representation (2.2)–(2.4). To formally obtain the Fourier integral representation of f from this series as $p \to \infty$, we begin by substituting the integral formulas for a_0, a_n, and b_n given by (2.3) and (2.4) into the Fourier series (2.2). This action leads to

$$f(x) = \frac{1}{2p}\int_{-p}^{p} f(t)dt + \sum_{n=1}^{\infty}\left[\frac{1}{p}\int_{-p}^{p} f(t)\cos\frac{n\pi t}{p}\cos\frac{n\pi x}{p}\,dt + \frac{1}{p}\int_{-p}^{p} f(t)\sin\frac{n\pi t}{p}\sin\frac{n\pi x}{p}\,dt\right]$$

or

$$f(x) = \frac{1}{2p}\int_{-p}^{p} f(t)dt + \frac{1}{p}\int_{-p}^{p} f(t)\sum_{n=1}^{\infty}\cos\frac{n\pi(t-x)}{p}\,dt \tag{2.5}$$

where we have interchanged the order of summation and integration and used the trigonometric identity

$$\cos A\cos B + \sin A\sin B = \cos(A - B)$$

We now wish to examine what happens as we let p tend to infinity. First, we must make the additional requirement that f is *absolutely integrable*, i.e.,

$$\int_{-\infty}^{\infty} |f(t)|dt < \infty \tag{2.6}$$

so that

$$\lim_{p\to\infty}\frac{1}{2p}\int_{-p}^{p} f(t)dt = 0 \tag{2.7}$$

For the remaining infinite sum in (2.5), it is convenient to let $\Delta s = \pi/p$ and then consider the equivalent limit

* By "formal," we mean a procedure that is not mathematically rigorous.

$$f(x) = \lim_{\Delta s \to 0} \frac{1}{\pi} \int_{-\pi/\Delta s}^{\pi/\Delta s} f(t) \sum_{n=1}^{\infty} \cos[n\Delta s(t - x)]\Delta s \, dt \tag{2.8}$$

(Observe that $\Delta s \to 0$ as $p \to \infty$.) When Δs is a small positive number, the points $n\Delta s$ are equally spaced along the s axis. In such a case we may expect the series in (2.8) to approximate the integral

$$\int_0^{\infty} \cos[s(t - x)]ds$$

in the limit as $\Delta s \to 0$. While this does *not* mean that the limit of the series in (2.8) is defined to be the above, we may take, under appropriate conditions on f, that (2.8) tends to the integral form

$$f(x) = \frac{1}{\pi} \int_{-\infty}^{\infty} f(t) \int_0^{\infty} \cos[s(t - x)]ds \; dt \tag{2.9}$$

Upon switching the order of integration, we get the equivalent form

$$f(x) = \frac{1}{\pi} \int_0^{\infty} \int_{-\infty}^{\infty} f(t)\cos[s(t - x)]dt \; ds \tag{2.10}$$

The purely formal procedure we just went through (since the passage to the limit cannot be rigorously justified) has led us to an important result known as *Fourier's integral theorem.** We will state the theorem here but not present its rather lengthy proof until Sec. 2.3.

Theorem 2.1 (*Fourier Integral Theorem*). If f and f' are piecewise continuous functions on every finite interval, and if

$$\int_{-\infty}^{\infty} |f(x)|dx < \infty$$

then

$$f(x) = \frac{1}{\pi} \int_0^{\infty} \int_{-\infty}^{\infty} f(t)\cos[s(t - x)]dt \; ds$$

at points x where f is continuous. If x is a point of discontinuity of f, the above integral converges to the average value $\frac{1}{2}[f(x^+) + f(x^-)]$ of the right-hand and left-hand limits.†

---

* For a rigorous discussion and a precise statement of the conditions under which (2.10) holds, see E. C. Titchmarsh, *Theory of Fourier Integrals*, Oxford: Clarendon Press, 1937.

† Right-hand and left-hand limits are defined, respectively, by $f(x^+) = \lim_{\varepsilon \to 0+} f(x + \varepsilon)$ and $f(x^-) = \lim_{\varepsilon \to 0+} f(x - \varepsilon)$. At points of continuity it follows that $f(x^-) = f(x^+) = f(x)$.

The conditions listed in Theor. 2.1 are only sufficient conditions, not necessary conditions. That is, there exist functions f that have valid integral representations but which do not satisfy the conditions of this theorem. Moreover, the conditions stated in Theor. 2.1 are not the most general set of sufficient conditions that have been established over the years. Nonetheless, these conditions are broad enough to cover most of the functions commonly occurring in practice.

To emphasize the analogy between Fourier series and the Fourier integral theorem, we rewrite (2.10) in the form

$$f(x) = \frac{1}{\pi}\int_0^{\infty}\int_{-\infty}^{\infty} f(t)(\cos st \cos sx + \sin st \sin sx)dt\, ds$$

or equivalently,

$$f(x) = \int_0^{\infty} [A(s)\cos sx + B(s)\sin sx]\, ds \tag{2.11}$$

where

$$A(s) = \frac{1}{\pi}\int_{-\infty}^{\infty} f(t)\cos st\, dt \tag{2.12}$$

and

$$B(s) = \frac{1}{\pi}\int_{-\infty}^{\infty} f(t)\sin st\, dt \tag{2.13}$$

In this setting we refer to (2.11) as the *Fourier integral representation* of the function f with coefficients defined by (2.12) and (2.13). The general theory concerning such representations closely parallels that of Fourier series.

Example 2.1: Find an integral representation of the form (2.11) for the rectangle function $f(x) = h(1 - |x|)$, where h is the Heaviside unit function (see Fig. 2.1).

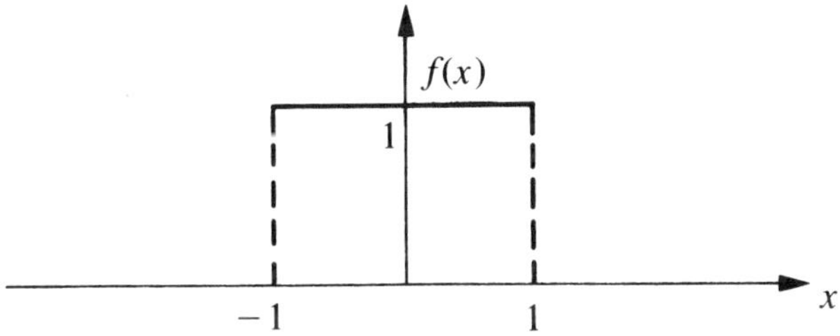

Figure 2.1 Graph of $f(x) = h(1 - |x|)$

Solution: The coefficients $A(s)$ and $B(s)$ are given by

$$A(s) = \frac{1}{\pi}\int_{-\infty}^{\infty} h(1 - |x|)\cos sx\, dx = \frac{1}{\pi}\int_{-1}^{1} \cos sx\, dx = \frac{2 \sin s}{\pi s}$$

and

$$B(s) = \frac{1}{\pi}\int_{-1}^{1} \sin sx\, dx = 0$$

Thus the Fourier integral representation becomes

$$f(x) = \frac{2}{\pi}\int_{0}^{\infty} \left(\frac{\sin s}{s}\right) \cos sx\, ds$$

Since $x = 0$ is a point of continuity of f in Exam. 2.1, we can use the Fourier integral theorem to deduce that

$$f(0) = 1 = \frac{2}{\pi}\int_{0}^{\infty} \frac{\sin s}{s}\, ds$$

which leads to the interesting result*

$$\int_{0}^{\infty} \frac{\sin s}{s}\, ds = \frac{\pi}{2} \tag{2.14}$$

Observe that at $x = \pm 1$ there is a jump discontinuity in the function f given above. At these points the Fourier integral converges to the average value of the left-hand and right-hand limits. Hence, it follows that

$$\frac{2}{\pi}\int_{0}^{\infty} \left(\frac{\sin s}{s}\right) \cos sx\, ds = \begin{cases} 1/2, & x = -1 \\ 1, & -1 < x < 1 \\ 1/2, & x = 1 \\ 0, & \text{otherwise} \end{cases} \tag{2.15}$$

Finally, it may be of interest to plot the "partial integral" of the function f, defined by

$$S_\mu(x) = \frac{2}{\pi}\int_{0}^{\mu} \left(\frac{\sin s}{s}\right) \cos sx\, ds \tag{2.16}$$

to see how it tends to $f(x)$ as $\mu \to \infty$. Recalling the identity

$$2 \sin A \cos B = \sin(A + B) + \sin(A - B)$$

we have

* This, of course, is a standard integral result that can also be derived by the use of complex variable theory. It is an important result that we will refer to on several occasions.

$$\begin{aligned} S_\mu(x) &= \frac{1}{\pi}\int_0^\mu \frac{\sin[s(1+x)]}{s}\,ds + \frac{1}{\pi}\int_0^\mu \frac{\sin[s(1-x)]}{s}\,ds \\ &= \frac{1}{\pi}\int_0^{\mu(1+x)} \frac{\sin t}{t}\,dt + \frac{1}{\pi}\int_0^{\mu(1-x)} \frac{\sin t}{t}\,dt \\ &= \frac{1}{\pi}\{\mathrm{Si}[\mu(1+x)] + \mathrm{Si}[\mu(1-x)]\} \end{aligned} \tag{2.17}$$

where $\mathrm{Si}(z)$ is the *sine integral* defined by

$$\mathrm{Si}(z) = \int_0^z \frac{\sin t}{t}\,dt \tag{2.18}$$

Equation (2.17) is plotted in Fig. 2.2 for values $\mu = 4,16,128$.

2.2.1 *Cosine and Sine Integral Representations*

If the function f is an *even function*, i.e., if $f(-x) = f(x)$, it follows from properties of integrals that

$$A(s) = \frac{1}{\pi}\int_{-\infty}^{\infty} f(x)\cos sx\,dx = \frac{2}{\pi}\int_0^{\infty} f(x)\cos sx\,dx \tag{2.19}$$

and

$$B(s) = \frac{1}{\pi}\int_{-\infty}^{\infty} f(x)\sin sx\,dx = 0 \tag{2.20}$$

from which we deduce

$$f(x) = \int_0^{\infty} A(s)\cos sx\,ds \tag{2.21}$$

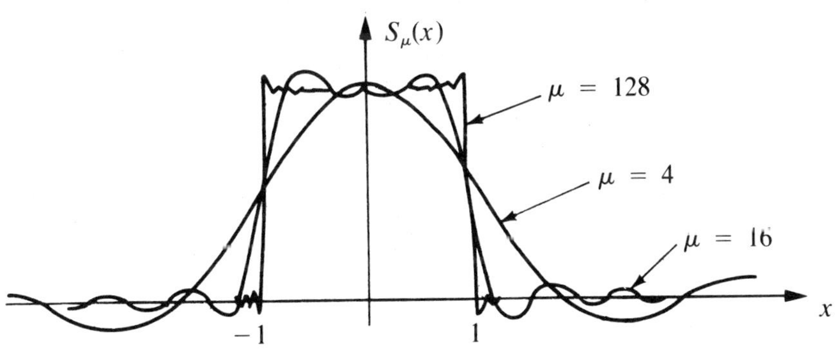

Figure 2.2 The partial integral of $f(x) = h(1 - |x|)$

We refer to (2.21) as a *Fourier cosine integral representation*. In a similar manner, if f is an *odd function*, i.e., $f(-x) = -f(x)$, we obtain the *Fourier sine integral representation*

$$f(x) = \int_0^\infty B(s)\sin sx\, ds \tag{2.22}$$

where $A(s) = 0$ and

$$B(s) = \frac{2}{\pi}\int_0^\infty f(x)\sin sx\, dx \tag{2.23}$$

Finally, if f should be a function defined only on the interval $0 < x < \infty$, we can represent it over this interval by either a Fourier cosine integral or a Fourier sine integral, analogous to the *half-range expansions* of Fourier series. Consider the following example.

Example 2.2: Find a Fourier cosine and Fourier sine integral representation of the function (see Fig. 2.3)

$$f(x) = \begin{cases} \cos x, & 0 < x < \pi/2 \\ 0, & x > \pi/2 \end{cases}$$

Solution: For a cosine integral representation, we compute

$$A(s) = \frac{2}{\pi}\int_0^{\pi/2} \cos x \cos sx\, dx = \frac{2\cos \pi s/2}{\pi(1 - s^2)}$$

and, therefore,

$$f(x) = \frac{2}{\pi}\int_0^\infty \left(\frac{\cos \pi s/2}{1 - s^2}\right) \cos sx\, ds$$

In the same manner

$$B(s) = \frac{2}{\pi}\int_0^{\pi/2} \cos x \sin sx\, dx = \frac{2}{\pi}\left(\frac{s - \sin \pi s/2}{s^2 - 1}\right)$$

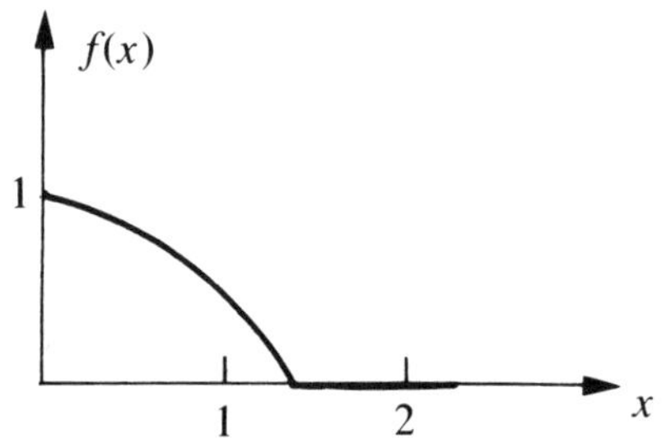

Figure 2.3 Graph of $f(x) = h(\pi/2 - x)\cos x$, $x > 0$

from which we deduce the sine integral representation

$$f(x) = \frac{2}{\pi}\int_0^\infty \left(\frac{s - \sin \pi s/2}{s^2 - 1}\right) \sin sx\, ds$$

EXERCISES 2.2

1. By using the result of Eq. (2.14), show directly that

(a) $\displaystyle\int_0^\infty \frac{\sin s \cos s}{s}\, ds = \frac{\pi}{4}$

(b) $\displaystyle\int_0^\infty \frac{\sin ax}{x}\, dx = \frac{\pi}{2}, \qquad a > 0$

2. If

$$f(x) = \begin{cases} 0, & x < 0 \\ e^{-x}, & x > 0 \end{cases}$$

(a) show that

$$f(x) = \frac{1}{\pi}\int_0^\infty \frac{\cos sx + s \sin sx}{s^2 + 1}\, ds$$

(b) Verify directly that the above integral representation converges to the value 1/2 at $x = 0$.

3. Find an integral representation for

$$f(x) = \begin{cases} 1 - x^2, & |x| < 1 \\ 0, & |x| > 1 \end{cases}$$

and deduce the value of the integral

$$I = \int_0^\infty \left(\frac{\sin x - x \cos x}{x^3}\right) \cos \frac{1}{2}x\, dx$$

4. Use the result of Prob. 3 to deduce that

(a) $\displaystyle\int_0^\infty \left(\frac{1 - \cos x}{x}\right)^2 dx = \frac{\pi}{2}$

(b) $\displaystyle\int_0^\infty \frac{\sin^4 x}{x^2}\, dx = \frac{\pi}{4}$

In Probs. 5–10, obtain the Fourier integral representation of the given function.

5. $f(x) = e^{-|x|}, \qquad -\infty < x < \infty$

6. $f(x) = \begin{cases} -1, & -1 < x < 0 \\ 1, & 0 < x < 1 \\ 0, & \text{otherwise} \end{cases}$

7. $f(x) = \begin{cases} \sin x, & |x| < \pi \\ 0, & |x| > \pi \end{cases}$ **8.** $f(x) = \begin{cases} \sin x, & 0 < x < \pi \\ 0, & \text{otherwise} \end{cases}$

9. $f(x) = \begin{cases} \cos x, & 0 < x < \pi \\ 0, & \text{otherwise} \end{cases}$ **10.** $f(x) = e^{-x^2}, \ -\infty < x < \infty$

11. Show that e^{-kx}, $k > 0$, has the half-range representations

(a) $e^{-kx} = \dfrac{2k}{\pi} \displaystyle\int_0^\infty \frac{\cos sx}{s^2 + k^2}\, ds, \qquad x > 0$

(b) $e^{-kx} = \dfrac{2}{\pi} \displaystyle\int_0^\infty \frac{s \sin sx}{s^2 + k^2}\, ds, \qquad x > 0$

12. Show that $e^{-x}\cos x$ has the half-range representations

(a) $e^{-x}\cos x = \dfrac{2}{\pi} \displaystyle\int_0^\infty \left(\frac{s^2 + 2}{s^4 + 4}\right) \cos sx\, ds, \qquad x > 0$

(b) $e^{-x}\cos x = \dfrac{2}{\pi} \displaystyle\int_0^\infty \frac{s^3 \sin sx}{s^4 + 4}\, ds, \qquad x > 0$

13. Using the results of Prob. 11, show that

$$e^{-x} - e^{-2x} = \frac{6}{\pi} \int_0^\infty \frac{s \sin sx}{(s^2 + 1)(s^2 + 4)}\, ds, \qquad x > 0$$

In Probs. 14–17, obtain the Fourier cosine and Fourier sine integral representations of the given function.

14. $f(x) = \begin{cases} K, & 0 < x < p,\ K > 0 \\ 0, & x > p \end{cases}$

15. $f(x) = \begin{cases} 1 - x, & 0 < x < 1 \\ 0, & x > 1 \end{cases}$

16. $f(x) = \begin{cases} \sin x, & 0 < x < \pi \\ 0, & x > \pi \end{cases}$

17. $f(x) = \begin{cases} x^2, & 0 < x < 1 \\ 0, & x > 1 \end{cases}$

18. Determine whether or not the following functions are absolutely integrable on the entire real axis.

(a) $f(x) = \begin{cases} |1 + x|, & |x| < 1 \\ 0, & |x| > 1 \end{cases}$ (b) $f(x) = \begin{cases} 1, & |x| < 1 \\ 1/x^2, & |x| > 1 \end{cases}$

(c) $f(x) = e^{-|x|}$ (d) $f(x) = \dfrac{\sin x}{x}$

(e) $f(x) = \left(\dfrac{\sin x}{x}\right)^2$

2.3 *Proof of the Fourier Integral Theorem*

In order to provide a more rigorous justification of the Fourier integral theorem (Theor. 2.1) we start with the following central result.

Lemma 2.1 (*Riemann–Lebesgue*). If f is piecewise continuous and absolutely integrable on the entire real axis, then

$$\lim_{\lambda\to\infty}\int_{-\infty}^{\infty} f(t)\cos\lambda t\,dt = \lim_{\lambda\to\infty}\int_{-\infty}^{\infty} f(t)\sin\lambda t\,dt = 0$$

or equivalently,

$$\lim_{\lambda\to\infty}\int_{-\infty}^{\infty} f(t)e^{i\lambda t}\,dt = 0$$

Proof: We will present the proof only for the case when f is continuous and has a bounded derivative f' on the real axis. A slight modification of the proof is required for the case when f has some finite discontinuities.

Using integration by parts over the finite interval $-p \le t \le p$, we get

$$\int_{-p}^{p} f(t)e^{i\lambda t}\,dt = \frac{f(t)e^{i\lambda t}}{i\lambda}\Bigg|_{-p}^{p} - \frac{1}{i\lambda}\int_{-p}^{p} f'(t)e^{i\lambda t}\,dt$$

Clearly, $f(t)e^{i\lambda t}$ is bounded on all finite intervals for any λ. Thus, the first term on the right-hand side vanishes in the limit as λ tends to infinity. Also, because we assume that f' is bounded, it follows that

$$\left|\int_{-p}^{p} f'(t)e^{i\lambda t}\,dt\right| \le \int_{-p}^{p} |f'(t)|dt < \infty$$

for all λ. Hence, we conclude that

$$\lim_{\lambda\to\infty}\int_{-p}^{p} f(t)e^{i\lambda t}\,dt = 0$$

and if we formally allow the limits of integration to become infinite in extent, i.e., $p \to \infty$, we get our intended result. ■

Remark: The assumption that f' is bounded in the proof of Lemma 2.1 is not required for the validity of the theorem. However, by adding this assumption, our proof is much simpler than would otherwise be required.

Lemma 2.2. If f is piecewise smooth and absolutely integrable on the entire real axis, and if x is a point of continuity of f, then

$$\lim_{\lambda\to\infty} \frac{1}{\pi}\int_{-\infty}^{\infty} f(x+t)\frac{\sin \lambda t}{t}\,dt = f(x)$$

Proof: We first note that

$$\begin{aligned}\lim_{\lambda\to\infty} \frac{1}{\pi}\int_{-p}^{p} \frac{\sin \lambda t}{t}\,dt &= \lim_{\lambda\to\infty} \frac{1}{\pi}\int_{-\lambda p}^{\lambda p} \frac{\sin t}{t}\,dt \\ &= \frac{1}{\pi}\int_{-\infty}^{\infty} \frac{\sin t}{t}\,dt \\ &= 1\end{aligned}$$

this last result following from (2.14). Hence, to prove the lemma we wish to show that

$$\lim_{\lambda\to\infty} \frac{1}{\pi}\int_{-p}^{p} \left[\frac{f(x+t)-f(x)}{t}\right]\sin \lambda t\,dt = 0$$

The function $[f(x+t) - f(x)]/t$ is piecewise continuous for all $t \neq 0$, and at $t = 0$, we find that

$$\lim_{t\to 0}\frac{f(x+t)-f(x)}{t} = f'(x)$$

which exists since f is piecewise smooth. Thus, the conditions of Lemma 2.1 are satisfied by this function, and the above integral necessarily vanishes, even in the limit as we formally allow p to become infinite. ■

2.3.1 *Convergence at a Point of Continuity*

We are now prepared to prove our main result, which is Theor. 2.1. We will present the proof only for points of continuity of the function f, leaving the proof for points of discontinuity to the exercises.

By changing the order of integration in the following iterated integral, we obtain

$$\begin{aligned}\frac{1}{\pi}\int_{0}^{\lambda}\int_{-\infty}^{\infty} f(t)\cos[s(t-x)]\,dt\,ds &= \frac{1}{\pi}\int_{-\infty}^{\infty} f(t)\int_{0}^{\lambda}\cos[s(t-x)]\,ds\,dt \\ &= \frac{1}{\pi}\int_{-\infty}^{\infty} f(t)\frac{\sin \lambda(t-x)}{t-x}\,dt \\ &= \frac{1}{\pi}\int_{-\infty}^{\infty} f(x+t)\frac{\sin \lambda t}{t}\,dt\end{aligned}$$

If we now allow $\lambda \to \infty$ and invoke Lemma 2.2, we obtain our desired result

$$\frac{1}{\pi}\int_0^{\infty}\int_{-\infty}^{\infty} f(t)\cos[s(t - x)]\,dt\,ds = f(x) \tag{2.24}$$

where x is a point of continuity of the function f.

EXERCISES 2.3

1. Prove that
   (a) $\lim\limits_{\lambda\to\infty}\displaystyle\int_0^p \frac{\sin\lambda t}{t}\,dt = \frac{\pi}{2}$
   (b) $\lim\limits_{\lambda\to\infty}\displaystyle\int_{-p}^0 \frac{\sin\lambda t}{t}\,dt = \frac{\pi}{2}$
2. Based on Lemma 2.1, show that if f and f' are piecewise continuous on $(0, p)$ and $(-p, 0)$, then
   (a) $\lim\limits_{\lambda\to\infty}\dfrac{2}{\pi}\displaystyle\int_0^p f(x + t)\frac{\sin\lambda t}{t}\,dt = f(x^+)$
   (b) $\lim\limits_{\lambda\to\infty}\dfrac{2}{\pi}\displaystyle\int_{-p}^0 f(x + t)\frac{\sin\lambda t}{t}\,dt = f(x^-)$
3. Prove Theor. 2.1 for points of finite discontinuity of f, i.e., prove that

$$\frac{1}{\pi}\int_0^{\infty}\int_{-\infty}^{\infty} f(t)\cos[s(t - x)]dt\,ds = \frac{1}{2}\Big[f(x^+) + f(x^-)\Big]$$

2.4 *Fourier Transform Pairs*

Fourier's integral theorem (Theor. 2.1) states that

$$f(x) = \frac{1}{\pi}\int_0^{\infty}\int_{-\infty}^{\infty} f(t)\cos[s(t - x)]\,dt\,ds. \tag{2.25}$$

Through the use of Euler's formula, $\cos x = \frac{1}{2}(e^{ix} + e^{-ix})$, we can express (2.25) in terms of complex exponential functions. That is,

$$\begin{aligned} f(x) &= \frac{1}{\pi}\int_0^{\infty}\int_{-\infty}^{\infty} f(t)\cos[s(t - x)]dt\,ds \\ &= \frac{1}{2\pi}\int_0^{\infty}\int_{-\infty}^{\infty} f(t)\left[e^{is(t-x)} + e^{-is(t-x)}\right]dt\,ds \\ &= \frac{1}{2\pi}\int_{-\infty}^{\infty}\int_{-\infty}^{\infty} f(t)e^{is(t-x)}\,dt\,ds \end{aligned}$$

or

$$f(x) = \frac{1}{2\pi}\int_{-\infty}^{\infty} e^{-isx}\int_{-\infty}^{\infty} e^{ist} f(t)\,dt\,ds \tag{2.26}$$

which is the *exponential form of Fourier's integral theorem.*

What we have established by the integral formula (2.26) is the pair of transform formulas*

$$F(s) = \frac{1}{\sqrt{2\pi}}\int_{-\infty}^{\infty} e^{ist} f(t)\,dt \tag{2.27}$$

and

$$f(t) = \frac{1}{\sqrt{2\pi}}\int_{-\infty}^{\infty} e^{-ist} F(s)\,ds \tag{2.28}$$

We define $F(s)$ as the *Fourier transform* of $f(t)$, also written as

$$F(s) = \mathscr{F}\{f(t);s\} \tag{2.29}$$

and $f(t)$ as the *inverse Fourier transform* of $F(s)$, which may be written as

$$f(t) = \mathscr{F}^{-1}\{F(s);t\} \tag{2.30}$$

The location of the constant $1/2\pi$ in the definition of the transform pairs is arbitrarily selected as long as (2.26) is satisfied. For reasons of symmetry we have split the constant between the transform pairs, but in the literature no universal agreement exists on the location of these constants. In some texts, the constant $1/2\pi$ is positioned in front of one of the transform pairs with no constant in front of the other. There is also some variation as to which integral represents the transform and which one represents the inverse transform. In practice, of course, these differences are of little consequence but the user should be aware of them when consulting different reference sources.

As an immediate consequence of (2.27), we observe that

$$|F(s)| \le \frac{1}{\sqrt{2\pi}}\int_{-\infty}^{\infty} |f(t)|\,dt \tag{2.31}$$

Hence, if f is absolutely integrable it follows that its transform function $F(s)$ is *bounded.* A similar argument applied to (2.28) shows that $f(t)$ is also bounded when $F(s)$ is absolutely integrable. Furthermore, the Riemann–Lebesgue Lemma (Lemma 2.1) shows that

$$\lim_{|s|\to\infty} F(s) = 0 \tag{2.32}$$

---

* Unless stated otherwise, we will generally assume that both t and s are real variables.

Since the transform function $F(s)$ associated with absolutely integrable functions that are also piecewise smooth must satisfy this last relation, it immediately rules out certain functions as possible transform functions. For example, sines, cosines, and polynomials do not satisfy this relation. Finally, it is a curious property that although the function $f(t)$ may have certain finite discontinuities, its transform $F(s)$ can be shown to be a *continuous* function. Because of this, the Fourier transform is sometimes called a "smoothing process."

2.4.1 *Fourier Cosine and Sine Transforms*

In Sec. 2.2.1 we found that when the function f is even, the Fourier integral representation of $f(x)$ reduces to

$$\begin{aligned} f(x) &= \int_0^\infty A(s)\cos sx\, ds \\ &= \frac{2}{\pi}\int_0^\infty \cos sx \int_0^\infty f(t)\cos st\, dt\, ds \end{aligned} \tag{2.33}$$

Based on this relation we introduce the *Fourier cosine transform*

$$\mathscr{F}_C\{f(t);s\} = \sqrt{\frac{2}{\pi}}\int_0^\infty f(t)\cos st\, dt = F_C(s), \qquad s > 0 \tag{2.34}$$

and *inverse cosine transform*

$$\mathscr{F}_C^{-1}\{F_C(s);t\} = \sqrt{\frac{2}{\pi}}\int_0^\infty F_C(s)\cos st\, ds = f(t), \qquad t > 0 \tag{2.35}$$

These results are interesting in that they imply the equivalence of the operators $\mathscr{F}_C$ and $\mathscr{F}_C^{-1}$. In other words, the cosine transform and its inverse are exactly the same in functional form.

Similarly, when f is an odd function its Fourier integral representation becomes

$$f(x) = \frac{2}{\pi}\int_0^\infty \sin sx \int_0^\infty f(t)\sin st\, dt\, ds \tag{2.36}$$

which leads to the *Fourier sine transform*

$$\mathscr{F}_S\{f(t);s\} = \sqrt{\frac{2}{\pi}}\int_0^\infty f(t)\sin st\, dt = F_S(s), \qquad s > 0 \tag{2.37}$$

and *inverse sine transform*

$$\mathscr{F}_S^{-1}\{F_S(s);t\} = \sqrt{\frac{2}{\pi}}\int_0^\infty F_S(s)\sin st\, ds = f(t), \qquad t > 0 \tag{2.38}$$

Hence, we see that the Fourier sine transform and its inverse are also exactly the same in functional form.

If the function f is neither even nor odd, but defined only for $t > 0$, then it may have both a cosine transform and a sine transform. Moreover, the even and odd extensions of f will then have exponential Fourier transforms. To see the relations between these various transforms, let us construct the even extension of f by setting

$$f_e(t) = f(|t|), \qquad -\infty < t < \infty \tag{2.39}$$

The Fourier transform of $f_e(t)$ leads to

$$\begin{aligned}\mathscr{F}\{f_e(t);s\} &= \frac{1}{\sqrt{2\pi}}\int_{-\infty}^{\infty} f_e(t)e^{ist}\,dt \\ &= \frac{1}{\sqrt{2\pi}}\int_{-\infty}^{\infty} f_e(t)\cos st\,dt + i\frac{1}{\sqrt{2\pi}}\int_{-\infty}^{\infty} f_e(t)\sin st\,dt \\ &= \sqrt{\frac{2}{\pi}}\int_0^{\infty} f(t)\cos st\,dt\end{aligned}$$

from which we deduce

$$\mathscr{F}\{f_e(t);s\} = \mathscr{F}_C\{f(t);s\}, \qquad -\infty < s < \infty \tag{2.40}$$

Based on (2.40), it is clear that the Fourier transform and cosine transform of an even function give identical results. In particular, their transforms are even functions of s (see Prob. 20 in Exer. 2.4). The odd extension of f is constructed by setting

$$f_0(t) = f(|t|)\text{sgn}(t), \qquad -\infty < t < \infty \tag{2.41}$$

where the *signum function* is defined by

$$\text{sgn}(t) = \begin{cases} -1, & t < 0 \\ 1, & t > 0 \end{cases} \tag{2.42}$$

In this case, we find

$$\begin{aligned}\mathscr{F}\{f_0(t);s\} &= \frac{1}{\sqrt{2\pi}}\int_{-\infty}^{\infty} f_0(t)e^{ist}\,dt \\ &= \frac{1}{\sqrt{2\pi}}\int_{-\infty}^{\infty} f_0(t)\cos st\,dt + i\frac{1}{\sqrt{2\pi}}\int_{-\infty}^{\infty} f_0(t)\sin st\,dt \\ &= i\sqrt{\frac{2}{\pi}}\int_0^{\infty} f(t)\sin st\,dt\end{aligned}$$

Because the Fourier transform of an odd function is also an odd function (see Prob. 20 in Exer. 2.4), we make the conclusion that the Fourier transform and sine transform are related by

$$\mathscr{F}\{f_0(t);s\} = i\mathscr{F}_S\{f(t);|s|\}\text{sgn}(s), \qquad -\infty < s < \infty \tag{2.43}$$

The practical use of (2.40) is that if we want to evaluate the Fourier transform of an even function, we can do so by simply calculating its cosine transform. If the function we wish to transform is odd, we first can find its sine transform and then use (2.43). These observations are very useful when using tables to find transforms, since most of the known transforms are either cosine or sine transforms. A short table of transforms is presented in Appendix B.

2.4.2 *Evaluating Transforms of Elementary Functions*

As already pointed out, many elementary functions like sines, cosines, polynomials, and in general any periodic function, do not have Fourier transforms (at least in the usual sense) because they are not absolutely integrable. A special class of elementary functions that do have Fourier transforms and can be calculated by basic methods are those involving exponential functions. Several such transforms are related to the integrals

$$I = \int_0^\infty e^{-at}\cos st\,dt = \frac{a}{s^2 + a^2}, \qquad a > 0 \tag{2.44}$$

and

$$J = \int_0^\infty e^{-at}\sin st\,dt = \frac{s}{s^2 + a^2}, \qquad a > 0 \tag{2.45}$$

One way to verify these integral formulas is to use integration by parts to obtain the relations

$$I = \frac{1}{a} - \frac{s}{a}\int_0^\infty e^{-at}\sin st\,dt = \frac{1}{a} - \frac{s}{a}J$$

and

$$J = (s/a)I$$

Solving these last two equations simultaneously for I and J yields the results given by (2.44) and (2.45).

Example 2.3: Find the Fourier transform of $e^{-a|t|}$, $a > 0$.

Solution: Because the function is even, we can use (2.40) to write

$$\begin{aligned}\mathscr{F}\{e^{-a|t|};s\} &= \mathscr{F}_C\{e^{-at};s\}\\ &= \sqrt{\frac{2}{\pi}}\int_0^\infty e^{-at}\cos st\,dt\end{aligned}$$

or, using (2.44), we see that

$$\mathscr{F}\{e^{-a|t|};s\} = \sqrt{\frac{2}{\pi}}\frac{a}{s^2 + a^2}, \qquad a > 0$$

Because transform relations occur in pairs as given by Eqs. (2.27) and (2.28), it follows that once we have established one transform or integral relation, the other one is automatically known. For instance, based on the result of Exam. 2.3, we have

$$\mathscr{F}\{e^{-a|t|};s\} = \frac{1}{\sqrt{2\pi}}\int_{-\infty}^{\infty} e^{ist}e^{-a|t|}\,dt = \sqrt{\frac{2}{\pi}}\frac{a}{s^2 + a^2}, \qquad a > 0 \tag{2.46}$$

and thus it immediately follows that

$$\mathscr{F}^{-1}\left\{\frac{1}{s^2 + a^2};t\right\} = \frac{1}{\sqrt{2\pi}}\int_{-\infty}^{\infty} \frac{e^{-ist}}{s^2 + a^2}\,ds = \frac{1}{a}\sqrt{\frac{\pi}{2}}\,e^{-a|t|}, \qquad a > 0 \tag{2.47}$$

Moreover, by interchanging the roles of t and s in (2.47), and taking the complex conjugate of the resulting expression (which is real in this example since we are dealing with even functions), we now deduce the additional Fourier transform

$$\mathscr{F}\left\{\frac{1}{t^2 + a^2};s\right\} = \frac{1}{a}\sqrt{\frac{\pi}{2}}\,e^{-a|s|}, \qquad a > 0 \tag{2.48}$$

In this fashion we see that the evaluation of a single Fourier transform has the effect of giving us two transform relations from each integral.

It can be shown that both (2.44) and (2.45) are uniformly converging integrals on any closed intervals for which a is positive and all closed intervals involving s. Related integrals which can be formally derived by differentiating or integrating (2.44) and (2.45) with respect to either parameter, a or s, can also be shown to converge uniformly. This means that we can formally differentiate or integrate both sides of (2.44) and (2.45) to produce new integral relations which can then be related to other integral transforms. Consider the following examples.

Example 2.4: Find the Fourier sine and cosine transforms of te^{-at}, $a > 0$.

Solution: Formal differentiation of both sides of (2.45), first with respect to a and then with respect to s, gives us, respectively,

$$-\int_0^{\infty} te^{-at}\sin st\,dt = -2as/(s^2 + a^2)^2$$

and

$$\int_0^\infty te^{-at}\cos st\, dt = (a^2 - s^2)/(s^2 + a^2)^2$$

Thus, we deduce that

$$\mathscr{F}_S\{te^{-at};s\} = \sqrt{\frac{2}{\pi}}\frac{2as}{(s^2 + a^2)^2}, \qquad a > 0$$

and

$$\mathscr{F}_C\{te^{-at};s\} = \sqrt{\frac{2}{\pi}}\frac{a^2 - s^2}{(s^2 + a^2)^2}, \qquad a > 0$$

Example 2.5: Find the Fourier sine transform of $(1/t)e^{-at}$, $a > 0$.

Solution: We begin by integrating both sides of (2.45) with respect to the parameter a from a to ∞, which leads to

$$\begin{aligned}\int_0^\infty \frac{1}{t} e^{-at}\sin st\, dt &= \int_a^\infty \frac{s}{s^2 + a^2}\, da \\ &= \frac{\pi}{2} - \tan^{-1}\frac{a}{s} \\ &= \tan^{-1}\frac{s}{a}\end{aligned}$$

Thus, it follows that

$$\mathscr{F}_S\left\{\frac{1}{t}e^{-at};s\right\} = \sqrt{\frac{2}{\pi}}\tan^{-1}\frac{s}{a}, \qquad a > 0$$

If we allow $a \to 0^+$ in the result of Exam. 2.5, we find

$$\mathscr{F}_S\{1/t;s\} = \sqrt{\frac{\pi}{2}} \tag{2.49}$$

This result is only a formal result since neither $1/t$ nor $\sqrt{\pi/2}$ satisfy the conditions of the Fourier integral theorem. Nonetheless, it can be useful to treat (2.49) as a limiting case of the transform relation given in Exam. 2.5. Using (2.43), we obtain the similar relation*

$$\mathscr{F}\{1/t;s\} = i\sqrt{\pi/2}\,\mathrm{sgn}(s) \tag{2.50}$$

* Formal results like (2.49) and (2.50) are discussed in more detail in Sec. 2.8.

Finally, as a bonus we see that (2.50) provides a generalization of (2.14), which is

$$\int_0^\infty \frac{\sin st}{t}\, dt = \frac{\pi}{2}\,\mathrm{sgn}(s) \tag{2.51}$$

Example 2.6: Find the Fourier transform of $e^{-a^2t^2}$, $a > 0$.

Solution: From definition, we have

$$\mathscr{F}\{e^{-a^2t^2};s\} = \frac{1}{\sqrt{2\pi}}\int_{-\infty}^{\infty} e^{ist-a^2t^2}\, dt$$

By writing

$$a^2t^2 - ist = (at - is/2a)^2 + s^2/4a^2$$

we find

$$\begin{aligned}\mathscr{F}\{e^{-a^2t^2};s\} &= \frac{1}{\sqrt{2\pi}}\, e^{-s^2/4a^2}\int_{-\infty}^{\infty} e^{-(at-is/a)^2}\, dt \\ &= \frac{1}{a\sqrt{2\pi}}\, e^{-s^2/4a^2}\int_{-\infty}^{\infty} e^{-x^2}\, dx \\ &= \frac{1}{a\sqrt{2\pi}}\, e^{-s^2/4a^2}\,\Gamma(1/2)\end{aligned}$$

where we have made the change of variable $x = at - is/a$ and used properties of the gamma function. Simplifying this last result leads to

$$\mathscr{F}\{e^{-a^2t^2};s\} = \frac{1}{a\sqrt{2}}\, e^{-s^2/4a^2}, \qquad a > 0$$

By setting $a = 1/\sqrt{2}$ in the result of Exam. 2.6, we obtain the interesting relation

$$\mathscr{F}\{e^{-t^2/2};s\} = e^{-s^2/2} \tag{2.52}$$

which says that the function $e^{-t^2/2}$ is *self-reciprocal*, i.e., it is its own transform.

EXERCISES 2.4

1. Given the following functions, develop the even and odd extensions, $f_e(t)$ and $f_0(t)$, respectively:
(a) $f(t) = e^{-at}$
(b) $f(t) = e^{-t^2}\sin t$
(c) $f(t) = (1 + t)e^{-at}$
(d) $f(t) = e^{-t} + \sinh t$

2. Determine the Fourier transform of the function $f(t)$, given that

$$\mathscr{F}_S\{f(t);s\} = \sqrt{\frac{2}{\pi}}\,(2e^{-bs} - 1), \qquad b > 0$$

In Probs. 3–10, determine the Fourier transform of each function.

3. $f(t) = \begin{cases} 0, & t < 0 \\ e^{-at}, & t > 0,\ a > 0 \end{cases}$

4. $f(t) = \begin{cases} 1, & 0 < t < 1 \\ 0, & \text{otherwise} \end{cases}$

5. $f(t) = \begin{cases} 1, & |t| < b \\ 0, & \text{otherwise} \end{cases}$

6. $f(t) = \begin{cases} \sin t, & 0 < t < \pi \\ 0, & \text{otherwise} \end{cases}$

7. $f(t) = \dfrac{1}{5t^2 + 1}$

8. $f(t) = \dfrac{t}{t^2 + 7}$

9. $f(t) = \dfrac{t}{(t^2 + 4)^2}$

10. $f(t) = \dfrac{1 - t^2}{(t^2 + 4)^2}$

11. Show that

$$\mathscr{F}\left\{\frac{\sin mt}{t};s\right\} = \sqrt{\frac{\pi}{2}}\,h(m - |s|), \qquad m > 0$$

12. Use the results of Exam. 2.4 to determine
(a) $\mathscr{F}\{te^{-a|t|};s\}, \quad a > 0$
(b) $\mathscr{F}\{|t|e^{-a|t|};s\}, \quad a > 0$

13. Given the triangle function $f(t) = (1 - |t|)h(1 - |t|)$, show that

(a) $\mathscr{F}\{f(t);s\} = 2\sqrt{\dfrac{2}{\pi}}\,\dfrac{\sin^2(s/2)}{s^2}$

(b) From (a), deduce that

$$\int_{-\infty}^{\infty} \left(\frac{\sin x}{x}\right)^2 dx = \pi$$

14. Letting $f(t) = 1/\sqrt{t}$ in the sine and cosine forms of Fourier's integral theorem given by (2.33) and (2.36), respectively, show that

(a) $\displaystyle\int_0^{\infty} \frac{\cos t}{\sqrt{t}}\,dt = \int_0^{\infty} \frac{\sin t}{\sqrt{t}}\,dt = \sqrt{\frac{\pi}{2}}$

(b) From (a), deduce that

$$\mathscr{F}_S\left\{\frac{1}{\sqrt{t}};s\right\} = \mathscr{F}_C\left\{\frac{1}{\sqrt{t}};s\right\} = \frac{1}{\sqrt{s}}$$

15. Use the result of Exam. 2.6 with $a = (1 - i)/2$ to derive

(a) $\mathscr{F}\{\cos(t^2/2);s\} = \dfrac{1}{\sqrt{2}}\,[\cos(s^2/2) + \sin(s^2/2)]$

(b) $\mathscr{F}\{\sin(t^2/2);s\} = \dfrac{1}{\sqrt{2}}[\cos(\tfrac{1}{2}s^2) - \sin(s^2/2)]$

(c) $\mathscr{F}_S\left\{\dfrac{1}{t}\cos(t^2/2);s\right\} = \sqrt{\dfrac{\pi}{2}}\,[C(s/\sqrt{\pi}) + S(s/\sqrt{\pi})]$*

16. Show that

(a) $\mathscr{F}_C\{e^{-at}\cos at;s\} = \sqrt{\dfrac{2}{\pi}}\,\dfrac{as^2 + 2a^3}{s^4 + 4a^4}, \qquad a > 0$

(b) $\mathscr{F}_C\{e^{-at}\sin at;s\} = \sqrt{\dfrac{2}{\pi}}\,\dfrac{2a^3 - as^2}{s^4 + 4a^4}, \qquad a > 0$

17. From the results of Prob. 16, deduce that

$$\mathscr{F}^{-1}\left\{\frac{1}{s^4 + k^4};t\right\} = \frac{\sqrt{\pi}}{2k^3}\,e^{-k|t|/\sqrt{2}}\left[\cos(kt/\sqrt{2}) + \sin(k|t|/\sqrt{2})\right], \qquad k > 0$$

18. Evaluate the sine transforms

(a) $\mathscr{F}_S\{e^{-at};s\}, \qquad a > 0$
(b) $\mathscr{F}_S\{e^{-at}\cos at;s\}, \qquad a > 0$
(c) $\mathscr{F}_S\{e^{-at}\sin at;s\}, \qquad a > 0$

19. From the results of Prob. 18, evaluate

(a) $\mathscr{F}_S\left\{\dfrac{t^3}{t^4 + k^4};s\right\}$ (b) $\mathscr{F}_S\left\{\dfrac{t}{t^4 + k^4};s\right\}$

(c) $\mathscr{F}_S\left\{\dfrac{t^3}{(t^4 + k^4)^2};s\right\}$ (d) $\mathscr{F}_S\left\{\dfrac{t}{(t^4 + k^4)^2};s\right\}$

20. Prove the following properties of the Fourier transform for real functions $f(t)$:

(a) If $f(t)$ is even, then $F(s)$ is real and even.
(b) If $f(t)$ is odd, then $F(s)$ is imaginary and odd.
(c) If $f(t)$ is neither even nor odd, then $F(s)$ has an even real part and an odd imaginary part.

2.5 *Properties of the Fourier Transform*

The calculation of integral transforms is often tedious and quite complex in some instances. However, once we have derived the transforms of some standard functions, we can deduce the transforms of many other

---

* $C(x)$ and $S(x)$ are the Fresnel integrals (see Sec. 1.3.2).

functions in a simple way through the use of certain *operational properties* associated with the transform. These operational properties are basically consequences of the properties of integrals.

If $f(t)$ satisfies the conditions of the Fourier integral theorem, its Fourier transform $F(s)$ is *uniquely* determined by the integral

$$F(s) = \frac{1}{\sqrt{2\pi}} \int_{-\infty}^{\infty} e^{ist} f(t)\, dt \tag{2.53}$$

Thus, there is only one transform function $F(s)$ associated with each function $f(t)$. However, if $f(t)$ and $g(t)$ are two functions that are identical everywhere except at certain isolated points, then both $f(t)$ and $g(t)$ will have the same transform, say $F(s)$. This means that the inverse transform of $F(s)$ can be either $f(t)$ or $g(t)$. Of course, the distinction between functions that differ only at isolated points is mostly of academic interest and has little effect in practical applications. If we agree to define a function at a point of finite discontinuity as the average of its left-hand and right-hand limits, then $f(t)$ is uniquely related to $F(s)$ by the inverse transform relation

$$\frac{1}{2}[f(t^+) + f(t^-)] = \frac{1}{\sqrt{2\pi}} \int_{-\infty}^{\infty} e^{-ist} F(s)\, ds \tag{2.54}$$

Another important property of the Fourier transform and inverse Fourier transform is the *linearity property*.

Theorem 2.2 (*Linearity property*). If $F(s)$ and $G(s)$ are the Fourier transforms, respectively, of $f(t)$ and $g(t)$, then for any constants C_1 and C_2, it follows that

$$\mathscr{F}\{C_1 f(t) + C_2 g(t);s\} = C_1 F(s) + C_2 G(s)$$
$$\mathscr{F}^{-1}\{C_1 F(s) + C_2 G(s);t\} = C_1 f(t) + C_2 g(t)$$

Proof: Directly from the defining integral, we have

$$\begin{aligned}\mathscr{F}\{C_1 f(t) + C_2 g(t);s\} &= \frac{1}{\sqrt{2\pi}} \int_{-\infty}^{\infty} e^{ist} [C_1 f(t) + C_2 g(t)]\, dt \\ &= \frac{C_1}{\sqrt{2\pi}} \int_{-\infty}^{\infty} e^{ist} f(t)\, dt + \frac{C_2}{\sqrt{2\pi}} \int_{-\infty}^{\infty} e^{ist} g(t)\, dt \\ &= C_1 F(s) + C_2 G(s)\end{aligned}$$

A similar argument proves the inverse transform linearity property. ■

If $a > 0$, then

$$\mathscr{F}\{f(at);s\} = \frac{1}{\sqrt{2\pi}}\int_{-\infty}^{\infty} e^{ist} f(at)\, dt$$

$$= \frac{1}{a\sqrt{2\pi}}\int_{-\infty}^{\infty} e^{iu(s/a)} f(u)\, du$$

where we have set $u = at$. Thus, if $F(s)$ is the Fourier transform of $f(t)$, we have just shown that

$$\mathscr{F}\{f(at);s\} = (1/a)F(s/a), \qquad a > 0 \tag{2.55a}$$

Similarly, if $a < 0$ it follows that

$$\mathscr{F}\{f(at);s\} = -(1/a)F(s/a), \qquad a < 0 \tag{2.55b}$$

so that in general we have the *scaling property*

$$\mathscr{F}\{f(at);s\} = (1/|a|)F(s/a), \qquad a \neq 0 \tag{2.56}$$

Example 2.7: Given that the Fourier transform of $f(t) = (\sin b\sqrt{1+t^2})/\sqrt{1+t^2}$ is $F(s) = \sqrt{\pi/2}J_0(\sqrt{b^2-s^2})$, $|s| < b$ and $F(s) = 0$, $|s| > b$, determine the Fourier transform of $(\sin b\sqrt{a^2+t^2})/\sqrt{a^2+t^2}$, $b > 0$.

Solution: We first observe that

$$\frac{\sin b\sqrt{a^2+t^2}}{\sqrt{a^2+t^2}} = \frac{\sin b|a|\sqrt{1+(t/a)^2}}{|a|\sqrt{1+(t/a)^2}}$$

and then using (2.56), we see that

$$\mathscr{F}\left\{\frac{\sin b\sqrt{a^2+t^2}}{\sqrt{a^2+t^2}};s\right\} = \begin{cases} \sqrt{\pi/2}\, J_0\,(\sqrt{b^2a^2-a^2s^2}), & |as| < b|a| \\ 0, & |as| > b|a| \end{cases}$$

$$= \begin{cases} \sqrt{\pi/2}\, J_0(|a|\sqrt{b^2-s^2}), & |s| < b \\ 0, & |s| > b \end{cases}$$

2.5.1 *Shift Properties*

Multiplication of either $f(t)$ or $F(s)$ by a complex exponential causes a shift in the transform variable upon completing the integration of the transform or inverse transform. More precisely, we have the following theorem.

Theorem 2.3 (*Shifting property*). If $f(t)$ and $F(s)$ are Fourier transform pairs, then

(a) $\mathscr{F}\{e^{iat}f(t);s\} = F(s+a)$
(b) $\mathscr{F}\{f(t-a);s\} = e^{ias}\, F(s)$

Proof: From definition,

$$\begin{aligned}\mathscr{F}\{e^{iat}f(t);s\} &= \frac{1}{\sqrt{2\pi}}\int_{-\infty}^{\infty} e^{ist}\, e^{iat}\, f(t)\, dt \\ &= \frac{1}{\sqrt{2\pi}}\int_{-\infty}^{\infty} e^{i(s+a)t}\, f(t)\, dt \\ &= F(s + a)\end{aligned}$$

In the same fashion,

$$\begin{aligned}\mathscr{F}^{-1}\{e^{ias}F(s);t\} &= \frac{1}{\sqrt{2\pi}}\int_{-\infty}^{\infty} e^{-its}\, e^{ias}\, F(s)\, ds \\ &= \frac{1}{\sqrt{2\pi}}\int_{-\infty}^{\infty} e^{-i(t-a)s}\, F(s)\, ds \\ &= f(t - a)\end{aligned}$$

from which we deduce

$$\mathscr{F}\{f(t - a);s\} = e^{ias}\, F(s) \qquad \blacksquare$$

Example 2.8: Find the Fourier inverse transform of $1/(s^2 + ias + b)$, $b > 0$.

Solution: By completing the square, we have

$$\frac{1}{s^2 + ias + b} = \frac{1}{[s + (ia/2)]^2 + [(a^2/4) + b]}$$

Then, using Theor. 2.3,*

$$\begin{aligned}\mathscr{F}^{-1}\left\{\frac{1}{s^2 + ias + b};t\right\} &= e^{-at/2}\mathscr{F}^{-1}\left\{\frac{1}{s^2 + [(a^2/4) + b]};t\right\} \\ &= \sqrt{\frac{2\pi}{a^2 + 4b}}\exp\left[-\frac{1}{2}(at + \sqrt{a^2 + 4b}|t|)\right]\end{aligned}$$

the last step of which follows from Eq. (2.47).

2.5.2 *Transforms of Derivatives and Derivatives of Transforms*

In applications involving differential equations it is important to know how the Fourier transform behaves on derivatives of a function. If f is continuous everywhere and f' is piecewise smooth, and both f and f' are absolutely integrable, then

---

* Note that Theor. 2.3 implies that $\mathscr{F}^{-1}\{F(s + a);t\} = e^{iat}\mathscr{F}^{-1}\{F(s);t\}$.

$$\mathcal{F}\{f'(t);s\} = \frac{1}{\sqrt{2\pi}} \int_{-\infty}^{\infty} e^{ist} f'(t)\, dt$$

$$= \frac{1}{\sqrt{2\pi}} f(t)e^{ist} \Big|_{-\infty}^{\infty} - \frac{is}{\sqrt{2\pi}} \int_{-\infty}^{\infty} e^{ist} f(t)\, dt$$

where we have employed an integration by parts. Now if f also satisfies

$$\lim_{|t|\to\infty} f(t) = 0$$

we then obtain

$$\mathcal{F}\{f'(t);s\} = -isF(s) \tag{2.57}$$

where $F(s)$ is the Fourier transform of $f(t)$. By repeated application of (2.57), we can prove the following more general result.

Theorem 2.4 (*Differentiation property*). If $f, f', \ldots, f^{(n-1)}$ are continuous everywhere and absolutely integrable, $f^{(n)}$ is piecewise smooth and absolutely integrable, and

$$\lim_{|t|\to\infty} f(t) = \lim_{|t|\to\infty} f'(t) = \ldots = \lim_{|t|\to\infty} f^{(n-1)}(t) = 0$$

then

$$\mathcal{F}\{f^{(n)}(t);s\} = (-is)^n F(s), \qquad n = 1,2,3,\ldots$$

where $F(s)$ is the Fourier transform of $f(t)$.

Remark: If $f(t)$ has a finite jump discontinuity at $t = a$, then $f'(t)$ contains an impulse at $t = a$ (see Sec. 1.5.2). In this case the Fourier transform of $f'(t)$ must also contain the Fourier transform of the impulse function. Such concepts, which involve the notion of generalized functions, will be discussed in Sec. 2.8.

In the case of the cosine and sine transforms, the above results are somewhat different. For example, in the case of the cosine transform we use integration by parts to obtain

$$\mathcal{F}_C\{f'(t);s\} = \sqrt{2/\pi} \int_0^{\infty} f'(t)\cos st\, dt$$

$$= -\sqrt{2/\pi}\, f(0) + s\sqrt{2/\pi} \int_0^{\infty} f(t)\sin st\, dt$$

from which we deduce

$$\mathcal{F}_C\{f'(t);s\} = sF_S(s) - \sqrt{2/\pi}\, f(0) \tag{2.58}$$

Similarly, it can be shown that

$$\mathscr{F}_S\{f'(t);s\} = -sF_C(s) \tag{2.59}$$

(see Prob. 9 in Exer. 2.5). For second derivatives, we are led to the relations

$$\mathscr{F}_C\{f''(t);s\} = -s^2F_C(s) - \sqrt{2/\pi}\,f'(0) \tag{2.60}$$

and

$$\mathscr{F}_S\{f''(t);s\} = -s^2F_S(s) + \sqrt{2/\pi}\,sf(0) \tag{2.61}$$

the verification of which is left to the exercises (see Prob. 10 in Exer. 2.5). These last two formulas give us some indication of which transform — cosine or sine — to use in a particular application. That is, in any problem in which $f(0)$ is known but $f'(0)$ is not known, we should use the Fourier sine transform of $f''(t)$. In the same way, if $f'(0)$ is known rather than $f(0)$, the Fourier cosine transform should be used.

If the transform of $f(t)$ is $F(s)$, then the transform of $t^m f(t)$, $m = 1,2,3,\ldots$, can be found by repeated differentiation of $F(s)$. To see this, let us start with the Fourier integral

$$F(s) = \frac{1}{\sqrt{2\pi}}\int_{-\infty}^{\infty} e^{ist} f(t)\,dt \tag{2.62}$$

and formally differentiate both sides with respect to s. This action yields

$$F'(s) = \frac{1}{\sqrt{2\pi}}\int_{-\infty}^{\infty} e^{ist}\,[itf(t)]\,dt$$

and thus we conclude that

$$\mathscr{F}\{tf(t);s)\} = -iF'(s) \tag{2.63}$$

Of course, the validity of (2.63) requires that the transform of $tf(t)$ exist. Continued differentiation of (2.62) with respect to s leads to

$$F^{(m)}(s) = \frac{1}{\sqrt{2\pi}}\int_{-\infty}^{\infty} e^{ist}\,[(it)^m f(t)]\,dt, \qquad m = 1,2,3,\ldots \tag{2.64}$$

which we now formulate as a theorem.

Theorem 2.5. If f is absolutely integrable and piecewise smooth, and if $t^m f(t)$ has a Fourier transform, then

$$\mathscr{F}\{t^m f(t);s\} = (-i)^m\,F^{(m)}(s), \qquad m = 1,2,3,\ldots$$

where $F(s)$ is the Fourier transform of $f(t)$.

By combining Theors. 2.4 and 2.5, we arrive at the result

$$\mathscr{F}\{t^m f^{(n)}(t);s\} = (-i)^{m+n} \frac{d^m}{ds^m} [s^n F(s)], \qquad m,n = 1,2,3,\ldots \tag{2.65}$$

Example 2.9: Find the Fourier transform of $te^{t-t^2/2}$.

Solution: Recalling Eq. (2.52), we have

$$\mathscr{F}\{e^{-t^2/2};s\} = e^{-s^2/2}$$

Then, using (2.63), it immediately follows that

$$\mathscr{F}\{te^{-t^2/2};s\} = -i(-s)e^{-s^2/2} = is\, e^{-s^2/2}$$

Finally, recalling Theor. 2.3, we deduce that

$$\begin{aligned}\mathscr{F}\{te^{t-t^2/2};s\} &= i(s - i)e^{\;(s-i)^2/2} \\ &= (1 + is)e^{is}\, e^{-(s^2-1)/2}\end{aligned}$$

EXERCISES 2.5

1. If $f(t)$ is a real function with transform $F(s)$, show that the complex conjugate of $F(s)$ satisfies

$$\overline{F(s)} = \mathscr{F}\{f(t);-s\}$$

2. If $f(t)$ is a complex function with transform $F(s)$, show that the complex conjugate of $F(s)$ satisfies

$$\overline{F(s)} = \mathscr{F}\{\overline{f(-t)};s\}$$

3. If $F(s)$ is the Fourier transform of $f(t)$, show that

$$\mathscr{F}\{e^{ibt/a}f(t/a);s\} = aF(as + b), \qquad a > 0$$

4. Show that

(a) $\mathscr{F}_C\{f(at);s\} = (1/a)F_C(s/a), \qquad a > 0$

(b) $\mathscr{F}_S\{f(at);s\} = (1/a)F_S(s/a), \qquad a > 0$

5. Show that

(a) $\mathscr{F}_C\{f(t)\cos at;s\} = \frac{1}{2}[F_C(s + a) + F_C(s - a)]$

(b) $\mathscr{F}_S\{f(t)\cos at;s\} = \frac{1}{2}[F_S(s + a) + F_S(s - a)]$

6. Show that

(a) $\mathscr{F}_C\{f(t)\sin at;s\} = \frac{1}{2}[F_S(s + a) - F_S(s - a)]$

(b) $\mathscr{F}_S\{f(t)\sin at;s\} = \frac{1}{2}[F_C(s - a) - F_C(s + a)]$

7. Use the results of Probs. 5 and 6 to show that

(a) $\mathscr{F}_C\{e^{-at}\cos at;s\} = \sqrt{\dfrac{2}{\pi}}\dfrac{as^2 + 2a^3}{s^4 + 4a^4}, \qquad a > 0$

(b) $\mathscr{F}_C\{e^{-at}\sin at;s\} = \sqrt{\dfrac{2}{\pi}}\dfrac{2a^3 - as^2}{s^4 + 4a^4}, \qquad a > 0$

8. Use the results of Probs. 5 and 6 to evaluate
(a) $\mathscr{F}_S\{e^{-at}\cos at;s\}, \quad a > 0$
(b) $\mathscr{F}_S\{e^{-at}\sin at;s\}, \quad a > 0$

9. Verify that the Fourier sine transform satisfies the relation

$$\mathscr{F}_S\{f'(t);s\} = -sF_C(s)$$

10. Derive the transform relations
(a) $\mathscr{F}_C\{f''(t);s\} = -s^2F_C(s) - \sqrt{2/\pi}f'(0)$
(b) $\mathscr{F}_S\{f''(t);s\} = -s^2F_S(s) + \sqrt{2/\pi}sf(0)$

In Probs. 11–15, evaluate the Fourier transform of the given function using known transforms and appropriate properties of the transform.

11. $f(t) = (1 - t)e^{-|t|}$

12. $f(t) = e^{bt-t^2}$

13. $f(t) = t^2e^{-t^2/2}$

14. $f(t) = e^{-t^2/2}\cos 2t$

15. $f(t) = t\cos t^2$

In Probs. 16–18, evaluate the Fourier inverse transform of the given function using known transform relations and appropriate properties of the transform.

16. $F(s) = \dfrac{1}{s^2 + 4s + 7}$

17. $F(s) = \dfrac{e^{-2is}}{s^2 + 4s + 7}$

18. $F(s) = \tan^{-1}(s/a)\operatorname{sgn}(s)$

Hint: Examine $F'(s)$.

19. Show that, under appropriate assumptions on f and its derivatives,

$$\mathscr{F}_C\{f^{(4)}(t);s\} = s^4F_C(s) + \sqrt{2/\pi}\,[s^2f'(0) - f^{(3)}(0)]$$

20. Show that, under appropriate assumptions on f and its derivatives,

$$\mathscr{F}_S\{f^{(4)}(t);s\} = s^4F_S(s) - \sqrt{2/\pi}\,[s^3f(0) - sf''(0)]$$

2.6 *Transforms of More Complicated Functions*

When the functions involved in a Fourier transform or inverse transform are of a more complicated nature, we usually must resort to techniques other than the standard integration methods of calculus. Sometimes it is useful to represent part of the integrand in a power series and perform termwise integration on the resulting expression.* In certain cases the resulting integrated series can be summed to yield the transform we are seeking. Other useful techniques are those involving the powerful methods of complex variables.

To illustrate the power series method mentioned above, let us consider the following example.

Example 2.10: Find the Fourier cosine transform of $(a^2 - t^2)^{p-1/2} \times h(a - t)$, $p > -1/2$, $a > 0$, where $h(t)$ is the Heaviside unit function.

Solution: From definition of the cosine transform, we have

$$\mathscr{F}_C\{(a^2 - t^2)^{p-1/2}h(a - t);s\} = \sqrt{\frac{2}{\pi}}\int_0^a (a^2 - t^2)^{p-1/2}\cos st\,dt$$
$$= \sqrt{\frac{2}{\pi}}\sum_{k=0}^{\infty}\frac{(-1)^k s^{2k}}{(2k)!}\int_0^a (a^2 - t^2)^{p-1/2}\,t^{2k}\,dt$$

where we have replaced the cosine function with its power series representation. The substitution $t = a\sin\theta$ in the above integral leads to

$$\int_0^a (a^2 - t^2)^{p-1/2}\,t^{2k}\,dt = a^{2p+2k}\int_0^{\pi/2}\cos^{2p}\theta\,\sin^{2k}\theta\,d\theta$$
$$= a^{2p+2k}\frac{\Gamma(p + \frac{1}{2})\Gamma(k + \frac{1}{2})}{2\,\Gamma(p + k + 1)}$$

by use of Eq. (1.11) (see Chap. 1). Next, employing the duplication formula of the gamma function [see (G7) in Sec. 1.2]

$$\sqrt{\pi}\Gamma(2z) = 2^{2z-1}\,\Gamma(z)\Gamma(z + \tfrac{1}{2})$$

we find that

$$\mathscr{F}_C\{(a^2 - t^2)^{p-1/2}h(a - t);s\} = \frac{a^{2p}\Gamma(p + \frac{1}{2})}{\sqrt{2\pi}}\sum_{k=0}^{\infty}\frac{(-1)^k\,\Gamma(k + \frac{1}{2})(as)^{2k}}{(2k)!\,\Gamma(k + p + 1)}$$
$$= 2^{p-1/2}\,\Gamma(p + \tfrac{1}{2})(a/s)^p\sum_{k=0}^{\infty}\frac{(-1)^k\,(as/2)^{2k+p}}{k!\Gamma(k + p + 1)}$$

This last power series is recognized as the Bessel function $J_p(as)$ (see Sec. 1.4), and thus we have our result

---

* The power series in such cases must converge everywhere.

$$\mathscr{F}_C\{(a^2 - t^2)^{p-1/2}h(a - t);s\}$$
$$= 2^{p-1/2}\Gamma(p + \tfrac{1}{2})(a/s)^p J_p(as), \qquad p > -1/2, a > 0$$

Because the Fourier cosine transform and inverse cosine transform are identical operations, we can use the result of Exam. 2.10 to deduce the additional cosine transform relation

$$\mathscr{F}_C\{t^{-p} J_p(at);s\} = \frac{(a^2 - s^2)^{p-1/2}h(a - s)}{2^{p-1/2}a^p\Gamma(p + \frac{1}{2})}, \qquad p > -1/2, a > 0 \quad (2.66)$$

Also, for $p = 0$ we obtain the special case

$$\mathscr{F}\{J_0(at);s\} = \mathscr{F}_C\{J_0(at);s\} = \begin{cases} \sqrt{\dfrac{2}{\pi}}\dfrac{1}{\sqrt{a^2 - s^2}}, & |s| < a \\ 0, & |s| > a \end{cases} \quad (2.67)$$

2.6.1 *The Use of Residue Theory*

The calculus of residues from complex variables is a powerful tool in the calculation of many transform formulas. To deal with some of the integrals that will arise, we will need to use the theorems from complex variables provided in Appendix A.

To begin, we wish to derive the pair of transform formulas:

$$\mathscr{F}_C\{t^{\alpha-1};s\} = \sqrt{\frac{2}{\pi}}\frac{\Gamma(\alpha)}{s^\alpha}\cos(\pi\alpha/2), \qquad s > 0, 0 < \alpha < 1 \quad (2.68a)$$

$$\mathscr{F}_S\{t^{\alpha-1};s\} = \sqrt{\frac{2}{\pi}}\frac{\Gamma(\alpha)}{s^\alpha}\sin(\pi\alpha/2), \qquad s > 0, 0 < \alpha < 1 \quad (2.68b)$$

Let us define the complex function $f(z) = z^{\alpha-1}e^{-sz}$ and integrate it around the closed contour shown in Fig. 2.4. From Cauchy's integral theorem (Theor. A.1 in Appendix A), it follows that

$$\oint_C f(z)dz = 0 \quad (2.69)$$

or

$$\int_{C\rho} f(z)dz + \int_\rho^R f(x)dx + \int_{C_R} f(z)dz + \int_R^\rho f(iy)d(iy) = 0 \quad (2.70)$$

Along the imaginary axis we have set $z = iy$, so that

$$f(iy) = i^{\alpha-1}y^{\alpha-1}e^{-isy}$$

If we now allow $\rho \to 0$ and $R \to \infty$, we have from Theors. A.3 and A.4 in Appendix A that

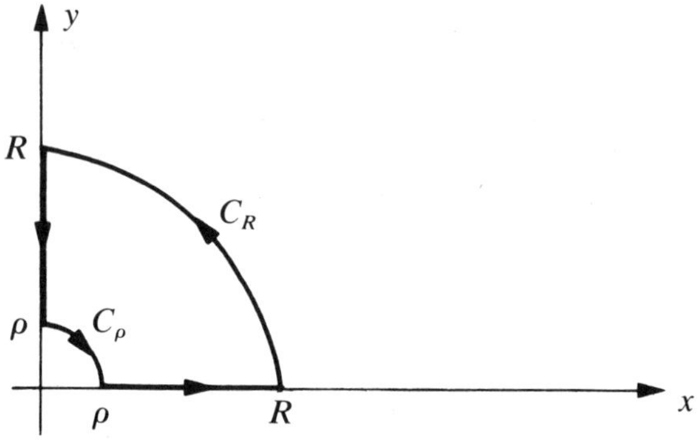

Figure 2.4 Contour of integration

$$\lim_{\rho\to 0}\int_{C_\rho} f(z)dz = 0, \qquad \lim_{R\to\infty}\int_{C_R} f(z)dz = 0 \tag{2.71}$$

and hence (2.70) reduces to

$$\int_0^\infty x^{\alpha-1}e^{-sx}dx + i^\alpha\int_\infty^0 y^{\alpha-1}e^{-isy}dy = 0$$

Now setting $i^{-\alpha} = e^{-i\pi\alpha/2}$, we can rewrite this last expression as

$$\int_0^\infty y^{\alpha-1}e^{-isy}dy = e^{-i\pi\alpha/2}\int_0^\infty x^{\alpha-1}e^{-sx}dx \tag{2.72}$$

In order to evaluate the integral on the right, we make the change of variable $u = sx$ to get

$$\begin{aligned}\int_0^\infty x^{\alpha-1}e^{-sx}dx &= s^{-\alpha}\int_0^\infty e^{-u}\,u^{\alpha-1}\,du\\ &= s^{-\alpha}\Gamma(\alpha)\end{aligned}$$

from which we deduce

$$\int_0^\infty y^{\alpha-1}e^{-isy}dy = \frac{\Gamma(\alpha)}{s^\alpha}\,e^{-i\pi\alpha/2} \tag{2.73}$$

Finally, multiplying both sides of (2.73) by the constant factor $\sqrt{2/\pi}$ and equating real and imaginary parts, we are led to the desired formulas given by (2.68a) and (2.68b).

If we let $\alpha = 1/2$ in (2.68a) and (2.68b), we get the special cases

$$\mathcal{F}_C\{1/\sqrt{t};s\} = 1/\sqrt{s} \tag{2.74a}$$

and

$$\mathcal{F}_S\{1/\sqrt{t};s\} = 1/\sqrt{s} \tag{2.74b}$$

which shows that $1/\sqrt{t}$ is self-reciprocal under cosine and sine transformations.

The residue calculus is especially helpful in finding transforms of rational functions. To illustrate the technique, let $f(z)$ denote a complex function with the following properties:

1. $f(z)$ has a finite number of poles $a_1, a_2, \ldots, a_n$ in the upper half-plane.
2. $f(z)$ is analytic along the real axis except at the points $b_1, b_2, \ldots, b_m$, which are simple poles.
3. $zf(z)e^{isz} \to 0$ and $z \to \infty$, $\mathrm{Im}(z) > 0$.

Suppose we integrate the complex function $f(z)e^{isz}$, $s > 0$, around the contour shown in Fig. 2.5. By use of the residue theorem of complex variables, we find that

$$\sum_{k=1}^{m+1} \int_{L_k} f(x)e^{isx}dx + \sum_{k=1}^{m} \int_{C_k} f(z)e^{isz}dz + \int_{C_R} f(z)e^{isz}dz = 2\pi i \sum_{k=1}^{n} \mathrm{Res}\{f(z)e^{isz};a_k\}$$

where $L_1, L_2, \ldots, L_{m+1}$ are the straight line segments along the x axis and $C_1, C_2, \ldots, C_m$ are small semicircles with centers at the simple poles $b_1, b_2, \ldots, b_m$. In the limit as $R \to \infty$ and the radii of the small semicircles tend to zero, we obtain

$$\int_{-\infty}^{\infty} f(x)e^{isx}dx - \pi i \sum_{k=1}^{m} \mathrm{Res}\{f(z)e^{isz};b_k\} = 2\pi i \sum_{k=1}^{n} \mathrm{Res}\{f(z)e^{isz};a_k\} \quad (2.75)$$

where (see Theor. A.3 in Appendix A)

$$\lim_{R\to\infty} \int_{C_R} f(z)e^{isz}dz = 0$$

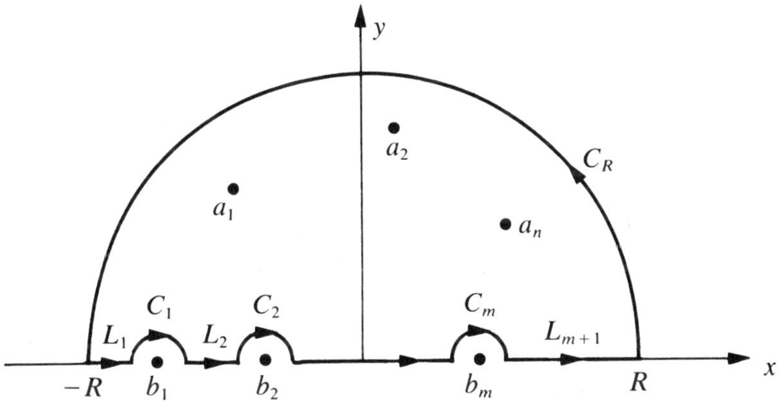

Figure 2.5 Contour of integration

Thus, we have developed the general transform formula for rational functions (changing x to t in the integral)

$$\mathscr{F}\{f(t);s\} = i\sqrt{2\pi}\left[\sum_{k=1}^{n} \operatorname{Res}\{f(z)e^{isz};a_k\} + \frac{1}{2}\sum_{k=1}^{m} \operatorname{Res}\{f(z)e^{isz};b_k\}\right] \tag{2.76}$$

for $s > 0$.

If f is either even or odd we can extend the result (2.76) to include $s < 0$ by utilizing the relations (2.40) and (2.43). For more general f we can set $s = -\sigma < 0$ and integrate the function $f(z)e^{-i\sigma z}$ around a contour similar to that in Fig. 2.5, but in the lower half-plane. Equivalently, we can replace z by $-z$ and integrate the function $f(-z)e^{-isz}$ around a contour in the upper half-plane. The result of this latter approach is the transform relation (see Prob. 9 in Exer. 2.6)

$$\mathscr{F}\{f(t);s\} = i\sqrt{2\pi}\left[\sum_{k=1}^{N} \operatorname{Res}\{f(-z)e^{-isz};\alpha_k\} + \frac{1}{2}\sum_{k=1}^{M} \operatorname{Res}\{f(-z)e^{-isz};\beta_k\}\right] \tag{2.77}$$

where $s < 0$. Here $\alpha_1, \alpha_2, \ldots, \alpha_N$ are the poles of $f(-z)$ in the upper half-plane and $\beta_1, \beta_2, \ldots, \beta_M$ are simple poles of $f(-z)$ along the real axis.

For calculating inverse Fourier transforms by this method, we simply observe that the cases corresponding to positive and negative t are the reverse of those for positive and negative s as a consequence of the fact that the kernel of the inverse transform is the complex conjugate of the kernel of the transform.

Example 2.11: Find the Fourier transform of $1/t(t^2 + k^2)$, $k > 0$.

Solution: The complex function

$$f(z) = 1/z(z^2 + k^2)$$

has simple poles at $z = 0$ and $z = \pm ik$. Hence, calculating residues at $z = 0$ and $z = ik$, we find

$$\operatorname{Res}\{f(z)e^{isz};0\} = \lim_{z\to 0} zf(z)e^{isz} = 1/k^2$$

and

$$\operatorname{Res}\{f(z)e^{isz};ik\} = \lim_{z\to ik} (z - ik)f(z)e^{isz} = -e^{-ks}/2k^2$$

From (2.76), we now have

$$\mathscr{F}\left\{\frac{1}{t(t^2+k^2)};s\right\} = i\sqrt{2\pi}\left(\frac{1}{2k^2} - \frac{e^{-ks}}{2k^2}\right)$$
$$= \sqrt{\frac{\pi}{2}}\frac{i}{k^2}(1 - e^{-ks}), \qquad s > 0$$

Because $f(t) = 1/t(t^2 + k^2)$ is an odd function, we can use (2.43) to deduce that, for all s,

$$\mathscr{F}\left\{\frac{1}{t(t^2+k^2)};s\right\} = \sqrt{\frac{\pi}{2}}\frac{i}{k^2}(1 - e^{-k|s|})\text{sgn}(s)$$

Example 2.12: Find the inverse Fourier transform of $F(s) = 1/(s^2 + ias + b)$, $a > 0$, $b > 0$.

Solution: By definition,

$$\mathscr{F}^{-1}\{F(s);t\} = \frac{1}{\sqrt{2\pi}}\int_{-\infty}^{\infty} e^{-its}F(s)\,ds$$
$$= \frac{1}{\sqrt{2\pi}}\int_{-\infty}^{\infty} e^{its}F(-s)\,ds$$

where we have replaced s by $-s$. The complex function

$$F(-z) = \frac{1}{z^2 - iaz + b}$$

has simple poles at $z = z_1$ and $z = z_2$, where

$$z_1 = \frac{i}{2}(a + \sqrt{a^2 + 4b})$$
$$z_2 = \frac{i}{2}(a - \sqrt{a^2 + 4b})$$

Clearly, z_1 lies in the upper half-plane while z_2 is in the lower half-plane. Calculating the residue at $z = z_1$ leads to*

$$\text{Res}\{F(-z)e^{itz};z_1\} = \left.\frac{e^{itz}}{2z - ia}\right|_{z=z_1} = \frac{e^{-(a+\sqrt{a^2+4b})t/2}}{i\sqrt{a^2+4b}}$$

and thus we have

* Recall that if $z = a$ is a simple pole of $f(z) = P(z)/Q(z)$, then

$$\text{Res}\{f(z);a\} = \lim_{z\to a}\frac{(z-a)P(z)}{Q(z)} = P(a)/Q'(a).$$

$$\mathscr{F}^{-1}\left\{\frac{1}{s^2 + ias + b};t\right\}$$
$$= \sqrt{\frac{2\pi}{a^2 + 4b}} \exp\left[-\frac{1}{2}(a + \sqrt{a^2 + 4b})t\right], \qquad t > 0$$

For $t < 0$, we consider the complex function

$$F(z) = \frac{1}{z^2 + iaz + b}$$

which has simple poles at $z = z_3$ and $z = z_4$, where

$$z_3 = -\frac{i}{2}(a + \sqrt{a^2 + 4b})$$

$$z_4 = -\frac{i}{2}(a - \sqrt{a^2 + 4b})$$

The pole at z_4 lies in the upper half-plane and

$$\text{Res}\{F(z)e^{-itz};z_4\} = \left.\frac{e^{-itz}}{2z + ia}\right|_{z=z_4} = \frac{e^{-(a-\sqrt{a^2+4b})t/2}}{i\sqrt{a^2 + 4b}}$$

Therefore,

$$\mathscr{F}^{-1}\left\{\frac{1}{s^2 + ias + b};t\right\}$$
$$= \sqrt{\frac{2\pi}{a^2 + 4b}} \exp\left[-\frac{1}{2}(a - \sqrt{a^2 + 4b})t\right], \qquad t < 0$$

and by combining results, we deduce that for all t,*

$$\mathscr{F}^{-1}\left\{\frac{1}{s^2 + ias + b};t\right\} = \sqrt{\frac{2\pi}{a^2 + 4b}} \exp\left[-\frac{1}{2}(at + \sqrt{a^2 + 4b}|t|)\right]$$

Example 2.13: Find the Fourier cosine transform of ($\cosh at/\cosh t$) and Fourier sine transform of ($\sinh at/\cosh t$), $|a| < 1$.

Solution: To start, let us integrate the complex function

$$f(z) = \frac{e^{(a+is)z}}{\cosh z}$$

around the rectangular contour shown in Fig. 2.6. The function $f(z)$ has simple poles at $z = (n + 1/2)\pi i$, $n = 0, \pm 1, \pm 2, \ldots$, but the

* Recall that we previously found this inverse transform relation through use of the shifting property (see Exam. 2.8).

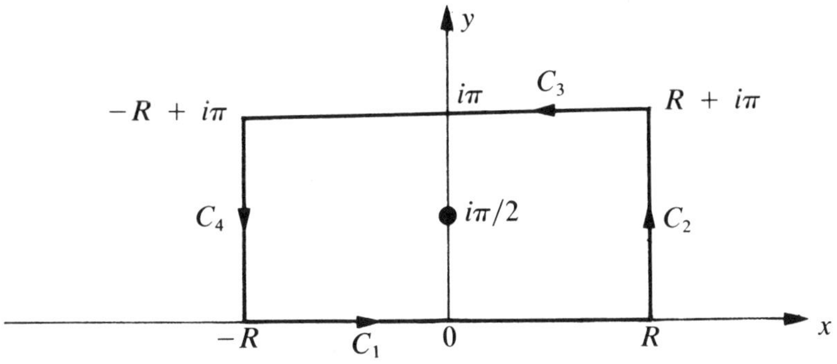

Figure 2.6 Contour of integration

only pole enclosed by the rectangular contour C is $i\pi/2$. Hence, from the residue theorem (Appendix A) it follows that

$$\oint_C f(z)dz = 2\pi i \operatorname{Res}\{f(z);i\pi/2\} = 2\pi e^{(ia-s)\pi/2}$$

This can also be written as

$$\int_{-R}^{R} f(x)dx + \int_{C_2} f(z)dz + \int_{C_3} f(z)dz + \int_{C_4} f(z)dz = 2\pi e^{(ia-s)\pi/2}$$

Along the line segment C_2 we have $z = R + iy$, and thus

$$\left|\int_{C_2} f(z)dz\right| = \left|\int_0^{\pi} \frac{e^{aR-sy}\,e^{i(sR+ay)}}{\cosh(R+iy)}\,dy\right|$$
$$\leq e^{aR}\int_0^{\pi} \frac{e^{-sy}}{|\cosh(R+iy)|}\,dy$$

It can be shown that the last integral vanishes in the limit as $R \to \infty$ (see Prob. 20 in Exer. 2.6). The same in true of the integral along the line segment C_4. We have $z = x + i\pi$ along the segment C_3, which leads to

$$\int_{C_3} f(z)dz = \int_R^{-R} \frac{e^{(a+is)(x+i\pi)}}{\cosh(x+i\pi)}\,dx$$
$$= e^{(ia-s)\pi}\int_{-R}^{R} \frac{e^{(a+is)x}}{\cosh x}\,dx$$

Therefore, as $R \to \infty$ we are left with

$$(1 + e^{(ia-s)\pi})\int_{-\infty}^{\infty} \frac{e^{(a+is)x}}{\cosh x}\,dx = 2\pi e^{(ia-s)\pi/2}$$

or

$$\int_{-\infty}^{\infty} \frac{e^{(a+is)x}}{\cosh x}\, dx = \frac{2\pi e^{(ia-s)\pi/2}}{1 + e^{(ia-s)\pi}}$$

$$= \frac{2\pi}{e^{(ia-s)\pi/2} + e^{-(ia-s)\pi/2}}$$

$$= \frac{\pi}{\cosh\left[\dfrac{(ia - s)\pi}{2}\right]}$$

By splitting this last expression into real and imaginary parts, we get

$$\int_{-\infty}^{\infty} \frac{\cosh ax}{\cosh x} \cos sx\, dx + i\int_{-\infty}^{\infty} \frac{\sinh ax}{\cosh x} \sin sx\, dx$$

$$= \frac{\pi}{\cos\dfrac{a\pi}{2} \cosh \dfrac{s\pi}{2} - i \sin \dfrac{a\pi}{2} \sinh \dfrac{s\pi}{2}}$$

$$= \frac{\pi \cos \dfrac{a\pi}{2} \cosh \dfrac{s\pi}{2}}{\cos^2\dfrac{a\pi}{2} \cosh^2 \dfrac{s\pi}{2} + \sin^2 \dfrac{a\pi}{2} \sinh^2 \dfrac{s\pi}{2}} + i\frac{\pi \sin \dfrac{a\pi}{2} \sinh \dfrac{s\pi}{2}}{\cos^2\dfrac{a\pi}{2} \cosh^2\dfrac{s\pi}{2} + \sin^2\dfrac{a\pi}{2} \sinh^2\dfrac{s\pi}{2}}$$

and by comparing real and imaginary parts, and simplifying the algebra, we deduce that

$$\mathscr{F}_C\left\{\frac{\cosh at}{\cosh t};s\right\} = \sqrt{2\pi}\,\frac{\cos\dfrac{a\pi}{2} \cosh\dfrac{s\pi}{2}}{\cos a\pi + \cosh s\pi}, \qquad |a| < 1$$

$$\mathscr{F}_S\left\{\frac{\sinh at}{\sinh t};s\right\} = \sqrt{2\pi}\,\frac{\sin\dfrac{a\pi}{2} \sinh\dfrac{s\pi}{2}}{\cos a\pi + \cosh s\pi}, \qquad |a| < 1$$

EXERCISES 2.6

1. Find the Fourier cosine transform of $h(a - t)$, $a > 0$ and compare this with Exam. 2.10 to deduce that

$$J_{1/2}(x) = \sqrt{\frac{2}{\pi x}} \sin x$$

2. Given that $f(t) = (t^2 - 1)^n h(1 - |t|)$, $n = 0,1,2,\ldots$, show that

$$\mathscr{F}\{f(t);s\} = \frac{(-1)^n n!}{\sqrt{2}} \left(\frac{2}{s}\right)^{n+1/2} J_{n+1/2}(s)$$

3. Use the result of Prob. 2 to deduce that

$$\mathscr{F}\{P_n(t)h(1 - |t|);s\} = \frac{i^n}{\sqrt{s}} J_{n+1/2}(s), \qquad n = 0,1,2,\ldots$$

where $P_n(t)$ is the nth *Legendre polynomial* defined by

$$P_n(t) = \frac{1}{2^n n!}\frac{d^n}{dt^n}\left[(t^2 - 1)^n\right], \qquad n = 0,1,2,\ldots$$

4. Show that

$$\mathscr{F}_S\{t(a^2 - t^2)^{p-3/2}h(a - t);s\}$$
$$= 2^{p-3/2}a^p s^{1-p}\Gamma(p - 1/2)J_p(as), \qquad a > 0,\ p > 1/2$$

5. Integrate both sides of Eq. (2.67) with respect to s to deduce the Fourier sine transform of $(1/t)J_0(at)$.

6. Show that

(a) $\mathscr{F}\{|t|^{-\alpha};s\} = \sqrt{\dfrac{2}{\pi}}\dfrac{\Gamma(1-\alpha)}{|s|^{1-\alpha}}\sin\dfrac{\pi\alpha}{2}, \qquad 0<\alpha<1$

(b) $\mathscr{F}\{|t|^{-\alpha}\mathrm{sgn}(t);s\} = i\sqrt{\dfrac{2}{\pi}}\dfrac{\Gamma(1-\alpha)}{|s|^{1-\alpha}}\cos\dfrac{\pi\alpha}{2}, \qquad 0<\alpha<1$

7. Given that $r = \sqrt{a^2 + s^2}$ and $\tan\theta = s/a$, show that

$$\mathscr{F}_C\{e^{-at}t^{p-1};s\} = \sqrt{\frac{2}{\pi}}\frac{\Gamma(p)}{r^p}\cos p\theta, \qquad a>0,\ p>0$$

$$\mathscr{F}_S\{e^{-at}t^{p-1};s\} = \sqrt{\frac{2}{\pi}}\frac{\Gamma(p)}{r^p}\sin p\theta, \qquad a>0,\ p>0$$

Hint: Consider the integral of $f(z) = z^{p-1}e^{-(s-ia)z}$ around the contour in Fig. 2.4.

8. Provide the details leading to Eq. (2.77).

In Probs. 9–14, use residue theory to find the Fourier transform of the given rational function.

9. $f(t) = \dfrac{1}{t^2 + 9}$

10. $f(t) = \dfrac{2it}{t^2 + 1}$

11. $f(t) = \dfrac{1}{t^4 + 1}$

12. $f(t) = \dfrac{1}{t^6 + 1}$

13. $f(t) = \dfrac{1}{(t^2 + 4)^2}$

14. $f(t) = \dfrac{t^2 - 1}{t(t^2 + 1)}$

In Probs. 15–17, use residue theory to find the inverse Fourier transform of the given rational function.

15. $F(s) = \dfrac{1}{s^2 + 1}$

16. $F(s) = \dfrac{2is + 1}{s^2 + 1}$

17. $F(s) = \dfrac{1 - s^2}{(s^2 + 4)^2}$

18. By expanding $J_0(at)$ in a power series, use residue theory to deduce that

$$\mathscr{F}_C\{(t^2 + 1)^{-1}J_0(at);s\} = \sqrt{\frac{\pi}{2}}\, I_0(a)e^{-s}, \qquad s > 1$$

Hint: Observe that $\displaystyle\int_{-\infty}^{\infty} \frac{t^{2k}}{t^2 + 1}\, e^{ist}\, dt = (-1)^k \frac{d^{2k}}{ds^{2k}} \int_{-\infty}^{\infty} \frac{e^{ist}}{t^2 + 1}\, dt$

19. Following the suggestion in Prob. 18, derive the Fourier sine transform of $t(t^2 + 1)^{-1}J_0(at)$.

20. Show that

(a) $|\cosh(R + iy)| \geq e^R/4, \qquad R \to \infty.$

(b) Use (a) to show that the integral along C_2 in Exam. 2.12 satisfies the inequality

$$\left| \int_{C_2} f(z)dz \right| \leq 4\pi e^{(a-1)R}, \qquad R \to \infty$$

and thus deduce that $\displaystyle\lim_{R\to\infty} \int_{C_2} f(z)dz = 0.$

21. By integrating the functions $\dfrac{\sinh(z/2)}{\sinh z}e^{isz}$ and $\dfrac{\cosh(z/2)}{\sinh z}e^{isz}$ around the contour shown in the accompanying figure, deduce that

$$\mathscr{F}_C\left\{\frac{\sinh(t/2)}{\sinh t};s\right\} = \sqrt{\frac{\pi}{2}}\operatorname{sech} \pi s$$

$$\mathscr{F}_S\left\{\frac{\cosh(t/2)}{\sinh t};s\right\} = \sqrt{\frac{\pi}{2}}\tanh \pi s$$

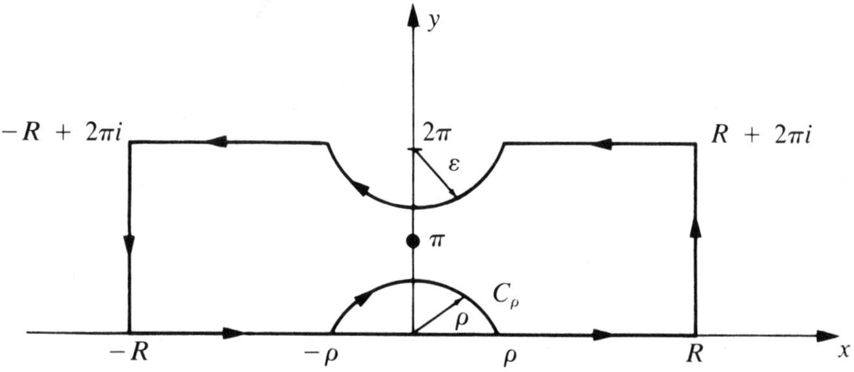

22. Given the transform function $F(s) = e^{i\pi/4}/\sqrt{2s}$,

(a) show that

$$\mathscr{F}^{-1}\{F(s);t\} = 0, \qquad t < 0$$

by integrating an appropriate complex function around a closed contour in the upper half-plane.

(b) By using the contour shown in the accompanying figure, show that

$$\mathscr{F}^{-1}\{F(s);t\} = \frac{1}{\sqrt{\pi}}\int_0^\infty \frac{e^{-\xi t}}{\sqrt{\xi}}\,d\xi, \qquad t > 0$$

and thus deduce that

$$\mathscr{F}^{-1}\left\{\frac{e^{i\pi/4}}{\sqrt{2s}};t\right\} = \frac{1}{\sqrt{t}}\,h(t)$$

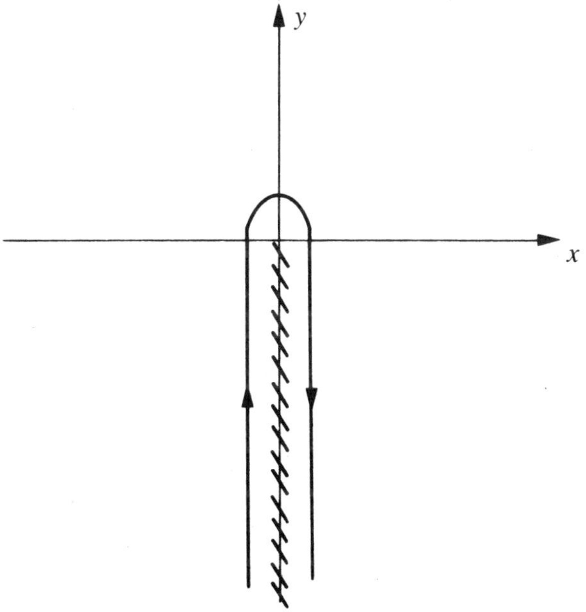

In Probs. 23–30, establish the given transform relation.

23. $\mathscr{F}_C\{(1 - t)h(1 - t);s\} = \sqrt{\frac{2}{\pi}}\frac{1}{s^2}(1 - \cos s)$

24. $\mathscr{F}_C\left\{\frac{\sinh at}{\sinh \pi t};s\right\} = \frac{1}{\sqrt{2\pi}}\frac{\sin a}{\cosh s + \cos a}, \qquad |a| < \pi$

25. $\mathscr{F}_S\left\{\frac{\cosh at}{\sinh \pi t};s\right\} = \frac{1}{\sqrt{2\pi}}\frac{\sinh s}{\cosh s + \cos a}, \qquad |a| < \pi$

26. $\mathscr{F}\left\{\frac{e^{-a|t|}}{\sqrt{t}};s\right\} = \left(\frac{\sqrt{s^2+a^2}+a}{2\pi(s^2+a^2)}\right)^{1/2}, \qquad a > 0$

27. $\mathscr{F}_S\left\{\frac{\sin t}{t};s\right\} = \frac{1}{\sqrt{2\pi}}\log\left|\frac{1+s}{1-s}\right|$

28. $\mathscr{F}_C\{t^{-1}(e^{-bt}-e^{-at});s\} = \frac{1}{\sqrt{2\pi}}\log\left(\frac{s^2+a^2}{s^2+b^2}\right), \qquad a > b$

29. $\mathscr{F}_S\{t^{-1}(e^{-bt}-e^{-at});s\} = \sqrt{\frac{2}{\pi}}\tan^{-1}\left[\frac{(a-b)s}{s^2+ab}\right], \qquad a > b$

30. $\mathscr{F}\left\{\frac{\sin b\sqrt{a^2+t^2}}{\sqrt{a^2+t^2}};s\right\} = \sqrt{\frac{\pi}{2}}\,J_0(a\sqrt{b^2-s^2})h(1-|s/b|)$

2.7 *The Convolution Integrals of Fourier*

One of the most important operational properties of the Fourier transform is the *convolution theorem*. To derive this important property, let us begin by defining the *convolution* of two integrable functions f and g by the expression

$$(f \circ g)(t) = \frac{1}{\sqrt{2\pi}}\int_{-\infty}^{\infty} f(u)g(t-u)\,du \tag{2.78}$$

The Fourier transform applied to this convolution integral leads to

$$\begin{aligned}\mathscr{F}\{(f\circ g)(t);s\} &= \frac{1}{2\pi}\int_{-\infty}^{\infty} e^{ist}\int_{-\infty}^{\infty} f(u)g(t-u)\,du\,dt\\ &= \frac{1}{2\pi}\int_{-\infty}^{\infty}\int_{-\infty}^{\infty} e^{ist}f(u)g(t-u)\,du\,dt \end{aligned}\tag{2.79}$$

We can interpret (2.79) as an iterated integral for which the order of integration can be interchanged. Hence, by changing the order of integration and making the change of variable $t = u + x$, we find

$$\begin{aligned}\mathscr{F}\{(f\circ g)(t);s\} &= \frac{1}{2\pi}\int_{-\infty}^{\infty}\int_{-\infty}^{\infty} e^{is(u+x)}f(u)g(x)\,dx\,du\\ &= \frac{1}{\sqrt{2\pi}}\int_{-\infty}^{\infty} e^{isu}f(u)du\cdot\frac{1}{\sqrt{2\pi}}\int_{-\infty}^{\infty} e^{isx}g(x)dx\end{aligned}$$

and thus conclude that

$$\mathscr{F}\{(f\circ g)(t);s\} = F(s)G(s) \tag{2.80}$$

where $F(s)$ and $G(s)$ are the Fourier transforms, respectively, of $f(t)$ and

$g(t)$. By applying the inverse Fourier transform to both sides of (2.80), we obtain the alternate form

$$\mathscr{F}^{-1}\{F(s)G(s);t\} = (f \circ g)(t) \tag{2.81}$$

This shows that the inverse transform of a product of transform functions can be found by convolving the inverse transforms of each product term.

Equation (2.81) is the *Fourier convolution theorem*, which sometimes is expressed in the form

$$\int_{-\infty}^{\infty} e^{-ist} F(s)G(s)\, ds = \int_{-\infty}^{\infty} f(u)g(t-u)\, du \tag{2.82}$$

An interesting consequence of (2.82) follows by first setting $t = 0$, which yields

$$\int_{-\infty}^{\infty} F(s)G(s)\, ds = \int_{-\infty}^{\infty} f(u)g(-u)\, du \tag{2.83}$$

For the special case where $g(-u) = \overline{f(u)}$, then*

$$G(s) = \mathscr{F}\{g(u);s\} = \mathscr{F}\{\overline{f(-u)};s\} = \overline{F(s)}$$

(see Prob. 2 in Exer. 2.5) and we obtain

$$\int_{-\infty}^{\infty} F(s)\overline{F(s)}\, ds = \int_{-\infty}^{\infty} f(u)\overline{f(u)}\, du$$

or

$$\int_{-\infty}^{\infty} |F(s)|^2\, ds = \int_{-\infty}^{\infty} |f(t)|^2\, dt \tag{2.84}$$

Equation (2.84) is called *Parseval's relation*. In physical applications the quantity on the right-hand side represents the total energy in a waveform, such as a sound wave or electrical signal. Thus, Parseval's relation states that the total energy is given by the area under the $|F(s)|^2$ curve. For this reason, the quantity $|F(s)|^2$ is called the *energy spectrum* or *energy spectral density function* of $f(t)$. Engineers are undoubtedly familiar with the Fourier series counterpart of Parseval's relation, which has the physical interpretation that the power associated with a periodic function equals the sum of the powers associated with its harmonic components.

Returning now to the convolution integral (2.78), we observe that it satisfies certain formal properties of ordinary products. For example, if C is a constant, it is immediately clear that

$$f \circ (Cg) = (Cf) \circ g = C(f \circ g) \tag{2.85}$$

---

* In most cases of interest to us the function $f(t)$ is real and thus $\overline{f(t)} = f(t)$.

Similarly, it follows that the *distributive law* holds, i.e.,

$$f \circ (g + k) = f \circ g + f \circ k \tag{2.86}$$

By definition,

$$(g \circ f)(t) = \frac{1}{\sqrt{2\pi}} \int_{-\infty}^{\infty} g(u)f(t - u)\, du$$

and by making the change of variable $v = t - u$, we have

$$\begin{aligned}(g \circ f)(t) &= -\frac{1}{\sqrt{2\pi}} \int_{\infty}^{-\infty} g(t - v)f(v)\, dv \\ &= \frac{1}{\sqrt{2\pi}} \int_{-\infty}^{\infty} f(v)g(t - v)\, dv\end{aligned}$$

from which we deduce the *commutative law*

$$f \circ g = g \circ f \tag{2.87}$$

Finally, the convolution integral (2.78) also satisfies the *associative law*

$$f \circ (g \circ k) = (f \circ g) \circ k \tag{2.88}$$

but we leave the proof of this result to the exercises (see Prob. 8 in Exer. 2.7).

Example 2.14: Use the convolution theorem to evaluate the inverse Fourier transform

$$\mathscr{F}^{-1}\left\{\frac{\sin s}{s(1 - is)};t\right\}$$

Solution: Let us define

$$F(s) = \frac{\sin s}{s} \text{ and } G(s) = \frac{1}{1 - is}$$

which have inverse Fourier transforms

$$\mathscr{F}^{-1}\left\{\frac{\sin s}{s};t\right\} = \sqrt{\frac{\pi}{2}}\, h(1 - |t|)$$

and

$$\mathscr{F}^{-1}\left\{\frac{1}{1 - is};t\right\} = \sqrt{2\pi} e^{-t}\, h(t)$$

Thus, using (2.81), we have

$$\mathscr{F}^{-1}\{F(s)G(s);t\} = \sqrt{\frac{\pi}{2}} \int_{-\infty}^{\infty} e^{-(t-u)} h(t-u)h(1-|u|)\,du$$

$$= \begin{cases} \sqrt{\frac{\pi}{2}}\, e^{-t} \int_{-1}^{t} e^{u}\,du, & |t| < 1 \\ \sqrt{\frac{\pi}{2}}\, e^{-t} \int_{-1}^{1} e^{u}\,du, & t > 1 \\ 0, & t < -1 \end{cases}$$

or

$$\mathscr{F}^{-1}\left\{\frac{\sin s}{s(1-is)};t\right\} = \begin{cases} \sqrt{\frac{\pi}{2}}\,(1 - e^{-(t+1)}), & |t| < 1 \\ \sqrt{\frac{\pi}{2}}\,(e - e^{-1})e^{-t}, & t > 1 \\ 0, & t < -1 \end{cases}$$

Example 2.15: Evaluate the integral

$$I = \int_{-\infty}^{\infty} \frac{dx}{(x^2 + a^2)(x^2 + b^2)}$$

Solution: The given integral has the form

$$I = \int_{-\infty}^{\infty} F(x)G(x)\,dx$$

where $F(x) = 1/(x^2 + a^2)$ and $G(x) = 1/(x^2 + b^2)$. Thus, by the use of Eq. (2.83), we immediately can relate this integral to

$$I = \int_{-\infty}^{\infty} f(t)g(-t)\,dt$$

where $f(t)$ and $g(t)$ are the inverse Fourier transforms of $F(x)$ and $G(x)$, respectively. Recalling Exam. 2.3, we see that

$$f(t) = \frac{1}{a}\sqrt{\frac{\pi}{2}}\, e^{-a|t|} \text{ and } g(t) = \frac{1}{b}\sqrt{\frac{\pi}{2}}\, e^{-b|t|}$$

and using the fact that these are even functions, we deduce that

$$I = 2\int_{0}^{\infty} f(t)g(t)\,dt$$

$$= \frac{\pi}{ab}\int_{0}^{\infty} e^{-(a+b)t}\,dt$$

or

$$\int_{-\infty}^{\infty} \frac{dx}{(x^2 + a^2)(x^2 + b^2)} = \frac{\pi}{ab(a + b)}$$

2.7.1 *Cosine and Sine Convolution Integrals*

Other convolution integrals involving the cosine and sine transforms can be derived similarly, but the resulting integrals are more complicated than (2.82). For example, if f and g are functions defined for $t > 0$ which have cosine transforms $F_C(s)$ and $G_C(s)$, respectively, then from previous results we know that

$$\mathscr{F}\{f(|t|);s\} = F_C(s)$$

and

$$\mathscr{F}\{g(|t|);s\} = G_C(s)$$

The substitution of these expressions into (2.82) yields

$$\int_{-\infty}^{\infty} e^{-ist} F_C(s)G_C(s)\, ds = \int_{-\infty}^{\infty} f(|u|)g(|t - u|)\, du \tag{2.89}$$

Since both $F_C(s)$ and $G_C(s)$ are necessarily even functions, we can write (2.89) as

$$\begin{aligned} 2\int_0^{\infty} \cos(st)F_C(s)G_C(s)\, ds &= \int_0^{\infty} f(u)g(|t - u|)\, du + \int_{-\infty}^{0} f(|u|)g(|t - u|)\, du \\ &= \int_0^{\infty} f(u)g(|t - u|)\, du + \int_0^{\infty} f(u)g(t + u)\, du, \qquad t \end{aligned}$$

and thus we have derived the convolution integral ($t > 0$)

$$\int_0^{\infty} \cos(st)F_C(s)G_C(s)\, ds = \frac{1}{2}\int_0^{\infty} f(u)\,[g(|t - u|) + g(t + u)]\, du \tag{2.90}$$

Without providing the details, we simply state that the following convolution integrals can be derived in a likewise manner (see Exer. 2.7):

$$\int_0^{\infty} \cos(st)F_S(s)G_S(s)\, ds = \frac{1}{2}\int_0^{\infty} f(u)\,[g(u + t) + g(u - t)]\, du \tag{2.91}$$

$$\int_0^{\infty} \sin(st)F_S(s)G_C(s)\, ds = \frac{1}{2}\int_0^{\infty} f(u)\,[g(|u - t|) - g(u + t)]\, du \tag{2.92}$$

$$\int_0^{\infty} \sin(st)F_C(s)G_S(s)\, ds = \frac{1}{2}\int_0^{\infty} f(u)\,[g(u + t) - g(u - t)]\, du \tag{2.93}$$

EXERCISES 2.7

1. Verify the convolution integral (2.82) for
(a) $f(t) = g(t) = h(1 - |t|)$
(b) $f(t) = g(t) = e^{-t^2/2}$

2. Use the convolution integral to find the inverse Fourier transform of

$$F(s) = \frac{1}{(1 - is)^2} = \frac{1}{1 - is} \cdot \frac{1}{1 - is}$$

3. Show that the convolution theorem can also be expressed in the form

$$\int_{-\infty}^{\infty} e^{ist} f(t)g(t)\, dt = \int_{-\infty}^{\infty} F(u)G(s - u)\, du$$

and use this result to evaluate the Fourier transform of $e^{-|t|} \dfrac{\sin t}{t}$.

4. For $a,b > 0$, show that

$$\int_{-\infty}^{\infty} \frac{x^2}{(x^2 + a^2)(x^2 + b^2)}\, dx = \frac{\pi}{a + b}$$

5. Use the Fourier transform relation

$$\mathscr{F}\{(a^2 - t^2)^{-1/2} h(a - |t|);s\} = \sqrt{\frac{\pi}{2}} J_0(as), \qquad a > 0$$

to show that

$$\int_0^{\infty} J_0(ax)J_0(bx)\, dx = \frac{2}{\pi b} K(a/b), \qquad 0 < a < b$$

where $K(m)$ denotes the *complete elliptic integral*

$$K(m) = \int_0^{\pi/2} (1 - m^2 \sin^2\theta)^{-1/2}\, d\theta$$

6. For $0 < a < b$, show that

$$\int_0^{\infty} \sin ax \sin bx \frac{dx}{x^2} = \frac{\pi a}{2}$$

7. Verify Parseval's relation (2.84) for
(a) $f(t) = e^{-a|t|}, \qquad a > 0$
(b) $f(t) = e^{-a^2t^2}, \qquad a > 0$

8. Show that the Fourier convolution (2.78) satisfies the associative property

$$f \circ (g \circ k) = (f \circ g) \circ k$$

9. For real $g(t) \equiv f(t)$, show that (2.90) leads to the Parseval relation

$$\int_0^\infty |F_C(s)|^2\,ds = \int_0^\infty |f(u)|^2\,du$$

10. Derive Eq. (2.91) by replacing $G_S(s)$ with its defining transform integral and interchanging the order of integration.

11. Derive Eq. (2.92).

Hint: See Prob. 10.

12. Derive Eq. (2.93).

Hint: See Prob. 10.

13. For real $g(t) \equiv f(t)$, show that (2.91) leads to the Parseval relation

$$\int_0^\infty |F_S(s)|^2\,ds = \int_0^\infty |f(u)|^2\,du$$

14. By substituting the functions

$$\begin{aligned} f(t) &= t^{-\alpha}, \qquad 0 < \alpha < 1 \\ g(t) &= (1 - t^2)^{\nu - 1/2} h(1 - t), \qquad \nu > -1/2 \end{aligned}$$

into Eq. (2.90).

(a) show that

$$\frac{2^\nu}{\sqrt{\pi}} \sin\left(\frac{\alpha\pi}{2}\right) \Gamma(1 - \alpha)\Gamma(\nu + \tfrac{1}{2}) \int_0^\infty s^{\alpha - \nu - 1} J_\nu(s)\,ds = \int_0^1 t^{-\alpha}(1 - t^2)^{\nu - 1/2}\,dt$$

(b) Using properties of the beta function (see Prob. 21 in Exer. 1.2), evaluate the integral on the right in (a) and deduce that

$$\int_0^\infty x^{\alpha - \nu - 1} J_\nu(x)\,dx = \frac{2^{\alpha - \nu - 1}\Gamma(\alpha/2)}{\Gamma(\nu + 1 - \alpha/2)}, \qquad 0 < \alpha < 1,\ \nu > -1/2$$

15. Using Eq. (2.90),
(a) show that

$$\mathscr{F}_C\left\{\frac{J_0(at)}{t^2 + 1}; s\right\} = \sqrt{\frac{2}{\pi}}\, e^{-s} \int_0^a \frac{\cosh u}{\sqrt{a^2 - u^2}}\,du, \qquad a > 0$$

(b) From (a), deduce that

$$\mathscr{F}_C\left\{\frac{J_0(at)}{t^2+1};s\right\} = \sqrt{\frac{\pi}{2}}\, e^{-s}\, I_0(a), \qquad a > 0$$

where $I_0(a)$ is the modified Bessel function of order zero.

Hint: Express cosh u in a power series and use properties of the beta function (see Prob. 21 in Exer. 1.2).

2.8 *Transforms Involving Generalized Functions*

In Sec. 2.4 we stated that the Fourier transform $F(s)$ is a bounded function provided the inverse transform $f(t)$ is *absolutely integrable*, i.e., provided

$$\int_{-\infty}^{\infty} |f(t)|\, dt < \infty \tag{2.94}$$

This is actually a sufficient condition of the Fourier integral theorem for the existence of the transform of a given function $f(t)$. Sinusoidal functions, the Heaviside unit function, polynomials, and so forth, do not satisfy the condition (2.94) and, therefore, do not have a Fourier transform in the usual sense. However, it is possible to extend our definition of Fourier transform to include such functions if we are willing to consider the notion of generalized functions such as the impulse function $\delta(t)$ and its derivatives $\delta^{(n)}(t)$, $n = 1,2,3,\ldots$.

The impulse function was first introduced in Sec. 1.5.2. It is a useful concept in a wide variety of physical problems involving the ideas of line spectrum, impulsive forces, or point sources. Because the rules of manipulation of impulse functions do not follow naturally from the methods of classical analysis, such functions are often referred to as *generalized functions*. The general theory of such functions has been put on a solid mathematical basis under the title of the *theory of distributions*.* Nonetheless, our treatment of impulse functions and their derivatives will continue to be based upon formal manipulations as developed in Sec. 1.5.2.

2.8.1 *Impulse Function*

By using the "sifting property" of the impulse function

* The most notable pioneering work on generalized functions is contained in L. Schwartz, *Théorie des Distributions*, Tomes 1 and 2, Paris: Hermann and Cie, 1950, 1951. See also A. H. Zemanian, *Distribution Theory and Transform Analysis*, New York: McGraw-Hill, 1965.

$$\int_{-\infty}^{\infty} \delta(t - a)f(t)\, dt = f(a) \tag{2.95}$$

we immediately obtain the formal result

$$\mathscr{F}\{\delta(t);s\} = \frac{1}{\sqrt{2\pi}} \int_{-\infty}^{\infty} e^{ist}\, \delta(t)\, dt = \frac{1}{\sqrt{2\pi}} \tag{2.96}$$

Hence, we say the Fourier transform of the impulse function $\delta(t)$ is the constant $1/\sqrt{2\pi}$. On the other hand, we can use this transform relation to deduce that

$$\mathscr{F}^{-1}\{1;t\} = \frac{1}{\sqrt{2\pi}} \int_{-\infty}^{\infty} e^{-ist}\, (1)\, ds = \sqrt{2\pi}\, \delta(s) \tag{2.97}$$

Of course, the integral in (2.97) has no meaning as an ordinary integral. Its interpretation lies strictly in the fact that it is used to define the generalized function $\delta(s)$.

To better understand how the concept of an impulse function might arise in a practical situation, let us briefly discuss the case where $f(t)$ represents a function in the time domain. For example, if $f(t)$ is a waveform — an electrical signal like a voltage or current, or an acoustic wave or optical wave, etc. — the Fourier transform of $f(t)$ describes the waveform in the *frequency domain*. It is customary in this setting to let $s = \omega$, where ω is the angular frequency variable. Although the waveform physically exists in the time domain, we can say that it consists of those components in its frequency domain description called its *spectrum*. Mathematically, we write*

$$F(\omega) = \frac{1}{\sqrt{2\pi}} \int_{-\infty}^{\infty} e^{i\omega t} f(t)\, dt = |F(\omega)| e^{i\phi(\omega)} \tag{2.98}$$

where $|F(\omega)|$ is the spectrum *amplitude* function and $\phi(\omega)$ describes the spectrum *phase*.

The sinusoidal functions $f(t) = \cos \omega_0 t$, which represents one of the simplest waveforms possible, does not satisfy the condition (2.94), and hence, does not have a Fourier transform in the strict sense. Yet, from a purely physical point of view we know this waveform has a single frequency component, or *line spectrum*, at $\omega = \omega_0$. This apparent contradiction can be explained by recognizing that the function $\cos \omega_0 t$ cannot exist for all time $-\infty < t < \infty$ as we assume in the formal definition of the Fourier transform. Like all waveforms, the cosine waveform exists for only some finite interval of time, and as such will

---

* In much of the engineering literature, the spectrum function is defined by a multiple of the complex conjugate of (2.98).

actually satisfy the condition of absolute integrability. For instance, let us assume that $f(t) = (\cos \omega_0 t)h(T - |t|)$, where $2T$ is the interval of time that the waveform is present. In this case, the Fourier transform of $f(t)$ leads to

$$\mathscr{F}\{f(t);\omega\} = \frac{1}{\sqrt{2\pi}}\int_{-T}^{T} e^{i\omega t}\cos \omega_0 t\, dt$$
$$= \sqrt{\frac{2}{\pi}}\int_{0}^{T}\cos \omega t \cos \omega_0 t\, dt$$

or

$$F(\omega) = \frac{1}{\sqrt{2\pi}}\left[\frac{\sin(\omega + \omega_0)T}{\omega + \omega_0} + \frac{\sin(\omega - \omega_0)T}{\omega - \omega_0}\right] \tag{2.99}$$

The graph of $|F(\omega)|$ is shown in Fig. 2.7 for a fixed value of T. By allowing T to become unbounded, it can be shown that the graph becomes more and more narrow and peaked around $\omega = \omega_0$ and $\omega = -\omega_0$. Hence, in a formal sense we may consider the Fourier transform of $\cos \omega_0 t$ for all time as the limit

$$\mathscr{F}\{\cos \omega_0 t;\omega\} = \lim_{T\to\infty}\frac{1}{\sqrt{2\pi}}\left[\frac{\sin(\omega + \omega_0)T}{\omega + \omega_0} + \frac{\sin(\omega - \omega_0)T}{\omega - \omega_0}\right] \tag{2.100}$$

provided this limit is meaningful.

In order to interpret the limit (2.100), let us recall that in the proof of Theor. 2.1 (see Sec. 2.3) we established the limit relation

$$\begin{aligned} f(x) &= \frac{1}{\pi}\int_{0}^{\infty}\int_{-\infty}^{\infty} f(t)\cos[s(t - x)]\, dt\, ds \\ &= \frac{1}{\pi}\int_{-\infty}^{\infty} f(t)\left[\lim_{\lambda\to\infty}\frac{\sin \lambda(t - x)}{t - x}\right] dt \end{aligned} \tag{2.101}$$

Based on the sifting property (2.95), it becomes clear that the limit in (2.101) leads to the formal definition

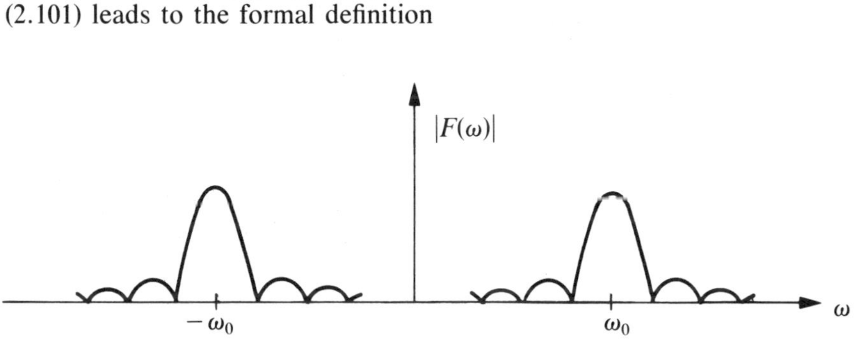

Figure 2.7 Spectrum of sinusoidal pulse

$$\frac{1}{\pi}\lim_{\lambda\to\infty}\frac{\sin\lambda(t-x)}{t-x}=\delta(t-x) \tag{2.102}$$

With this interpretation of the limit, we see that (2.100) gives the expected result*

$$\mathscr{F}\{\cos\omega_0 t;\omega\}=\sqrt{\pi/2}\,[\delta(\omega+\omega_0)+\delta(\omega-\omega_0)] \tag{2.103}$$

Hence, the notion of impulse functions is consistent with our previous claim that the spectrum of $\cos\omega_0 t$ should be a single line at frequency $\omega=\omega_0$ (and from symmetry of the transform, a line also at $\omega=-\omega_0$).

Example 2.16: Find the Fourier transform of sgn(t).

Solution: Recalling the transform relation [see Eq. (2.50)]

$$\mathscr{F}\left\{\frac{1}{t};s\right\}=i\sqrt{\frac{\pi}{2}}\,\mathrm{sgn}(s)$$

it therefore follows from inverse transform relations that

$$\frac{1}{\sqrt{2\pi}}\int_{-\infty}^{\infty}e^{-ist}\,\mathrm{sgn}(s)\,ds=\sqrt{\frac{2}{\pi}}\frac{1}{it}$$

Interchanging the roles of t and s, and taking the complex conjugate of the result, leads to

$$\frac{1}{\sqrt{2\pi}}\int_{-\infty}^{\infty}e^{ist}\,\mathrm{sgn}(t)\,dt=-\sqrt{\frac{2}{\pi}}\frac{1}{is}$$

and thus we deduce that

$$\mathscr{F}\{\mathrm{sgn}(t);s\}=\sqrt{\frac{2}{\pi}}\frac{i}{s}$$

Example 2.17: Find the Fourier transform of the Heaviside unit function $h(t)$.

Solution: We first note that

$$h(t)=\frac{1}{2}[1+\mathrm{sgn}(t)]$$

Hence,

---

* We could also obtain (2.103) directly by writing $\cos\omega_0 t=(e^{i\omega_0 t}+e^{-i\omega_0 t})/2$ and using the formal relation (2.97).

$$\mathscr{F}\{h(t);s\} = \frac{1}{2}\left[\mathscr{F}\{1;s\} + \mathscr{F}\{\text{sgn}(t);s\}\right]$$
$$= \sqrt{\frac{\pi}{2}}\left[\delta(s) + \frac{i}{\pi s}\right]$$

where we are using (2.97)* and the result of Exam. 2.16.

Observe that, in Exam. 2.17, a superficial application of Eq. (2.57) to $\delta(t) = h'(t)$ would have resulted in

$$\mathscr{F}\{\delta(t);s\} = -is\mathscr{F}\{h(t);s\}$$

leading to the *incorrect* transform relation

$$\mathscr{F}\{h(t);s\} = -\frac{1}{is\sqrt{2\pi}}$$

That is, if $sF(s) = sG(s)$, it does not follow that $F(s) = G(s)$, but rather that

$$F(s) = G(s) + k\delta(s)$$

where k is a constant. This is a consequence of the property $s\delta(s) = 0$ (recall Prob. 4a in Exer. 1.5).

EXERCISES 2.8

1. Show formally that

(a) $\mathscr{F}\{\delta(t - a);s\} = \frac{1}{\sqrt{2\pi}} e^{ias}$

(b) $\mathscr{F}\{e^{iat};s\} = \sqrt{2\pi}\,\delta(s + a)$

2. Find the Fourier transform of $\sin \omega_0 t$.

3. Show that for any real function of time, the amplitude spectrum $|F(\omega)|$ is necessarily an even function of ω. Use this result to explain why the Fourier transform of $\cos \omega_0 t$ leads to two impulse functions.

4. Using the property

$$\mathscr{F}\{f^{(n)}(t);s\} = (-is)^n F(s), \qquad n = 1,2,3,\ldots$$

show formally that

(a) $\mathscr{F}\{\delta'(t);s\} = -is.$

(b) $\mathscr{F}\{\delta^{(n)}(t);s\} = (-is)^n, \qquad n = 1,2,3,\ldots$

* Observe that $\mathscr{F}\{1;s\} = \mathscr{F}^{-1}\{1;s\}$ due to the *even* property of $f(t) = 1$.

5. Using the property

$$\mathscr{F}\{t^n f(t);s\} = (-i)^n F^{(n)}(s), \qquad n = 1,2,3,\ldots$$

show formally that
(a) $\mathscr{F}\{t;s\} = -i\sqrt{2\pi}\,\delta'(s)$
(b) $\mathscr{F}\{t^n;s\} = (-i)^n\sqrt{2\pi}\,\delta^{(n)}(s), \qquad n = 1,2,3,\ldots$

6. Based on the Fourier transform relation

$$\mathscr{F}\left\{\frac{1}{t};s\right\} = i\sqrt{\frac{\pi}{2}}\,\operatorname{sgn}(s)$$

show that by writing

$$\frac{1}{t^n} = \frac{(-1)^{n-1}}{(n-1)!}\frac{d^{n-1}}{dt^{n-1}}\left(\frac{1}{t}\right)$$

it can be formally deduced that

$$\mathscr{F}\left\{\frac{1}{t^n};s\right\} = i\sqrt{\frac{\pi}{2}}\frac{(is)^{n-1}}{(n-1)!}\operatorname{sgn}(s), \qquad n = 1,2,3,\ldots$$

7. If

$$\int_{-\infty}^{\infty} f(t)\,dt = F(0) \neq 0$$

where $F(s)$ is the Fourier transform of $f(t)$, show that

$$\mathscr{F}\left\{\int_{-\infty}^{t} f(x)\,dx;s\right\} = \frac{i}{s}F(s) + \pi F(0)\delta(s)$$

Hint: Write $\int_{-\infty}^{t} f(x)dx = \int_{-\infty}^{\infty} h(t-x)f(x)\,dx$ and use the convolution theorem.

8. Starting with the identity $t\delta(t) = 0$,
(a) show formally that

$$\delta'(t) = -\frac{1}{t}\delta(t)$$

(b) Use the result in (a) to deduce that

$$\delta^{(n)}(t) = (-1)^n\,(n!/t^n)\,\delta(t), \qquad n = 1,2,3,\ldots$$

9. If m and n are positive integers such that $m < n$, use Prob. 8 to formally deduce that

$$t^{n-m}\,\delta^{(n)}(t) = (-1)^{n-m}(n!/m!)\,\delta^{(m)}(t)$$

10. Based on the result of Prob. 9, formally show that

(a) $e^{-xt}\,\delta^{(n)}(t) = \sum_{k=0}^{n} \binom{n}{k} t^{n-k}\,\delta^{(k)}(t)$

(b) $f(t)\delta^{(n)}(t) = \sum_{k=0}^{n} \binom{n}{k} (-1)^k f^{(n-k)}(0)\delta^{(k)}(t)$

In Probs. 11–15, verify the given Fourier transform relation.

11. $\mathscr{F}\{t^n \operatorname{sgn}(t);s\} = \sqrt{\frac{2}{\pi}} \frac{n!}{(-is)^{n+1}}, \qquad n = 0,1,2,\ldots$

12. $\mathscr{F}\{t^n h(t);s\} = \sqrt{\frac{\pi}{2}}\, i^n \left[\delta^{(n)}(s) + \frac{(-1)^n n!}{is^{n+1}}\right], \qquad n = 0,1,2,\ldots$

13. $\mathscr{F}\{t^{-\alpha} h(t);s\} = \frac{\Gamma(1-\alpha)}{\sqrt{2\pi}} |s|^{\alpha-1} e^{i\pi(1-\alpha)\operatorname{sgn}(s)}, \qquad \alpha > 1$ (α nonintegral)

14. $\mathscr{F}\{|t|^{-\alpha};s\} = \sqrt{\frac{2}{\pi}}\, \Gamma(1-\alpha)|s|^{\alpha-1} \cos[\tfrac{1}{2}\pi(1-\alpha)], \qquad \alpha > 1$ (α nonintegral)

15. $\mathscr{F}\{|t|^{-\alpha}\operatorname{sgn}(t);s\} = i\sqrt{\frac{2}{\pi}}\, \Gamma(1-\alpha)|s|^{\alpha-1} \sin[\tfrac{1}{2}\pi(1-\alpha)], \qquad \alpha > 1$ (α nonintegral)

2.9 *Hilbert Transforms*

In applications involving systems analysis to electrical networks, one often finds the need to determine the frequency response function (i.e., the Fourier transform of a waveform) when only its real or imaginary part is known. Basically the mathematical problem is to determine the complex function

$$F(\omega) = R(\omega) + iX(\omega) \tag{2.104}$$

from only knowledge of the real functions $R(\omega)$ or $X(\omega)$. If the function $f(t)$, which is the inverse Fourier transform of $F(\omega)$, is a *causal function* [i.e., $f(t) = 0$ for $t < 0$], then this mathematical problem can be solved by means of Hilbert transforms, which fundamentally provide integral relations between the real functions $R(\omega)$ and $X(\omega)$.

A related problem involving Hilbert transforms is to find a complex representation of a real signal. For instance, if $x(t)$ is a real signal and $\hat{x}(t)$ is its Hilbert transform, then the *analytic signal*

$$u(t) = x(t) + i\hat{x}(t) \tag{2.105}$$

has a Fourier transform that leads to a one-sided frequency spectrum, i.e., the spectrum vanishes for negative frequencies. Analytic signal representations are particularly useful in the analysis of narrowband waveforms.

For consistency of notation with that normally used in the discussion of Hilbert transforms, let us redefine the Fourier transform of a given function $f(t)$ by

$$\mathcal{F}\{f(t);\omega\} = \int_{-\infty}^{\infty} e^{-i\omega t} f(t)\, dt = F(\omega) \tag{2.106}$$

The corresponding inversion formula then takes the form

$$\mathcal{F}^{-1}\{F(\omega);t\} = \frac{1}{2\pi}\int_{-\infty}^{\infty} e^{i\omega t}\, F(\omega)\, d\omega = f(t) \tag{2.107}$$

If $f(t)$ is a real function, it follows that

$$R(\omega) = \int_{-\infty}^{\infty} f(t)\cos \omega t\, dt \tag{2.108}$$

and

$$X(\omega) = -\int_{-\infty}^{\infty} f(t)\sin \omega t\, dt \tag{2.109}$$

where $R(\omega)$ and $X(\omega)$ are the real and imaginary parts of $F(\omega)$ as given in Eq. (2.104). We can immediately deduce from this that $R(\omega)$ is an *even function* and that $X(\omega)$ is an *odd function*.

In the special case when $f(t)$ is an even function, we see that

$$R(\omega) = 2\int_{0}^{\infty} f(t)\cos \omega t\, dt \tag{2.110}$$

and $X(\omega) = 0$, and through the inverse cosine transform relation, it can be shown that

$$f(t) = \frac{1}{\pi}\int_{0}^{\infty} R(\omega)\cos \omega t\, d\omega \tag{2.111}$$

Similarly, when $f(t)$ is an odd function we have $R(\omega) = 0$ and

$$X(\omega) = -2\int_{0}^{\infty} f(t)\sin \omega t\, dt \tag{2.112}$$

Thus,

$$f(t) = -\frac{1}{\pi}\int_{0}^{\infty} X(\omega)\sin \omega t\, d\omega \tag{2.113}$$

We see, therefore, that if $f(t)$ is an even (odd) function, it can be recovered entirely from the real (imaginary) part of its Fourier transform $F(\omega)$.

Regardless of whether a function is even or odd, it can always be expressed as the sum of an even and an odd function. That is, by writing

$$f(t) = \tfrac{1}{2}[f(t) + f(-t)] + \tfrac{1}{2}[f(t) - f(-t)]$$

we have

$$f(t) = f_e(t) + f_o(t) \tag{2.114}$$

where

$$f_e(t) = \tfrac{1}{2}[f(t) + f(-t)] \tag{2.115}$$

is an even function and

$$f_o(t) = \tfrac{1}{2}[f(t) - f(-t)] \tag{2.116}$$

is an odd function. From our discussion above, it now follows that $R(\omega)$ is the Fourier transform of $f_e(t)$ while $X(\omega)$ is the Fourier transform of $f_o(t)$. Hence, we have the transform pairs

$$R(\omega) = 2\int_0^\infty f_e(t)\cos \omega t \, dt \tag{2.117}$$

$$f_e(t) = \frac{1}{\pi}\int_0^\infty R(\omega)\cos \omega t \, d\omega \tag{2.118}$$

and

$$X(\omega) = -2\int_0^\infty f_o(t)\sin \omega t \, dt \tag{2.119}$$

$$f_o(t) = \frac{1}{\pi}\int_0^\infty X(\omega)\sin \omega t \, d\omega \tag{2.120}$$

In the case when $f(t)$ is causal, it happens that $f(-t) = 0$ for $t > 0$, and therefore we deduce from (2.115) and (2.116) that

$$f(t) = 2f_e(t) = 2f_o(t), \qquad t > 0 \tag{2.121}$$

Under this condition, Eqs. (2.118) and (2.120) can be written as

$$f(t) = \frac{2}{\pi}\int_0^\infty R(\omega)\cos \omega t \, d\omega \tag{2.122}$$

and

$$f(t) = -\frac{2}{\pi}\int_0^\infty X(\omega)\sin \omega t \, d\omega \tag{2.123}$$

Hence, a causal function $f(t)$ can always be recovered from knowledge of either $R(\omega)$ or $X(\omega)$, which suggests a possible relation between the functions $R(\omega)$ and $X(\omega)$.

To obtain the relation between $R(\omega)$ and $X(\omega)$, we first note that $f_e(t)$ and $f_o(t)$ are related by

$$f_e(t) = f_o(t)\mathrm{sgn}(t) \tag{2.124}$$

$$f_o(t) = f_e(t)\mathrm{sgn}(t) \tag{2.125}$$

Hence, using (2.124) and properties of even and odd functions, we can rewrite (2.117) as

$$\begin{aligned} R(\omega) &= \int_{-\infty}^{\infty} e^{-i\omega t} f_e(t)\,dt \\ &= \int_{-\infty}^{\infty} e^{-i\omega t} f_o(t)\mathrm{sgn}(t)\,dt \end{aligned} \tag{2.126}$$

Treating this last expression as the Fourier transform of a product, we can use the convolution integral to evaluate it. Recalling that

$$\mathscr{F}\{f_o(t);\omega\} = iX(\omega) \tag{2.127}$$

and

$$\mathscr{F}\{\mathrm{sgn}(t);\omega\} = 2/i\omega \tag{2.128}$$

we find that (2.126) is equivalent to

$$R(\omega) = \frac{1}{2\pi}\int_{-\infty}^{\infty} iX(y)\left[\frac{2}{i(\omega - y)}\right] dy$$

or

$$R(\omega) = \frac{1}{\pi}\int_{-\infty}^{\infty} \frac{X(y)}{\omega - y}\,dy \tag{2.129}$$

Similarly, since

$$\begin{aligned} iX(\omega) &= \int_{-\infty}^{\infty} e^{-i\omega t} f_o(t)\,dt \\ &= \int_{-\infty}^{\infty} e^{-i\omega t} f_e(t)\mathrm{sgn}(t)\,dt \end{aligned}$$

we likewise deduce that

$$X(\omega) = -\frac{1}{\pi}\int_{-\infty}^{\infty} \frac{R(y)}{\omega - y}\,dy \tag{2.130}$$

Equations (2.129) and (2.130) form what we call a *Hilbert transform pair*. They are also called *Kramers–Krönig relations* in electromagnetic theory and are basically expressions of the causality condition, namely, the effect must not precede in time the cause producing it.

What we have shown by (2.129) and (2.130) is that the real and imaginary parts of the Fourier transform of a causal function satisfy Hilbert transform relations. Conversely, it can also be shown (although we do not give the proof here) that if the real and imaginary parts of the Fourier transform $F(\omega)$ satisfy Hilbert transform relations (2.129) and (2.130), the inverse Fourier transform $f(t)$ is a causal function.

Finally, there is an alternate form of the Hilbert transform pair that is sometimes more convenient to use. We obtain this alternate form by first recognizing that, since $R(\omega)$ is even and $X(\omega)$ is odd,

$$R(-\omega) = -\frac{1}{\pi}\int_{-\infty}^{\infty} \frac{X(y)}{\omega + y}\,dy = R(\omega) \tag{2.131}$$

Summing (2.129) and (2.131) leads to

$$2R(\omega) = \frac{1}{\pi}\int_{-\infty}^{\infty} X(y)\left(\frac{1}{\omega - y} - \frac{1}{\omega + y}\right) dy$$

from which we deduce

$$R(\omega) = \frac{2}{\pi}\int_{0}^{\infty} \frac{yX(y)}{\omega^2 - y^2}\,dy \tag{2.132}$$

In the same manner, we can show that

$$X(\omega) = -\frac{2\omega}{\pi}\int_{0}^{\infty} \frac{R(y)}{\omega^2 - y^2}\,dy \tag{2.133}$$

EXERCISES 2.9

In Probs. 1–4, verify that the given functions satisfy the Hilbert transform relations (2.129) and (2.130).

1. $R(\omega) = -\dfrac{a}{\omega^2 + a^2}, \qquad X(\omega) = -\dfrac{\omega}{\omega^2 + a^2}$

2. $R(\omega) = \dfrac{1 - \cos\omega - \omega\sin\omega}{-\omega^2}, \qquad X(\omega) = \dfrac{\sin\omega - \omega\cos\omega}{-\omega^2}$

3. $R(\omega) = \dfrac{a(a^2 + \omega^2 + 1)}{(a^2 + \omega^2)^2 + 2(a^2 - \omega^2) + 1}$

$X(\omega) = \dfrac{\omega(1 - a^2 - \omega^2)}{(a^2 + \omega^2)^2 + 2(a^2 - \omega^2) + 1}, \quad a > 0$

4. $R(\omega) = \dfrac{a\cos\omega - a\sin\omega}{\omega^2 + a^2}, \qquad X(\omega) = \dfrac{-\omega\cos\omega - a\sin\omega}{\omega^2 + a^2}, \qquad a > 0$

5. Prove that the Hilbert transform of a constant is zero, i.e., show that

$$\int_{-\infty}^{\infty} \frac{dy}{\omega - y} = 0$$

6. Given that $R(\omega) = -a/(\omega^2 + a^2)$, determine the causal function $f(t)$.

7. Given that $X(\omega) = -\omega/(\omega^2 + a^2)$, determine the causal function $f(t)$.

8. Show that

(a) $$X(\omega) = \frac{1}{\pi}\int_{-\infty}^{\infty} \frac{R(\omega) - R(y)}{\omega - y}\,dy$$

(b) $$X(\omega) = \frac{2\omega}{\pi}\int_{0}^{\infty} \frac{R(\omega) - R(y)}{\omega^2 - y^2}\,dy$$

(c) $$X(\omega) = \frac{1}{\pi}\int_{0}^{\infty} R'(y)\log\left|\frac{\omega + y}{\omega - y}\right|\,dy$$

Hint: Integrate by parts in the result of (b).

9. Let $F(\omega)$ be a complex function of ω that is analytic in the upper-half ω-plane and tends to zero as $\omega \to \infty$. [The condition that $F(\omega)$ has no poles in the upper-half ω-plane is equivalent to the causality condition that the inverse transform of $F(\omega)$ vanishes for $t < 0$.] Then show that (2.129) and (2.130) follow by applying Cauchy's integral theorem to the integral

$$I(\omega_1) = \int_C \frac{F(\omega)}{\omega - (\omega_1 - i\varepsilon)}\,d\omega, \qquad \varepsilon \to 0^+$$

where ω_1 is real, and the contour C is shown in the following figure as the sum of C_1 along the real axis and C_2 at infinity.

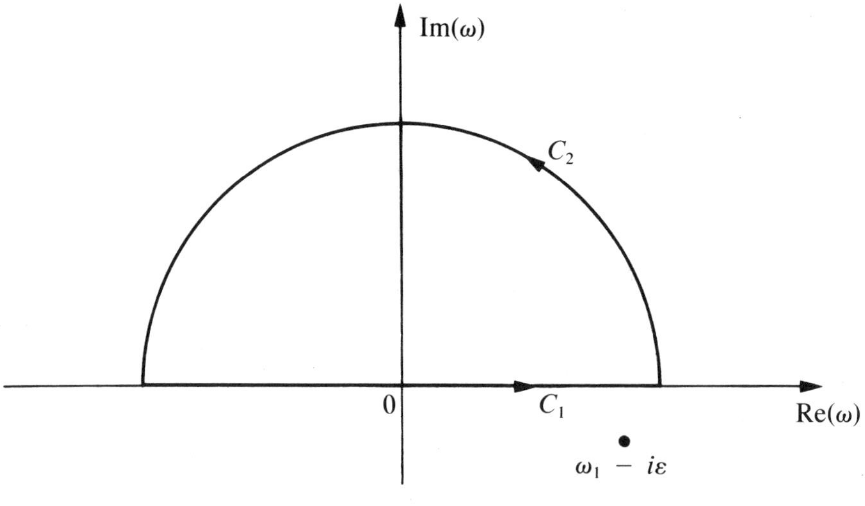

10. If $\hat{x}(t)$ is the Hilbert transform of $x(t)$, defined by

$$\hat{x}(t) = \frac{1}{\pi}\int_{-\infty}^{\infty} \frac{x(\tau)}{t-\tau}\, d\tau$$

show that $x(t)$ and $\hat{x}(t)$ satisfy the *orthogonality relation*

$$\int_{-\infty}^{\infty} x(t)\hat{x}(t)\, dt = 0$$

11. Let $x(t)$ be a real waveform which has no constant component. Show that $x(t)$ and its Hilbert transform $\hat{x}(t)$ satisfy the *Parseval relation*

$$\int_{-\infty}^{\infty} |x(t)|^2 dt = \int_{-\infty}^{\infty} |\hat{x}(t)|^2 dt$$

2.10 *Additional Topics*

In this final section on Fourier transforms we wish to touch briefly upon some related topics.

2.10.1 *Method of Stationary Phase*

In the use of Fourier transforms or inverse transforms we often have to deal with integrals whose evaluation by standardized methods is either difficult or impossible. In such cases we may have to use a numerical or approximation method to obtain the desired results. Sometimes we may only need to know the asymptotic behavior of the integral for large values of a parameter. In such instances Kelvin's *method of stationary phase* can be very useful.

To illustrate the technique, let us consider the integral

$$I(t) = \int_a^b F(s)e^{itG(s)}\, ds \tag{2.134}$$

whose asymptotic behavior for $t \to \infty$ is that which we seek. It is assumed that a and b are real constants and that $F(s)$ and $G(s)$ are both twice continuously differentiable real functions of s. Kelvin argued that the integrand in (2.134), regarded as a function of s, oscillates with increasing rapidity as $t \to \infty$ so that the contributions to $I(t)$ of adjacent portions of the integrand cancel one another except in the immediate vicinity of the end points of the interval and in the vicinity of those points at which $G(s)$ is *stationary*, i.e., points where $G'(s) = 0$. Moreover, in the first approximation the contribution of stationary points is more significant than that of the end points. Hence, if s_o is a point on the interval $a < s_o < b$ for which $G'(s_o) = 0$ and $G''(s_o) \neq 0$, then $G(s)$ may be approximated by the finite Taylor series

$$G(s) \simeq G(s_o) + \tfrac{1}{2} G''(s_o)(s - s_o)^2, \qquad |s - s_o| < \varepsilon \tag{2.135}$$

Similarly, we approximate $F(s)$ by $F(s_o)$, the first term in its Taylor expansion. Under the assumption that the major contribution to the integral occurs in the neighborhood of s_o, these approximations lead to

$$I(t) \sim F(s_o)e^{itG(s_o)} \int_{s_0-\varepsilon}^{s_0+\varepsilon} e^{it\frac{1}{2}G''(s_o)(s-s_o)^2} ds \tag{2.136}$$

By formally extending the limits of integration over $(-\infty,\infty)$ and using the result (see Prob. 11 in Exer. 2.10)

$$\int_{-\infty}^{\infty} e^{\pm ia^2u^2}\, du = \frac{\sqrt{2\pi}}{|a|}\, e^{\pm i\pi/4} \tag{2.137}$$

we deduce that

$$I(t) \sim F(s_o)\sqrt{2\pi/t|G''(s_o)|}\, \exp[itG(s_o) \pm i\pi/4], \qquad t \to \infty \tag{2.138}$$

where the plus sign goes with $G''(s_o) > 0$ and the minus sign with $G''(s_o) < 0$.

Example 2.18: Find the asymptotic behavior for $t \to \infty$ of

$$I(t) = \int_0^1 \cos[t(s^2 - s)]\, ds$$

Solution: Here we first write

$$I(t) = \mathrm{Re}\left\{\int_0^1 e^{it(s^2-s)}\, ds\right\}$$

from which we identify $F(s) = 1$ and $G(s) = s^2 - s$. Thus, calculating the derivatives

$$G'(s) = 2s - 1,\ G''(s) = 2$$

we find $s_o = 1/2$, and from (2.138) it immediately follows that

$$I(t) \sim \frac{\pi}{t} \cos\frac{1}{4}(t - \pi),\ t \to \infty$$

For a more thorough investigation of the method of stationary phase, the reader is advised to consult A. Erdelyi, *Asymptotic Expansions*, New York: Dover, 1956.

2.9.2 *Multiple Fourier Transforms*

The Fourier transform pairs (2.27) and (2.28) may be extended to functions of two or more variables. For example, if $f(x,y)$ is a piecewise continuous function such that

$$\int_{-\infty}^{\infty}\int_{-\infty}^{\infty} |f(x,y)|\, dx\, dy < \infty \tag{2.139}$$

we may consider the double Fourier transform pair

$$F(\xi,\eta) = \frac{1}{2\pi}\int_{-\infty}^{\infty}\int_{-\infty}^{\infty} e^{i(\xi x+\eta y)} f(x,y)\, dx\, dy \tag{2.140}$$

and

$$f(x,y) = \frac{1}{2\pi}\int_{-\infty}^{\infty}\int_{-\infty}^{\infty} e^{-i(\xi x+\eta y)} F(\xi,\eta) d\xi d\eta \tag{2.141}$$

We can interpret (2.140) as an iterated Fourier transform applied sequentially to the variable x and then to y. That is, we define

$$f^*(\xi,y) = \frac{1}{\sqrt{2\pi}}\int_{-\infty}^{\infty} e^{i\xi x} f(x,y)\, dx$$

followed by

$$F(\xi,\eta) = \frac{1}{\sqrt{2\pi}}\int_{-\infty}^{\infty} e^{i\eta y} f^*(\xi,y)\, dy$$

The inverse transform (2.141) follows in a similar manner. The *double Fourier transform* and *inverse double Fourier transform* are also denoted, respectively, by the notation

$$\mathscr{F}_{(2)}\{f(x,y);x\to\xi,\ y\to\eta\} = F(\xi,\eta) \tag{2.142}$$

and

$$\mathscr{F}_{(2)}^{-1}\{F(\xi,\eta);\xi\to x,\ \eta\to y\} = f(x,y) \tag{2.143}$$

Operational properties of the Fourier transform developed in Sec. 2.5 carry over in a natural sort of way to the double Fourier transform defined here. For example, the analogues of the *shifting properties* (Theor. 2.3) are

$$\mathscr{F}_{(2)}\{e^{i(ax+by)}f(x,y);x\to\xi,\ y\to\eta\} = F(\xi + a,\ \eta + b) \tag{2.144}$$

and

$$\mathscr{F}_{(2)}\{f(x - a,\ y - b);x\to\xi,\ y\to\eta\} = e^{i(a\xi+b\eta)}F(\xi,\eta) \tag{2.145}$$

The double Fourier *transform of derivatives* leads to the results

$$\mathscr{F}_{(2)}\left\{\frac{\partial f}{\partial x}(x,y);x\to\xi,\ y\to\eta\right\} = -i\xi F(\xi,\eta) \tag{2.146}$$

$$\mathscr{F}_{(2)}\left\{\frac{\partial^2 f}{\partial x\partial y}(x,y);x\to\xi,\ y\to\eta\right\} = -\xi\eta F(\xi,\eta) \tag{2.147}$$

and so on, and the *convolution theorem* assumes the form

$$\mathscr{F}_{(2)}^{-1}\{F(\xi,\eta)G(\xi,\eta);x\to\xi,\ y\to\eta\}$$
$$= \frac{1}{2\pi}\int_{-\infty}^{\infty}\int_{-\infty}^{\infty} f(x-u,\ y-v)g(u,v)du\ dv \quad (2.148)$$

The verification of these properties is left to the reader.

Generalizations of the double Fourier transform and its properties to n-dimensional Fourier transforms is obvious. For example, in three variables we define

$$F(\xi,\eta,\zeta) = (2\pi)^{-3/2}\int_{-\infty}^{\infty}\int_{-\infty}^{\infty}\int_{-\infty}^{\infty} e^{i(\xi x+\eta y+\zeta z)}f(x,y,z)dx\ dy\ dz \quad (2.149)$$

with a similar expression for the inverse transform. For reasons of theoretical tractability, one normally does not use Fourier transforms in dimensions greater than two. Actually, even though physical systems all have three-dimensions, we are often able to ignore one or two dimensions in our analysis by using some reasonable simplifying assumption, such as symmetry, and so on. The classical example where simplification to one or two dimensions is not possible, however, is the problem of diffraction of X-rays by crystals, which must be analyzed in three dimensions.

EXERCISES 2.10

In Probs. 1–3, verify the given asymptotic behavior of each integral.

1. $\displaystyle\int_0^1 e^{its^2}\,ds \sim \frac{1}{2}\sqrt{\frac{\pi}{t}}\,e^{i\pi/4}, \qquad t\to\infty$

2. $\displaystyle\int_0^1 e^{it\cos s}\,ds \sim \sqrt{\frac{\pi}{2t}}\,e^{i(t-\pi/4)}, \qquad t\to\infty$

3. $\displaystyle\int_0^\infty \cos\,(ts^2 - s)\,ds \sim \frac{1}{2}\sqrt{\frac{\pi}{2t}}, \qquad t\to\infty$

4. Using properties of the Fresnel integrals (Sec. 1.3.2), show that

(a) $\displaystyle\int_0^1 \cos\,[t(s^2 - s)]ds = \sqrt{\frac{2\pi}{t}}\left[\cos\frac{t}{4}C\left(\sqrt{\frac{t}{2\pi}}\right) + \sin\frac{t}{4}S\left(\sqrt{\frac{t}{2\pi}}\right)\right]$

(b) As $t\to\infty$, show that the result on the right-hand side in (a) approaches that given in Exam. 2.18.

5. If $G'(s_o) = G''(s_o) = \ldots = G^{(m-1)}(s_o) = 0$, $G^{(m)}(s_o) \neq 0$, show that Eq. (2.138) is generalized to

$$\int_a^b F(s)e^{itG(s)}ds \sim \frac{\Gamma(1/m)}{m}F(s_o)\left[\frac{m!}{t|G^{(m)}(s_o)|}\right]^{1/m}\exp\left[itG(s_o) \pm \frac{i\pi}{2m}\right]$$

6. Given the integral representation

$$J_n(x) = \frac{1}{\pi}\int_0^{\pi} \cos(x \sin s - ns)\, ds, \qquad n = 0,1,2,\ldots$$

show that large-order Bessel functions have the asymptotic behavior

$$J_n(x) \sim \frac{\Gamma(1/3)}{2^{2/3}3^{1/6}\pi n^{1/3}}, \quad n \to \infty$$

Hint: Use Prob. 5.

In Probs. 7–10, use the method of stationary phase to find an asymptotic expression for the given integral as $t \to \infty$. In some cases it may be necessary to use the result of Prob. 5.

7. $\int_0^1 e^{its^2}\cosh s^2\, ds$

8. $\int_0^1 \cos(ts^4)\tan s\, ds$

9. $\int_0^1 e^{it(s-\sin s)}\, ds$

10. $\int_0^1 \sin[t(s + \frac{1}{6}s^3 - \sinh s)]\cos s\, ds$

11. Show that

(a) $$\frac{1}{\sqrt{2\pi}}\int_{-\infty}^{\infty} e^{\pm ia^2u^2}\, du = \frac{1}{|a|}\sqrt{\frac{2}{\pi}}\left[\int_0^{\infty}\frac{\cos v}{\sqrt{v}}\, dv \pm i\int_0^{\infty}\frac{\sin v}{\sqrt{v}}\, dv\right]$$

(b) From (a), deduce that

$$\frac{1}{\sqrt{2\pi}}\int_{-\infty}^{\infty} e^{\pm ia^2u^2}\, du = \frac{1}{|a|}e^{\pm i\pi/4}$$

12. Show that

$$\mathscr{F}_{(2)}\{f(ax,by);x\to\xi,\ y\to\eta\} = \frac{1}{|ab|}F(\xi/a,\ \eta/b)$$

13. Show that

$$\mathscr{F}_{(2)}\{f(x-a,\ y-b);x\to\xi,\ y\to\eta\} = e^{i(a\xi+b\eta)}F(\xi,\eta)$$

14. Show that

$$\mathscr{F}_{(2)}\{f(x,y)\cos\omega x;x\to\xi,\ y\to\eta\} = \tfrac{1}{2}[F(\xi+\omega,\eta) + F(\xi-\omega,\eta)]$$

15. Verify the convolution formula

$$\mathscr{F}_{(2)}^{-1}\{F(\xi,\eta)G(\xi,\eta);\xi\to x,\ \eta\to y\} = (f*g)(x,y)$$

where

$$(f*g)(x,y) = \frac{1}{2\pi}\int_{-\infty}^{\infty}\int_{-\infty}^{\infty} f(u,v)g(u-x,\ v-y)\, du\, dv$$

3

Applications Involving Fourier Transforms

3.1 *Introduction*

The use of Fourier integrals and Fourier transforms in applications is quite extensive. Therefore, we will make no attempt to consider all the various applications involving these integrals, but rather briefly discuss how they are used in several representative areas of application.

The basic aim of the transform method is to transform the given problem into one that is easier to solve. In the case of an ordinary differential equation (ODE) with constant coefficients, the transformed problem is algebraic. The effect of applying an integral transform to a partial differential equation (PDE) is to exclude temporarily a chosen independent variable and to leave for solution a PDE in one less variable. The solution of the transformed problem in either case will be a function of the transform variable s and any remaining independent variables. Inverting this solution produces the solution of the original problem.

The exponential Fourier transform may be applied to derivatives of all orders that may occur in the formulation of a given problem. However, since it incorporates no boundary conditions in transforming these derivatives, it is best suited for solving DEs on infinite domains where the boundary conditions usually only require bounded solutions. On the other hand, the Fourier cosine and Fourier sine transforms are well suited for solving certain problems on semiinfinite domains where the governing DE involves only even-order derivatives. That is, because the cosine (sine) transform of an odd-order derivative involves the sine (cosine) transform of the original function, these transforms are generally not useful in solving DEs containing odd-ordered derivatives.

3.2 *Boundary Value Problems*

Fourier integrals and Fourier transforms are very useful in solving boundary value problems on *infinite domains*. Such problems, however, are classified as *singular* since they contain no finite boundaries.* In this case we normally prescribe boundary conditions of the form

$$y(x),\ y'(x) \text{ finite as } |x| \to \infty \tag{3.1}$$

Nonetheless, the use of Fourier transforms often forces us (at least initially) to impose the more stringent requirements

$$y(x) \to 0,\ y'(x) \to 0 \text{ as } |x| \to \infty \tag{3.2}$$

These stronger requirements are necessary to ensure that the Fourier transforms of $y'(x)$ and $y''(x)$ exist. Even so, the formal solutions that we generate by the transform method may not satisfy (3.2). In such cases we normally require $y(x)$ to at least satisfy (3.1). Because of this, we often take the point of view in practice that the transform method produces a "tentative solution" which must be independently verified that it is indeed a proper solution of the problem.

Example 3.1: Use the Fourier transform to solve

$$y'' - y = -h(1 - |x|), \qquad -\infty < x < \infty$$
$$y(x) \to 0,\ y'(x) \to 0 \text{ as } |x| \to \infty$$

Solution: Let us introduce the Fourier transforms

$$\mathscr{F}\{y(x);s\} = Y(s)$$
$$\mathscr{F}\{y''(x);s\} = -s^2Y(s)$$

and

$$\mathscr{F}\{h(1 - |x|);s\} = \frac{1}{\sqrt{2\pi}}\int_{-1}^{1} e^{isx}\,dx = \sqrt{\frac{2}{\pi}}\,\frac{\sin s}{s}$$

Thus, by applying the Fourier transform to each term in the equation,

* For a discussion of singular boundary value problems, see Sec. 1.6.4 in L. C. Andrews, *Elementary Partial Differential Equations with Boundary Value Problems*, Orlando: Academic Press, 1986.

the DE is converted into the algebraic equation

$$-s^2Y(s) - Y(s) = -\sqrt{\frac{2}{\pi}}\frac{\sin s}{s}$$

with solution

$$Y(s) = \sqrt{\frac{2}{\pi}}\frac{\sin s}{s(s^2+1)}$$

Finally, the solution $y(x)$ that we seek is simply the inverse Fourier transform of $Y(s)$, which we can obtain through use of the convolution theorem. That is, we have

$$\mathscr{F}^{-1}\left\{\sqrt{\frac{2}{\pi}}\frac{\sin s}{s};x\right\} = h(1 - |x|)$$

and

$$\mathscr{F}^{-1}\left\{\frac{1}{s^2+1};x\right\} = \sqrt{\frac{\pi}{2}}e^{-|x|}$$

from which we deduce

$$\begin{aligned} y(x) &= \mathscr{F}^{-1}\{Y(s);x\} \\ &= \frac{1}{2}\int_{-\infty}^{\infty} e^{-|u|}\, h(1 - |x - u|)du \\ &= \frac{1}{2}\int_{x-1}^{x+1} e^{-|u|}\, du \end{aligned}$$

The evaluation of this last integral leads to

$$y(x) = \begin{cases} \sinh(1)e^x, & -\infty < x < -1 \\ 1 - e^{-1}\cosh x, & -1 \le x \le 1 \\ \sinh(1)e^{-x}, & 1 < x < \infty \end{cases}$$

(see Prob. 1 in Exer. 3.2).

This one example illustrates the basic technique used in solving boundary value problems with constant coefficients by the transform method. That is, the transform applied term-by-term to the DE leads to an algebraic equation in the transform function $Y(s)$. This algebraic equation can be readily solved for $Y(s)$, the inversion of which yields the desired solution $y(x)$. It is perhaps the final step of inverting $Y(s)$ that is generally the most difficult part of the process. In many cases, we simply represent our solution as an integral whose evaluation is left to numerical procedures.

If the domain of interest in Ex. 3.1 were modified to the semiinifinite region $0 < x < \infty$, then either the sine or cosine transform would be used to solve the problem. The decision as to which of these two transforms to use will depend upon the type of boundary condition specified at the finite boundary $x = 0$. That is, from Eqs. (2.60) and (2.61) in Sec. 2.5, we recall

$$\mathscr{F}_C\{y''(x);s\} = -s^2 Y_C(s) - \sqrt{\frac{2}{\pi}}\, y'(0) \tag{3.3}$$

$$\mathscr{F}_S\{y''(x);s\} = -s^2 Y_S(s) + \sqrt{\frac{2}{\pi}}\, sy(0) \tag{3.4}$$

and so if $y(0)$ is specified we use the sine transform, whereas the cosine transform is called for when $y'(0)$ is specified. Problems involving these transforms are taken up in the exercises.

Finally, we remark that the technique illustrated in Ex. 3.1 can readily be generalized to boundary value problems of the more general form

$$\begin{gathered} y'' + ay' - by = f(x), \qquad -\infty < x < \infty \\ y(x) \to 0,\ y'(x) \to 0 \text{ as } |x| \to \infty \end{gathered} \tag{3.5}$$

where a and b ($b > 0$) are constants. Proceeding as before, the Fourier transform applied to (3.5) leads to the algebraic problem

$$-(s^2 + ias + b)Y(s) = F(s)$$

where $F(s)$ is the Fourier transform of $f(x)$. Solving for $Y(s)$, we get

$$Y(s) = -\frac{F(s)}{s^2 + ias + b} \tag{3.6}$$

The inversion of (3.6) by use of the convolution theorem [see Eq. (2.81) in Sec. 2.7] gives us the solution formula

$$y(x) = -\int_{-\infty}^{\infty} f(u)g(x - u)\, du \tag{3.7}$$

where we define

$$g(x) = \frac{1}{\sqrt{2\pi}}\mathscr{F}^{-1}\left\{\frac{1}{s^2 + ias + b};x\right\} \tag{3.8}$$

Recalling Exam. 2.8 in Chap. 2, we have

$$g(x) = \frac{1}{\sqrt{a^2 + 4b}}\exp\left[-\frac{1}{2}\left(ax + \sqrt{a^2 + 4b}\,|x|\right)\right] \tag{3.9}$$

and our solution process is complete.

The use of Fourier integral representations to solve boundary value problems is very similar to that of the Fourier transform. Let us illustrate with the following example.

Example 3.2: Let us consider an "infinitely long" beam resting on an elastic foundation as shown in Fig. 3.1. Such a problem might physically correspond to a railroad track on a road bed. If $f(x)$ represents a prescribed distributed load on this beam, the classical *beam theory of Euler* predicts that the vertical deflections $y(x)$ will satisfy the differential equation*

$$EIy^{(4)} + ky = f(x), \quad -\infty < x < \infty$$

where E and I are physical constants and k is the foundation modulus (i.e., a stiffness factor associated with the road bed). Solve for the deflection $y(x)$ when $f(x) = F_0 h(1 - |x|)$, where F_0 is a constant.

Solution: We begin by assuming the solution y has the integral representation

$$y(x) = \int_0^\infty [A(s)\cos sx + B(s)\sin sx]\, ds$$

where $A(s)$ and $B(s)$ are unknown functions. Differentiating this expression with respect to x yields

$$y^{(4)}(x) = \int_0^\infty s^4 [A(s)\cos sx + B(s)\sin sx]\, ds$$

The function $f(x) = F_0 h(1 - |x|)$ has the Fourier integral representation (see Exam. 2.1 in Chap. 2)

$$f(x) = \frac{2F_0}{\pi} \int_0^\infty \left(\frac{\sin s}{s}\right) \cos sx\, ds$$

and by substituting these integral representations directly into the governing DE, we find

$$\int_0^\infty (EIs^4 + k)[A(s)\cos sx + B(s)\sin sx]\, dx = \frac{2F_0}{\pi} \int_0^\infty \left(\frac{\sin s}{s}\right) \cos sx\, ds$$

Next, we equate like coefficients of $\cos sx$ and $\sin sx$ under the integrals, which leads to the system of algebraic equations

$$(EIs^4 + k)A(s) = \frac{2F_0}{\pi} \frac{\sin s}{s}$$

$$(EIs^4 + k)B(s) = 0$$

---

* See the discussion in Sec. 3.4.2 on vibrating beams.

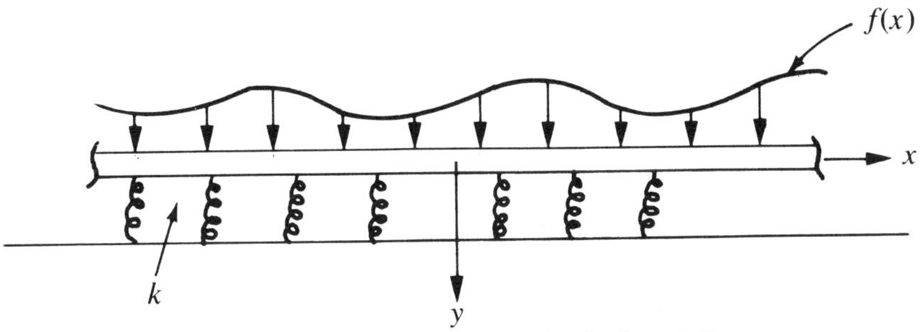

Figure 3.1 Infinite beam on elastic foundation

Solving, we deduce that $B(s) = 0$ and

$$A(s) = \frac{2F_0}{\pi} \frac{\sin s}{s(EIs^4 + k)}$$

Therefore, our solution becomes

$$y(x) = \frac{2F_0}{\pi} \int_0^\infty \frac{\sin s}{s(EIs^4 + k)} \cos sx \, ds$$

The final evaluation of this integral by using residue theory is left to the exercises (see Prob. 8 in Exer. 3.2).

3.2.1 *Linear Integral Equations*

Another type of boundary-value problem includes equations of the form

$$f(x) = \int_a^b k(x,t)u(t) \, dt \tag{3.10}$$

where $f(x)$ and $k(x,t)$ are known functions and the function $u(x)$ is to be determined. Such equations are called *linear integral equations of the first kind*. A familiar example of this kind of equation is the integral transform

$$F(s) = \frac{1}{\sqrt{2\pi}} \int_{-\infty}^\infty e^{ist} f(t) \, dt \tag{3.11}$$

where $F(s)$ is known and we want $f(t)$. The characteristic feature of an integral equation is that the unknown function appears under the integral — hence, the name *integral equation*.

Linear integral equations of the first kind are essentially integral transforms where $k(x,t)$ is the kernel. They are solved by finding an inverse transform relation. For example, Eq. (3.11) has the solution (under appropriate conditions)

$$f(t) = \frac{1}{\sqrt{2\pi}} \int_{-\infty}^{\infty} e^{-its} F(s)\, ds \tag{3.12}$$

Another variety of integral equation has the more general form

$$f(x) = u(x) - \lambda \int_a^b k(x,t)u(t)\, dt \tag{3.13}$$

called a *linear integral equation of the second kind*. Again it is assumed that $f(x)$ and $k(x,t)$ are known functions and the unknown function is $u(x)$. If $f(x)$ is identically zero on the interval $a \leq x \leq b$, the integral equation is called *homogeneous*, and *nonhomogeneous*, otherwise.

Integral equations like (3.13) with $f(x) \equiv 0$ always possess the *trivial solution* $u = 0$. Nontrivial solutions of the homogeneous equation are possible only for certain values of the parameter λ, called *eigenvalues*. The corresponding nontrivial solutions are then called *eigenfunctions*. The theory concerning such eigenvalue problems closely parallels the corresponding theory involving ordinary differential equations.* When the function $f(x)$ is not identically zero, the resulting nonhomogeneous equation usually has a solution only if λ is *not* one of the eigenvalues.

In some instances the upper limit of integration in (3.13) is variable, i.e.,

$$f(x) = u(x) - \lambda \int_a^x k(x,t)u(t)\, dt \tag{3.14}$$

and we refer to such an equation as a *Volterra equation*. If the kernel $k(x,t)$ in either (3.13) or (3.14) satisfies the inequality

$$\int_a^b \int_a^b |k(x,t)|^2\, dt\, dx < \infty$$

then it is called a *Hilbert–Schmidt kernel* and (3.13) is further known as a *Fredholm integral equation*.

Our approach here will not be to discuss the general theory or solution techniques applicable to solving either (3.13) or (3.14). Rather, we will single out a few examples of integral equations for which the transform method is particularly well suited.

Perhaps the simplest integral equations to tackle by means of the Fourier transform are those of *convolution type*,

$$\frac{1}{\sqrt{2\pi}} \int_{-\infty}^{\infty} u(t)k(x-t)\, dt = f(x), \quad -\infty < x < \infty \tag{3.15}$$

---

* See L. C. Andrews, *Elementary Partial Differential Equations with Boundary Value Problems*, Orlando: Academic Press, 1986. For a discussion of the theory of integral equations, see W. L. Lovitt, *Linear Integral Equations*, New York: Dover, 1950.

in which the functions $f(x)$ and $k(x)$ are prescribed for all real values of x and are assumed to possess Fourier transforms $F(s)$ and $K(s)$, respectively. Taking the Fourier transform of each side of (3.15), we have formally

$$U(s)K(s) = F(s)$$

which may be written in the form

$$U(s) = F(s)\cdot\frac{1}{K(s)} = F(s)L(s) \tag{3.16}$$

Now, if the inverse Fourier transform

$$\ell(x) = \mathscr{F}^{-1}\{L(s);x\} \tag{3.17}$$

exists, then by applying the convolution theorem we obtain the solution of (3.15) in the form

$$u(x) = \frac{1}{\sqrt{2\pi}}\int_{-\infty}^{\infty} f(t)\ell(x - t)\,dt \tag{3.18}$$

In some cases it may turn out that $L(s) = 1/K(s)$ does not have the inverse transform given by (3.17), but the related inverse transform

$$m(x) = \mathscr{F}^{-1}\left\{\frac{1}{(-is)^n K(s)};x\right\} = \mathscr{F}^{-1}\{M(s);x\} \tag{3.19}$$

does exist for some n ($n = 1,2,3,\ldots$). Thus, by writing (3.16) in the form

$$U(s) = [(-is)^n F(s)]M(s) \tag{3.20}$$

we obtain the formal solution

$$u(x) = \frac{1}{\sqrt{2\pi}}\int_{-\infty}^{\infty} f^{(n)}(t)m(x - t)\,dt \tag{3.21}$$

Example 3.3: Solve

$$f(x) = \frac{1}{\sqrt{2\pi}}\int_{-\infty}^{\infty} |x - \xi|^{-1/2}\,u(\xi)\,d\xi$$

Solution: In this case we apply the Fourier transform to each side of the given integral equation to get

$$F(s) = |s|^{-1/2}\,U(s)$$

and thus,

$$U(s) = |s|^{1/2}\,F(s)$$

Because the function $|s|^{1/2}$ does not have an inverse Fourier transform, we choose to rewrite $U(s)$ in the equivalent form

$$U(s) = i\,\mathrm{sgn}(s)|s|^{-1/2}\cdot(-is)F(s)$$

By recalling the transform relations

$$\mathscr{F}^{-1}\{i\,\mathrm{sgn}(s)|s|^{-1/2};x\} = |x|^{-1/2}\mathrm{sgn}(x)$$

and

$$\mathscr{F}^{-1}\{(-is)F(s);x\} = f'(x)$$

we deduce by use of the convolution theorem that

$$u(x) = \frac{1}{\sqrt{2\pi}}\int_{-\infty}^{x} |x - \xi|^{-1/2}f'(\xi)d\xi - \frac{1}{\sqrt{2\pi}}\int_{x}^{\infty} |\xi - x|^{-1/2}f'(\xi)d\xi$$

Ex. 3.3 involves an integral equation of the first kind. To illustrate the integral transform method for integral equations of the second kind, consider the following example.

Example 3.4: Solve the integral equation

$$u(x) - \frac{1}{2}\int_{-\infty}^{\infty} e^{-2|x - t|}\, u(t)\, dt = f(x), \qquad -\infty < x < \infty$$

Solution: Applying the Fourier transform to each side of the equation, we find

$$U(s) - \frac{2U(s)}{s^2 + 4} = F(s)$$

where $U(s)$ and $F(s)$ are the Fourier transforms, respectively, of $u(x)$ and $f(x)$. This algebraic equation has solution

$$U(s) = F(s) + 2F(s)/(s^2 + 2)$$

the inverse Fourier transform of which leads to

$$u(x) = f(x) + \frac{1}{\sqrt{2}}\int_{-\infty}^{\infty} e^{-\sqrt{2}|x - t|}\, f(t)\, dt$$

EXERCISES 3.2

1. Given the integral

$$I = \int_{x-1}^{x+1} e^{-|u|}\, du$$

show that

(a) $I = \int_{x-1}^{x+1} e^{u}\, du, \quad -\infty < x < -1$

(b) $I = \int_{x-1}^{0} e^{u}\, du + \int_{0}^{x+1} e^{-u}\, du, \qquad -1 \le x \le 1$

(c) $I = \int_{x-1}^{x+1} e^{-u}\, du, \quad 1 < x < \infty$

(d) From (a), (b), and (c), deduce the value of I.

In Probs. 2–6, use an appropriate integral transform to solve the given problem.

2. $y'' - y = e^{-|x|}, \qquad -\infty < x < \infty,$
$y(x) \to 0,\ y'(x) \to 0$ as $|x| \to \infty$

3. $y'' - y = e^{-x}, \qquad 0 < x < \infty$
$y(0) = 0,\ y(x) \to 0,\ y'(x) \to 0$ as $x \to \infty$

4. $y'' - y = e^{-x}, \qquad 0 < x < \infty$
$y'(0) = 0,\ y(x) \to 0,\ y'(x) \to 0$ as $x \to \infty$

5. $y'' - k^2 y = -h(1 - x), \qquad k > 0,\ 0 < x < \infty$
$y'(0) = 0,\ y(x) \to 0,\ y'(x) \to 0$ as $x \to \infty$

6. $y'' - y = xe^{-x}, \qquad 0 < x < \infty$
$y(0) = 0,\ y(x) \to 0,\ y'(x) \to 0$ as $x \to \infty$

Hint: $\mathscr{F}_S\{t^2 e^{-t}; s\} = \sqrt{2/\pi}\, 2s(3 - s^2)/(s^2 + 1)^3$

7. Using Fourier transforms, show that

$$y^{(4)} + K^4 y = f(x), \qquad -\infty < x < \infty$$

has the formal solution

$$y(x) = \frac{1}{\sqrt{2\pi}} \int_{-\infty}^{\infty} f(\xi) g(x - \xi)\, d\xi$$

where

$$g(x) = \frac{\sqrt{\pi}}{2K^3} e^{-K|x|/\sqrt{2}} [\cos(Kx/\sqrt{2}) + \sin(K|x|/\sqrt{2})]$$

8. Use residue theory to show that the integral solution in Exam. 3.2 can be reduced to

$$y(x) = \frac{F_0}{2k}\left[e^{-b(1+x)/\sqrt{2}} \sin\frac{b(1 + x)}{\sqrt{2}} + e^{-b(1-x)/\sqrt{2}} \sin\frac{b(1-x)}{\sqrt{2}} \right]$$

where $b^4 = k/EI$.

In Probs. 9–14, solve the given integral equation of the first kind.

9. $\int_{-\infty}^{\infty} u(t)e^{ixt}\,dt = \frac{\sin x}{x}$ **10.** $\int_0^{\infty} u(t)\cos xt\,dt = \frac{\pi}{2}e^{-x}$

11. $\frac{1}{\sqrt{2\pi}}\int_{-\infty}^{\infty} e^{-|x-\xi|}u(\xi)\,d\xi = e^{-x^2/2}$

12. $\frac{1}{2a\sqrt{\pi t}}\int_{-\infty}^{\infty} u(\xi)e^{-(x-\xi)^2/4a^2t}\,d\xi = \frac{1}{\sqrt{1+t}}e^{-x^2/4a^2(1+t)}$

13. $\int_{-\infty}^{\infty} \frac{u(\xi)}{(x-\xi)^2+a^2}\,d\xi = \frac{1}{x^2+b^2},\quad 0 < a < b$

14. $\int_{-\infty}^{\infty} u(\xi)u(x-\xi)\,d\xi = e^{-x^2}$

15. Show that

$$u(x) = e^{-|x|} + \lambda\int_x^{\infty} e^{x-\xi}\,u(\xi)\,d\xi, \qquad 0 < \lambda < 1$$

has the solution

$$u(x) = \begin{cases} \dfrac{2}{2-\lambda}e^{(1-\lambda)x}, & x < 0 \\ \dfrac{2}{2-\lambda}e^{-x}, & x \geq 0 \end{cases}$$

16. Given the homogeneous integral equation

$$u(x) = \frac{1}{\pi}\int_{-\infty}^{\infty} \frac{tu(t)dt}{[1+(x-t)^2](1+t^2)}$$

(a) show that

$$u(x) = \sqrt{2\pi}\int_0^{\infty} \frac{\sin xt}{e^t - 1}\,dt$$

(b) Using the relation

$$\coth \pi x = \frac{1}{\pi x} + \frac{2x}{\pi}\sum_{n=1}^{\infty} \frac{1}{x^2+n^2}$$

show that (a) reduces to

$$u(x) = \sqrt{2\pi}\left(\frac{\pi}{2}\coth \pi x - \frac{1}{2x}\right)$$

3.3 *Heat Conduction in Solids*

It is well known that if the temperature u in a solid body is not constant, heat energy flows in the direction of the gradient $-\nabla u$ with magnitude $k|\nabla u|$. The quantity k is called the *thermal conductivity* of the material and the above principle is called *Fourier's law of heat conduction*. This law combined with the *law of conservation of thermal energy*, which states that

> "... the rate of heat entering a region plus that which is generated inside the region equals the rate of heat leaving the region plus that which is stored...,"

leads to the PDE*

$$\nabla^2 u = a^{-2}u_t - q(x,y,z,t) \tag{3.22}$$

where a^2 is another physical constant called the *diffusivity*. Eq. (3.22) is commonly called the *heat equation* or *diffusion equation*. In addition to being the governing equation in determining the temperature distribution in homogeneous solids, it also occurs in problems involving electromagnetic theory, diffusion processes, and the propagation of current in transmission lines. When present, the term q is proportional to a heat source distributed throughout the solid body.

The quantity $\nabla^2 u$ in (3.22) is called the *Laplacian* and is a measure of the difference between the value of u at a point and the average value of u in a small neighborhood of the point. In rectangular coordinates, the Laplacian takes the form

$$\nabla^2 u = u_{xx} + u_{yy} + u_{zz} \tag{3.23}$$

Remark: We are adopting the standard notation of writing partial derivatives as subscripted variables. Thus, we will write u_x for $\partial u/\partial x$, u_{xx} for $\partial^2 u/\partial x^2$, u_{xy} for $\partial^2 u/\partial y \partial x$, and so on.

The fundamental problem in the mathematical theory of heat conduction is to solve Eq. (3.22) for the temperature u in a homogeneous solid when the distribution of temperature throughout the solid is known at time $t = 0$ and a certain boundary condition is prescribed at each exposed point of the solid. There are three distinct kinds of boundary conditions that might ordinarily be prescribed for such problems. The first is to specify

---

* For details, see J. Fourier, *The Analytical Theory of Heat*, New York: Dover, 1955.

the temperature u along each finite surface of the solid (*Dirichlet condition*); the second condition is to prescribe the flux of heat across the surface, which is accomplished by specifying the *normal derivative* of u at the surface (*Neumann condition*); and the third boundary condition is to prescribe the rate at which heat is lost from the solid due to surface radiation into the surrounding medium (*Newton's law of cooling*). This last boundary condition is sometimes called *Robin's condition*.

When the unknown function u depends only upon the spatial variable x and on time t, then (3.22) becomes

$$u_{xx} = a^{-2}u_t - q(x,t) \tag{3.24}$$

and when no heat source is present, (3.24) reduces further to

$$u_{xx} = a^{-2}u_t \tag{3.25}$$

Both (3.24) and (3.25) are called *one-dimensional* heat equations. Among other areas of application, these one-dimensional heat equations govern the temperature distribution in a long rod or wire whose lateral surface is impervious to heat (i.e., insulated). For modeling purposes we assume in this case that the rod coincides with a portion of the x axis, is made of homogeneous material, and has uniform cross section. Further, we will assume that the temperature $u(x,t)$ is the same at any pont in a particular cross section of the rod, but may change from cross section to cross section.

3.3.1 *Heat Equation on an Infinite Line*

Let us first consider the flow of heat in the infinite medium $-\infty < x < \infty$ when the initial temperature distribution $f(x)$ is known and the region is free of any heat sources. Physically, this problem might represent the linear flow of heat in a very long slender rod whose lateral surface is insulated. In such cases, the solution will represent the temperature distribution in the middle portion of the infinitely long rod, prior to the time when such temperatures are greatly influenced by the actual boundary conditions of the rod. The problem is mathematically characterized by

$$\begin{aligned} & u_{xx} = a^{-2}u_t, \qquad -\infty < x < \infty,\ t > 0 \\ \text{B.C.:}\quad & u(x,t) \to 0,\ u_x(x,t) \to 0 \text{ as } |x| \to \infty \\ \text{I.C.:}\quad & u(x,0) = f(x), \qquad -\infty < x < \infty \end{aligned} \tag{3.26}$$

Remark: Since both initial conditions and boundary conditions are prescribed in (3.26), we have designated the boundary conditions by "B.C." and the initial conditon by "I.C." to clearly identify each type of auxiliary condition. Also, the limiting boundary conditions are based primarily upon physical considerations of heat flow, but in some cases are prescribed merely to ensure that the relevant integral transforms exist.

The infinite range on the spatial variable x suggests use of the exponential Fourier transform rather than either the cosine or sine transform. However, because the function we wish to transform depends upon more than one independent variable, it is useful to adopt a special kind of notation to desigate the variable being transformed. For example, the Fourier transform of $u(x,t)$ with respect to x is defined by

$$\mathscr{F}\{u(x,t);x\to s\} = \frac{1}{\sqrt{2\pi}}\int_{-\infty}^{\infty} e^{isx}u(x,t)\,dx = U(s,t) \tag{3.27}$$

Similarly, we have that

$$\mathscr{F}\{u_x(x,t);x\to s\} = -\,isU(s,t) \tag{3.28}$$

$$\mathscr{F}\{u_{xx}(x,t);x\to s\} = -\,s^2U(s,t) \tag{3.29}$$

and

$$\begin{aligned}\mathscr{F}\{u_t(x,t);x\to s\} &= \frac{1}{\sqrt{2\pi}}\int_{-\infty}^{\infty} e^{isx}\,u_t(x,t)\,dx\\ &= \frac{1}{\sqrt{2\pi}}\frac{\partial}{\partial t}\int_{-\infty}^{\infty} e^{isx}\,u(x,t)\,dx\end{aligned}$$

or

$$\mathscr{F}\{u_t(x,t);x\to s\} = U_t(s,t) \tag{3.30}$$

Using the above results together with the relation $F(s) = \mathscr{F}\{f(x);s\}$, we find that the Fourier transform applied to the problem described by (3.26) leads to the transformed problem

$$\begin{aligned} &U_t + a^2s^2U = 0, \qquad t > 0\\ \text{I.C.}\quad &U(s,0) = F(s), \qquad -\infty < s < \infty\end{aligned} \tag{3.31}$$

We recognize (3.31) as a first-order initial-value problem whose solution is readily found to be

$$U(s,t) = F(s)e^{-a^2s^2t} \tag{3.32}$$

The solution of the original problem is now found by taking the inverse Fourier transform of (3.32). The product form of (3.32) suggests use of the convolution integral, which yields

$$u(x,t) = \frac{1}{\sqrt{2\pi}}\int_{-\infty}^{\infty} f(\xi)g(x-\xi,t)\,d\xi \tag{3.33}$$

where

$$g(x,t) = \mathscr{F}^{-1}\{e^{-a^2s^2t};s\to x\} = \frac{1}{a\sqrt{2t}}\,e^{-x^2/4a^2t} \tag{3.34}$$

Hence, we have obtained the formal solution

$$u(x,t) = \frac{1}{2a\sqrt{\pi t}}\int_{-\infty}^{\infty} f(\xi)e^{-(x-\xi)^2/4a^2t}\,d\xi \tag{3.35}$$

The integral (3.35) expresses the solution $u(x,t)$ as a sum of the effects of the initial temperature distribution $f(\xi)$ at various points along the infinite rod. The heat kernel

$$\frac{1}{2a\sqrt{\pi t}}\,e^{-(x-\xi)^2/4a^2t}$$

physically represents the temperature distribution at point x and time t due to a concentrated "hot spot" (heat source) at point $x = \xi$ and time $t = 0$. That this heat kernel depends only on the difference $x - \xi$ is a consequence of the fact that the medium is homogeneous, i.e., the coefficients in the heat equation do not depend on x.

By making the change of variable $z = (\xi - x)/2a\sqrt{t}$, we can express (3.35) in the equivalent form

$$u(x,t) = \frac{1}{\sqrt{\pi}}\int_{-\infty}^{\infty} f(x + 2az\sqrt{t})e^{-z^2}\,dz \tag{3.36}$$

This form of the solution is particularly useful if the initial temperature distribution is constant over various intervals of the rod. For example, if $f(x) = T_0$, $-\infty < x < \infty$, then (3.36) leads to the trivial result

$$u(x,t) = \frac{T_0}{\sqrt{\pi}}\int_{-\infty}^{\infty} e^{-z^2}\,dz = T_0 \tag{3.37}$$

by using properties of the gamma function or error function. It is interesting, however, that (3.37) does not satisfy the limiting boundary condition

$$u(x,t) \to 0 \text{ as } |x| \to \infty$$

prescribed in (3.26). Moreover, neither the function $f(x) = T_0$ nor the solution $u(x,t) = T_0$ have a Fourier transform in the usual sense. In spite of this, the solution (3.37) is valid!*

---

* Recall our discussion in Sec. 3.2. Also, since solutions given by (3.35) and (3.36) can be derived independent of the Fourier transform method, their validity may extend beyond that of the latter.

Example 3.5: Solve the problem described by (3.26) when the initial temperature distribution in the rod is given by

$$f(x) = e^{-x^2/4a^2}, \qquad -\infty < x < \infty$$

Solution: Using the transform relation

$$\mathscr{F}\{e^{-x^2/4a^2};s\} = \sqrt{2}a\, e^{-a^2s^2}$$

we find that the solution (3.32) of the transformed problem becomes

$$U(s,t) = \sqrt{a}e^{-a^2s^2(1+t)}$$

Rather than use the convolution theorem, we can invert this solution directly to obtain

$$u(x,t) = \frac{1}{\sqrt{1+t}}\, e^{-x^2/4a^2(1+t)}$$

Suppose now we consider the problem of heat flow in an infinite rod when a heat source is present. The problem is characterized by

$$u_{xx} = a^{-2}u_t - q(x,t), \qquad -\infty < x < \infty, \qquad t > 0$$

$$\text{B.C.:} \quad u(x,t) \to 0,\ u_x(x,t) \to 0 \text{ as } |x| \to \infty, \qquad t > 0 \tag{3.38}$$

$$\text{I.C.:} \quad u(x,0) = 0, \quad -\infty < x < \infty$$

where $q(x,t)$ is proportional to the heat source. For simplicity, we are assuming the initial temperature is zero. By using the Fourier transforms

$$\mathscr{F}\{u(x,t);x\to s\} = U(s,t)$$

$$\mathscr{F}\{q(x,t);x\to s\} = Q(s,t)$$

we are led to the nonhomogeneous first-order initial-value problem

$$U_t + a^2s^2U = a^2Q(s,t), \qquad t > 0 \tag{3.39}$$

$$\text{I.C.:} \quad U(s,0) = 0, \qquad -\infty < s < \infty$$

the solution of which is given by

$$U(s,t) = a^2 \int_0^t e^{-a^2s^2(t-\tau)}Q(s,\tau)\, d\tau \tag{3.40}$$

Inverting (3.40) by use of the convolution theorem, we have

$$u(x,t) = \frac{a^2}{\sqrt{2\pi}} \int_{-\infty}^{\infty} \int_0^t g(x-\xi,\ t-\tau)q(\xi,\tau)\, d\tau\, d\xi \tag{3.41}$$

where $g(x,t)$ is defined by (3.34).

3.3.2 *Heat Equation on a Semiinfinite Line*

Let us consider the problem of finding the temperature distribution near the end of a long rod which is insulated. In such a case we might model the rod as if it were extended over the interval $0 < x \leq \infty$. If the initial temperature distribution in the rod is $f(x)$, the problem we wish to solve is mathematically described by

$$u_{xx} = a^{-2}u_t, \qquad 0 < x < \infty,\ t > 0$$

$$\text{B.C.:} \quad u_x(0,t) = 0,\ u(x,t) \to 0,\ u_x(x,t) \to 0 \text{ as } x \to \infty \tag{3.42}$$

$$\text{I.C.:} \quad u(x,0) = f(x), \qquad 0 < x < \infty$$

The fact that the interval is semiinfinite, together with the prescribed boundary condition at $x = 0$, suggests that the Fourier cosine transform be used in this case. Hence, if we define*

$$\mathscr{F}_C\{u(x,t);x\to s\} = U(s,t) \tag{3.43}$$

it follows from properties of the cosine transform that

$$\begin{aligned}\mathscr{F}_C\{u_{xx}(x,t);x\to s\} &= -\, s^2U(s,t) - \sqrt{2/\pi}\, u_x(0,t)\\ &= -\, s^2U(s,t)\end{aligned} \tag{3.44}$$

Also, by setting $F(s) = \mathscr{F}_C\{f(x);s\}$, the transformed problem becomes

$$\begin{aligned} &U_t + a^2s^2U = 0, \qquad t > 0\\ \text{I.C.:}\quad &U(s,0) = F(s), \qquad 0 < s < \infty\end{aligned} \tag{3.45}$$

The solution of (3.45) is

$$U(s,t) = F(s)e^{-a^2s^2t} \tag{3.46}$$

and by applying the inverse cosine transform, we get the formal solution

$$u(x,t) = \sqrt{2/\pi}\int_0^\infty e^{-a^2s^2t}\, F(s)\cos sx\, ds \tag{3.47}$$

A more convenient form of the solution (3.47) can be developed by first replacing $F(s)$ with its integral representation

$$F(s) = \sqrt{2/\pi}\int_0^\infty f(\xi)\cos s\xi\, d\xi$$

---

* To avoid confusion with partial derivatives, we have chosen not to subscript the transformed functions by C to denote a cosine transform.

which yields

$$u(x,t) = \frac{2}{\pi}\int_0^\infty \int_0^\infty e^{-a^2s^2t} f(\xi) \cos s\xi \cos sx \, d\xi \, ds$$

$$= \frac{1}{\pi}\int_0^\infty \int_0^\infty e^{-a^2s^2t} f(\xi) \{\cos[s(x-\xi)] + \cos[s(x+\xi)]\} \, d\xi \, ds \qquad (3.48)$$

Finally, interchanging the order of integration and using the result (see Prob. 7 in Exer. 3.3)

$$\int_0^\infty e^{-bx^2} \cos cx \, dx = \frac{1}{2}\sqrt{\frac{\pi}{b}}\, e^{-c^2/4b}, \qquad b > 0 \qquad (3.49)$$

the above expression becomes*

$$u(x,t) = \frac{1}{2a\sqrt{\pi t}} \int_0^\infty f(\xi) \left\{ \exp\left[-\frac{(x-\xi)^2}{4a^2t} \right] + \exp\left[-\frac{(x+\xi)^2}{4a^2t} \right] \right\} d\xi \qquad (3.50)$$

As our final example of one-dimensional heat conduction problems, let us consider the problem where one end of a very long rod is exposed to a time-varying heat reservoir. We will assume the initial temperature distribution is zero, and thus our problem reads

$$u_{xx} = a^{-2}u_t, \qquad 0 < x < \infty, \; t > 0$$

B.C.: $u(0,t) = f(t), \; u(x,t) \to 0, \; u_x(x,t) \to 0$ as $x \to \infty$ (3.51)

I.C.: $u(x,0) = 0, \qquad 0 < x < \infty$

This time we will use the Fourier sine transform, which leads to the transformed problem

$$U_t + a^2s^2U = \sqrt{2/\pi}\, a^2 s f(t), \qquad t > 0 \qquad (3.52)$$

I.C.: $U(s,0) = 0, \qquad 0 < s < \infty$

where $U(s,t)$ denotes the Fourier sine transform of $u(x,t)$. The solution of this initial-value problem is

$$U(s,t) = \sqrt{2/\pi}\, a^2 s \int_0^t f(\tau) e^{-a^2s^2(t-\tau)} \, d\tau \qquad (3.53)$$

* Equation (3.50) could also be derived directly from the convolution integral relation given by Eq. (2.90) in Sec. 2.7.1.

and by using the transform relation [see Prob. 7(b) in Exer. 3.3]

$$\mathscr{F}_S^{-1}\{se^{-a^2s^2(t-\tau)};s\to x\} = \frac{x}{a^3\sqrt{8}}(t-\tau)^{-3/2}e^{-x^2/4a^2(t-\tau)} \tag{3.54}$$

we obtain from (3.53)

$$u(x,t) = \frac{x}{2a\sqrt{\pi}}\int_0^t \frac{f(\tau)}{(t-\tau)^{3/2}} e^{-x^2/4a^2(t-\tau)}\,d\tau \tag{3.55}$$

Remark: The problem described by (3.51) can also be solved by using the Laplace transform (see Sec. 5.4.1).

Example 3.6: Solve (3.51) for the special case when $f(t) = T_1$ (constant).

Solution: The solution for any $f(t)$ is that given by Eq. (3.55). By making the change of variable

$$z = x/2a\sqrt{t-\tau}$$

we find that (3.55) becomes

$$u(x,t) = \frac{2}{\sqrt{\pi}}\int_{x/2a\sqrt{t}}^{\infty} f(t-x^2/4a^2z^2)e^{-z^2}\,dz$$

which for $f(t) = T_1$ reduces to

$$u(x,t) = T_1\frac{2}{\sqrt{\pi}}\int_{x/2a\sqrt{t}}^{\infty} e^{-z^2}\,dz$$

Recalling the definition of the *complementary error function* (see Sec. 1.3.1)

$$\operatorname{erfc}(x) = \frac{2}{\sqrt{\pi}}\int_x^{\infty} e^{-z^2}\,dz$$

we can express our solution as

$$u(x,t) = T_1\operatorname{erfc}(x/2a\sqrt{t})$$

The physical interpretation of the solution suggests that for any fixed value of x, the temperature in the rod at that point will eventually approach T_1 if we wait long enough ($t \to \infty$). However, at any particular instant of time t, the temperature $u(x,t) \to 0$ as $x \to \infty$, in agreement with our prescribed boundary condition. Finally, we recognize that the temperature $u(x,t)$ is constant along any member of the family of parabolas in the xt plane defined by

$$x/2a\sqrt{t} = \text{constant}$$

3.3.3 *Heat Flow in an Infinite Rectangle*

Let us now consider the flow of heat in a rectangular plate that is so large we consider it mathematically unbounded in both the x and y directions. If the initial temperature distribution is $f(x,y)$ and the flat surfaces of the plate are insulated, the problem we wish to solve is described by

$$u_{xx} + u_{yy} = a^{-2}u_t, \qquad -\infty < x < \infty,\ -\infty < y < \infty,\ t > 0$$

$$\text{I.C.:}\quad u(x,y,0) = f(x,y), \quad -\infty < x < \infty,\ -\infty < y < \infty \tag{3.56}$$

We also assume that u and its normal derivatives vanish near the infinite edges of the plate.

By introducing the double Fourier transforms (see Sec. 2.9.2)

$$\mathscr{F}_{(2)}\{u(x,y,t);x\rightarrow\xi,\ y\rightarrow\eta\} = U(\xi,\eta,t) \tag{3.57}$$

$$\mathscr{F}_{(2)}\{f(x,y);x\rightarrow\xi,\ y\rightarrow\eta\} = F(\xi,\eta) \tag{3.58}$$

we find that (3.56) reduces to

$$\begin{aligned} &U_t + a^2(\xi^2 + \eta^2)U = 0, \qquad t > 0 \\ \text{I.C.:}\quad &U(\xi,\eta,0) = F(\xi,\eta) \end{aligned} \tag{3.59}$$

The solution of this transformed problem is

$$U(\xi,\eta,t) = F(\xi,\eta)e^{-a^2(\xi^2 + \eta^2)t} \tag{3.60}$$

and, therefore, we can write

$$u(x,y,t) = \frac{1}{2\pi}\int_{-\infty}^{\infty}\int_{-\infty}^{\infty} F(\xi,\eta)e^{-a^2(\xi^2+\eta^2)t}\,e^{-i(\xi x+\eta y)}\,d\xi\,d\eta \tag{3.61}$$

This solution can also be expressed in the equivalent form (see Prob. 23 in Exer. 3.3)

$$u(x,y,t) = \frac{1}{2\pi}\int_{-\infty}^{\infty}\int_{-\infty}^{\infty} f(x',y')g(x-x',\ y-y',\ t)\,dx'\,dy' \tag{3.62}$$

where

$$g(x,y,t) = \frac{1}{2a^2t}\,e^{-(x^2+y^2)/4a^2t} \tag{3.63}$$

EXERCISES 3.3

In Probs. 1–5, solve the heat conduction problem described by (3.26) when the initial temperature distribution $f(x)$ is prescribed as given.

1. $f(x = \begin{cases} T_0, & |x| < c \\ 0, & |x| > c \end{cases}$ **2.** $f(x) = \begin{cases} 0, & x < 0 \\ T_0, & x > 0 \end{cases}$

3. $f(x) = \begin{cases} T_0, & 0 < x < c \\ 0, & \text{otherwise} \end{cases}$ **4.** $f(x) = \begin{cases} 0, & x < 0 \\ e^{-x}, & x > 0 \end{cases}$

5. $f(x) = e^{-|x|}, \quad -\infty < x < \infty$ *Hint:* See Prob. 6

6. Show that

$$\int_a^\infty e^{-z^2-2bz}\,dz = \frac{\sqrt{\pi}}{2}\,e^{b^2}\operatorname{erfc}(a+b)$$

7. Show that

(a) $$\int_0^\infty e^{-bx^2}\cos cx\,dx = \frac{1}{2}\sqrt{\frac{\pi}{b}}\,e^{-c^2/4b}, \qquad b > 0$$

Hint: Expand cos cx in a Maclaurin series and use properties of the gamma function.

(b) By differentiating the result of (a) with respect to c, deduce that

$$\int_0^\infty xe^{-b^2x^2/2}\sin cx\,dx = \frac{c}{b^3}\sqrt{\frac{\pi}{2}}\,e^{-c^2/2b^2}, \qquad b > 0$$

8. If $f(x) = T_0$ (constant) and $q(x,t) = \delta(x)\delta(t)$, where δ is the delta function, show that the problem described by (3.38) leads to

$$u(x,t) = T_0 + \frac{1}{2a\sqrt{\pi t}}\,e^{-x^2/4a^2t}$$

9. A heat source of strength $q(t)h(t)$, where h is the Heaviside unit function, appears at the origin of a long rod at time $t = 0$ and moves along the positive x axis with constant speed v. The problem is characterized by

$$u_{xx} = u_t - \delta(x - vt)q(t)h(t), \qquad -\infty < x < \infty,\ t > 0$$

B.C.: $u(x,t) \to 0,\ u_x(x,t) \to 0$ as $|x| \to \infty$

I.C.: $u(x,0) = 0, \quad -\infty < x < \infty$

where δ denotes the delta function. Show that

$$u(x,t) = \frac{1}{2\sqrt{\pi}}\int_0^t \frac{q(\tau)}{\sqrt{t-\tau}}\,e^{-(x-v\tau)^2/4(t-\tau)}\,d\tau$$

10. (*Duhamel's principle*) If $u(x,t;\tau)$ denotes the solution of

$$u_{xx} = a^{-2}u_t, \quad u(x,0;\tau) = f(x,\tau), \quad -\infty < x < \infty$$

show that

$$v(x,t) = \int_0^t u(x,t-\tau;\tau)\,d\tau$$

is a solution of

$$a^2 v_{xx} = v_t - f(x,t), \qquad v(x,0) = 0, \ -\infty < x < \infty$$

(Duhamel's principle basically points out the equivalence between solving nonhomogeneous PDEs with a homogeneous initial condition and solving homogeneous PDEs with a nonhomogeneous initial condition.)

11. Show that a formal solution of

$$u_{xx} = a^{-2}u_t, \qquad 0 < x < \infty, \ t - 0$$

B.C.: $\quad u(0,t) = 0, \qquad u(x,t) \to 0 \text{ as } x \to \infty$

I.C.: $\quad u(x,0) = f(x), \qquad 0 < x < \infty$

(a) is given by

$$u(x,t) = \sqrt{\frac{2}{\pi}} \int_0^\infty e^{-a^2s^2t}\, F(s) \sin sx\, ds$$

(b) Show that the solution in (a) is equivalent to

$$u(x,t) = \frac{1}{2a\sqrt{\pi t}} \int_0^\infty f(\xi) \left\{ \exp\left[-\frac{(x-\xi)^2}{4a^2t} \right] - \exp\left[-\frac{(x+\xi)^2}{4a^2t} \right] \right\} d\xi$$

Hint: See Prob. 7(a).

12. Solve Prob. 11 when

$$f(x) = \begin{cases} 0, & 0 < x < c \\ T_0, & x > c \end{cases}$$

13. Solve the problem described by (3.42) when

$$f(x) = \begin{cases} 0, & 0 < x < c \\ T_0, & x > c \end{cases}$$

14. Solve the problem described by (3.42) when $f(x) = e^{-x}$, $0 < x < \infty$.

15. Given the boundary-value problem

$$u_{xx} = a^{-2}u_t, \qquad 0 < x < \infty, \ t > 0$$

B.C.: $\quad u(0,t) = 0, \qquad u(x,t) \to 0 \text{ as } x \to \infty$

I.C.: $\quad u(x,0) = T_0 x e^{-x^2/4}, \qquad 0 < x < \infty$

show that

$$u(x,t) = T_0 x(1 + a^2 t)^{-3/2} \exp[-x^2/4(1 + a^2 t)]$$

Hint: See Prob. 7(b).

16. If the boundary condition in (3.51) is $f(t) = T_1/\sqrt{t}$, where T_1 is constant, show that

$$u(x,t) = \frac{T_1}{\sqrt{t}} e^{-x^2/4a^2 t}$$

17. If the boundary condition in (3.51) is

$$f(t) = \begin{cases} T_1, & 0 < t < b \\ 0, & t \geq b \end{cases}$$

show that

$$u(x,t) = \begin{cases} T_1 \operatorname{erfc}(x/2a\sqrt{t}), & 0 < t < b \\ T_1[\operatorname{erf}(x/2a\sqrt{t-b}) - \operatorname{erf}(x/2a\sqrt{t})], & t \geq b \end{cases}$$

18. Given the boundary value problem

$$u_{xx} = a^{-2} u_t, \qquad 0 < x < \infty,\ t > 0$$

B.C.: $\quad u_x(0,t) = -f(t), \qquad u(x,t) \to 0$ as $x \to \infty$

I.C.: $\quad u(x,0) = 0, \qquad 0 < x < \infty$

show that

$$u(x,t) = \frac{a}{\sqrt{\pi}} \int_0^t \frac{f(\tau)}{\sqrt{t-\tau}} e^{-x^2/4a^2(t-\tau)}\, d\tau$$

19. Show that the solution in Prob. 18 can also be expressed in the form

$$u(x,t) = \frac{x}{\sqrt{\pi}} \int_{x/2a\sqrt{t}}^{\infty} f(t - x^2/4a^2 z^2) z^{-2} e^{-z^2}\, dz$$

20. For the special case of $f(t) = K$ (constant),
(a) show that the solution of Prob. 18 is

$$u(x,t) = K\left[2a\sqrt{\frac{t}{\pi}}\, e^{-x^2/4a^2 t} - x \operatorname{erfc}\left(\frac{x}{2a\sqrt{t}}\right)\right]$$

(b) What is the temperature at the end $x = 0$?

21. Find a solution in the form of Eq. (3.41) for

$$u_{xx} = a^{-2} u_t - q(x,t), \qquad 0 < x < \infty,\ t > 0$$

B.C.: $u(0,t) = 0, \quad u(x,t) \to 0$ as $x \to \infty$

I.C.: $u(x,0) = 0, \quad 0 < x < \infty$

22. Find a solution in the form of Eq. (3.41) for

$$u_{xx} = a^{-2}u_t - q(x,t), \qquad 0 < x < \infty, \; t > 0$$

B.C.: $u_x(0,t) = 0, \quad u(x,t) \to 0$ as $x \to \infty$

I.C.: $u(x,0) = 0, \quad 0 < x < \infty$

23. Show that

(a) $\mathscr{F}_{(2)}^{-1}\{e^{-a^2(\xi^2+\eta^2)t}; \xi \to x, \; \eta \to y\} = (1/2a^2t)\, e^{-(x^2+y^2)/4a^2t}$

(b) Use the result of (a) to deduce the solution formula (3.62) of the problem described by (3.56).

3.4 *Mechanical Vibrations*

The *wave equation*

$$\nabla^2 u = c^{-2}u_{tt} - q(x,y,z,t) \tag{3.64a}$$

where c is a constant having the dimension of velocity, describes various wave motions in nature and mechanical systems such as sound waves emanating from a struck bell, surface waves propagating radially outward when a pebble is dropped into a pool, and the deflections of a membrane set in motion. The term q is proportional to an external "force" acting on the system under investigation. A properly-posed problem involving the wave equation consists of *two* initial conditions and *one* boundary condition at each boundary point.

The *one-dimensional* wave equation

$$u_{xx} = c^{-2}u_{tt} - q(x,t) \tag{3.64b}$$

is the governing equation for such rudimentary problems as the transverse oscillations of a tightly stretched string or the longitudinal vibrations of a beam. The transverse vibrations of an elastic beam involve a fourth-order PDE having similar characteristics (see Sec. 2.4.2).

3.4.1 *Wave Equation on an Infinite Line*

As our first illustrative example, we seek the transverse deflections of an infinite string which is given an initial deflection $f(x)$ and initial velocity $g(x)$. Assuming there are no external forces acting on the string, the problem is characterized by

$$u_{xx} = c^{-2} u_{tt}, \qquad -\infty < x < \infty, t > 0$$

B.C.: $u(x,t) \to 0$ as $|x| \to \infty$

I.C.: $u(x,0) = f(x), \quad u_t(x,0) = g(x), \; -\infty < x < \infty$ (3.65)

The infinite extent on x once again suggests the use of the Fourier exponential transform. Hence, by introducing

$$\mathscr{F}\{u(x,t);x \to s\} = U(s,t) \tag{3.66}$$

we get the transformed problem

$$U_{tt} + c^2 s^2 U = 0, \qquad t > 0$$

I.C.: $U(s,0) = F(s), \quad U_t(s,0) = G(s)$ (3.67)

where $F(s) = \mathscr{F}\{f(x);s\}$ and $G(s) = \mathscr{F}\{g(x);s\}$. By standard solution techniques, we find

$$U(s,t) = F(s) \cos cst + \frac{G(s)}{cs} \sin cst \tag{3.68}$$

the inverse transform of which leads to the integral representation

$$u(x,t) = \frac{1}{\sqrt{2\pi}} \int_{-\infty}^{\infty} e^{-isx} \left[F(s) \cos cst + \frac{G(s)}{cs} \sin cst \right] ds \tag{3.69}$$

Although (3.69) is a formal solution of (3.65), an interesting result emerges if we choose to write the sine and cosine appearing in (3.69) in terms of complex exponentials by use of the Euler formulas

$$\cos x = \frac{1}{2}(e^{ix} + e^{-ix}), \; \sin x = \frac{1}{2i}(e^{ix} - e^{-ix})$$

Making the appropriate substitutions, we find

$$u(x,t) = \frac{1}{2\sqrt{2\pi}} \int_{-\infty}^{\infty} [e^{-is(x-ct)} + e^{-is(x+ct)}] F(s)\, ds$$
$$+ \frac{1}{2\sqrt{2\pi}} \int_{-\infty}^{\infty} [e^{-is(x-ct)} - e^{-is(x+ct)}] \frac{G(s)}{ics}\, ds \tag{3.70}$$

The first integral in (3.70) is recognized as being equal to the expression

$$\tfrac{1}{2}[f(x - ct) + f(x + ct)]$$

The second integral can be similarly identified if we start with

$$g(z) = \frac{1}{\sqrt{2\pi}} \int_{-\infty}^{\infty} e^{-isz} G(s)\, ds \tag{3.71}$$

and integrate from $x - ct$ to $x + ct$. Hence,

$$\int_{x-ct}^{x+ct} g(z)dz = \frac{1}{\sqrt{2\pi}} \int_{-\infty}^{\infty} [e^{-is(x-ct)} - e^{-is(x+ct)}] \frac{G(s)}{is} ds$$

which implies that (3.70) is equivalent to

$$u(x,t) = \frac{1}{2}\left[f(x - ct) + f(x + ct)\right] + \frac{1}{2c} \int_{x-ct}^{x+ct} g(z)dz \tag{3.72}$$

Equation (3.72) is the well-known *d'Alembert solution* of the wave equation on the infinite line. J. d'Alembert (1717–1783) introduced his solution of the wave equation in 1747, six years before Daniel Bernoulli (1700–1782) provided his series solution through separation of variables. The method used by d'Alembert was to make a transformation of variables to put the equation in what we now call canonical form. From this special form he was able to solve the wave equation and finally produce the solution (3.72).

3.4.2 *Transverse Vibrations of an Elastic Beam*

Consider a uniform elastic beam lying along the x axis in its equilibrium position as shown in Fig. 3.2. If the beam is initially displaced from this equilibrium position in the vertical plane it will cause the beam to vibrate freely in the transverse direction. From the elementary theory of small beam deflections it is known that the bending moment M at any cross section of the beam satisfies the relation*

$$M = EI/R \simeq EIu_{xx}, \qquad |u_x| << 1 \tag{3.73}$$

where E is the modulus of elasticity of the beam, I is the moment of inertia of the cross section of the beam with respect to the neutral axis passing through the centroid of each cross section, R is the radius of curvature of the bent beam, and u is the deflection in the transverse direction. Also, the balance of moments acting on an infinitesimal element bounded by two adjacent cross sections of the beam requires that

$$M_{xx} = -\frac{\gamma A}{g} u_{tt} \tag{3.74}$$

where γ is the specific weight of the beam, A is the area of the cross section, and g is the acceleration of gravity. The term on the right-hand side in (3.74) representing an inertia force actually behaves like a load intensity along the entire length of the beam. Combining (3.73) and (3.74),

* See S. Timoshenko, *Strength of Materials*, 3rd ed., New York: Van Nostrand, 1955.

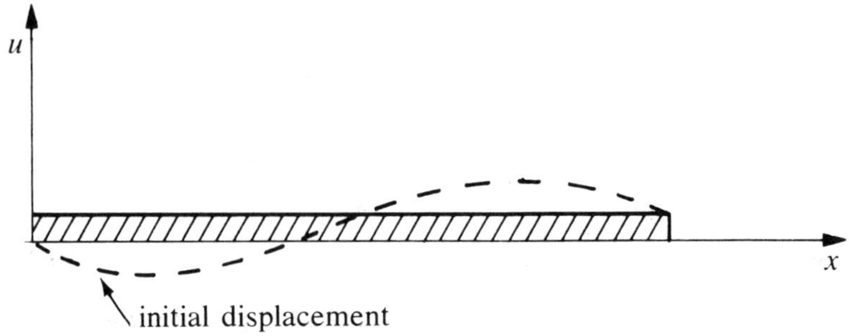

Figure 3.2 Freely vibrating beam in transverse direction

we obtain the differential equation of motion

$$u_{xxxx} + a^{-2}u_{tt} = 0 \tag{3.75}$$

where $a^2 = EIg/\gamma A$.

For illustrative purposes, let us consider the case involving the free vibrations of a uniform beam of infinite length where the motion is produced by distorting the beam initially in the shape $f(x)$ with zero velocity. Assuming units such that $a^2 = 1$, the problem is formulated by

$$\begin{aligned} &u_{xxxx} + u_{tt} = 0, \quad -\infty < x < \infty,\ t > 0 \\ \text{B.C.:}\quad &u(x,t) \to 0, \quad u_x(x,t) \to 0 \text{ as } |x| \to \infty \\ \text{I.C.:}\quad &u(x,0) = f(x), \quad u_t(x,0) = 0,\ -\infty < x < \infty \end{aligned} \tag{3.76}$$

Using the Fourier exponential transform, we convert (3.76) into the initial value problem

$$\begin{aligned} &U_{tt} + s^4U = 0, \quad t > 0 \\ \text{I.C.:}\quad &U(s,0)) = F(s), \quad U_t(s,0) = 0 \end{aligned} \tag{3.77}$$

with solution

$$U(s,t) = F(s)\cos ts^2 \tag{3.78}$$

By taking the inverse Fourier transform of (3.78), we find the motions of the beam are described by

$$u(x,t) = \frac{1}{\sqrt{2\pi}} \int_{-\infty}^{\infty} e^{-isx} F(s)\cos ts^2\, ds \tag{3.79}$$

An alternate form of the solution can be obtained by first noting the inverse transform relation (see Prob. 15 in Exer. 2.4)

$$\mathscr{F}^{-1}\{\cos(ts^2);s\to\xi\} = \frac{1}{2\sqrt{t}}[\cos(\xi^2/4t) + \sin(\xi^2/4t)] \tag{3.80}$$

and then using the convolution theorem on (2.78) to get

$$u(x,t) = \frac{1}{2\sqrt{2\pi t}} \int_{-\infty}^{\infty} f(x-\xi)\,[\cos(\xi^2/4t) + \sin(\xi^2/4t)]\,d\xi \qquad (3.81)$$

Finally, making the change of variable $\eta = x - \xi$ we can also express this last relation in the more compact form

$$u(x,t) = \frac{1}{2\sqrt{\pi t}} \int_{-\infty}^{\infty} f(\eta) \sin\left[\frac{(x-\eta)^2}{4t} + \frac{1}{4}\pi\right] d\eta \qquad (3.82)$$

EXERCISES 3.4

In Probs. 1–4, solve the vibrating string problem described by (3.65) when the initial conditions are prescribed by the given functions.

1. $f(x), = e^{-|x|}, \quad g(x) = 0, \ -\infty < x < \infty$

2. $f(x) = 0, \quad g(x) = e^{-|x|}, \ -\infty < x < \infty$

3. $f(x) = \begin{cases} F_0, & |x| < 1 \\ 0, & |x| > 1 \end{cases} \quad g(x) = V_0, \ -\infty < x < \infty$ (F_0, V_0 constants)

4. $f(x) = \begin{cases} F_0, & 0 < x < a \\ 0, & \text{otherwise} \end{cases} \quad g(x) = \begin{cases} V_0, & 0 < x < a \\ 0, & \text{otherwise} \end{cases}$

(F_0, V_0 constants)

5. Solve Prob. 1 above by using the solution formula (3.69).

6. Show that

$$u_{xx} = c^{-2}u_{tt}, \qquad 0 < x < \infty,\ t > 0$$

B.C.: $u(0,t) = 0, \quad u(x,t) \to 0$ as $x \to \infty$

I.C.: $u(x,0) = f(x), \quad u_t(x,0) = 0,\ 0 < x < \infty$

has the solution

$$u(x,t) = \tfrac{1}{2}[f_o(x + ct) + f_o(x - ct)]$$

where f_o denotes the odd extension of f over the entire axis.

7. Show that

$$u_{xx} = u_{tt}, \qquad 0 < x < \infty,\ t > 0$$

B.C.: $u_x(0,t) = 0, \quad u(x,t) \to 0$ as $x \to \infty$

I.C.: $u(x,0) = e^{-x}, \quad u_t(x,0) = 0,\ 0 < x < \infty$

has the solution

$$u(x,t) = \begin{cases} e^{-t}\cosh x, & x < t \\ e^{-x}\cosh t, & x > t \end{cases}$$

8. Solve

$$u_{xx} = c^{-2}u_{tt}, \qquad 0 < x < \infty,\ t > 0$$

B.C.: $u(0,t) = f(t), \qquad u(x,t) \to 0 \text{ as } x \to \infty$

I.C.: $u(x,0) = 0, \qquad u_t(x,0) = 0,\ 0 < x < \infty$

9. Solve

$$u_{xx} = u_{tt}, \qquad 0 < x < \infty,\ t > 0$$

B.C.: $u_x(0,t) = -1, \qquad u(x,t) \to 0 \text{ as } x \to \infty$

I.C.: $u(x,0) = 0, \qquad u_t(x,0) = 0,\ 0 < x < \infty$

10. Solve

$$u_{xx} = u_{tt}, \qquad 0 < x < \infty,\ t > 0$$

B.C.: $u_x(0,t) = A\sin\omega t, \qquad u(x,t) \to 0 \text{ as } x \to \infty$

I.C.: $u(x,0) = 0,\ u_t(x,0) = 0,\ 0 < x < \infty$

11. Solve the problem described by (3.76) when

(a) $f(x) = \begin{cases} F_0, & 0 < x < \infty \\ 0, & -\infty < x < 0 \end{cases}$

(b) $f(x) = \begin{cases} F_0, & |x| < 1 \\ 0, & |x| > 1 \end{cases}$

12. If $f(x) = e^{-x^2/4}$ in the problem described by (3.76),

(a) show that

$$u(x,t) = \frac{1}{2}\left[\frac{e^{-x^2/4(1+it)}}{\sqrt{1+it}} + \frac{e^{-x^2/4(1-it)}}{\sqrt{1-it}}\right]$$

(b) By setting $1 + it = Re^{i\phi}$, where

$$R = \sqrt{1+t^2}, \qquad \tan\phi = t$$

show that the solution in (a) becomes

$$u(x,t) = R^{-1/2}e^{-x^2(\cos\phi)/4R}\cos\left(\frac{x^2\sin\phi}{4R} - \frac{\phi}{2}\right)$$

13. Given the vibrating beam problem

$$u_{xxxx} + u_{tt} = 0, \qquad 0 < x < \infty,\ t > 0$$

B.C.: $u(0,t) = 1, \quad u_{xx}(0,t) = 0, \quad u(x,t) \to 0 \text{ as } x \to \infty$

I.C.: $u(x,0) = 0, \qquad u_t(x,0) = 0,\ 0 < x < \infty$

use the Fourier sine transform to show that
(a) the solution of the transformed problem is

$$U(s,t) = 2\sqrt{\frac{2}{\pi}}\left(\frac{1 - \cos ts^2}{s}\right)$$

(b) the solution of the original problem is

$$u(x,t) = 1 - C\left(\frac{x}{\sqrt{2\pi t}}\right) - S\left(\frac{x}{\sqrt{2\pi t}}\right)$$

where $C(x)$ and $S(x)$ are the *Fresnel integrals* (see Sec. 1.3.2).

Hint: Recall Prob. 20 in Exer. 2.5 and Prob. 15 in Exer. 2.4.

14. Given the vibrating beam problem

$$u_{xxxx} + u_{tt} = 0, \qquad 0 < x < \infty,\ t > 0$$

B.C.: $u(0,t) = f(t), \quad u_{xx}(0,t) = 0, \quad u(x,t) \to 0$ as $x \to \infty$

I.C.: $u(x,0) = 0, \quad u_t(x,0) = 0,\ 0 < x < \infty$

show that

(a) the solution of the transformed problem is

$$U(s,t) = \sqrt{\frac{2}{\pi}}\, s \int_0^t f(t - \tau)\sin \tau s^2 \, d\tau$$

(b) the solution of the original problem is

$$u(x,t) = \frac{1}{\sqrt{\pi}} \int_{x/\sqrt{2t}}^{\infty} f(t - x^2/2v^2)\,[\cos(\tfrac{1}{2}v^2) + \sin(\tfrac{1}{2}v^2)]\, dv$$

3.5 *Potential Theory*

Perhaps the single most important PDE in mathematical physics is the *equation of Laplace*, or *potential equation*. In two and three dimensions, respectively, we have the rectangular coordinate representations

$$u_{xx} + u_{yy} = 0 \tag{3.83a}$$

$$u_{xx} + u_{yy} + u_{zz} = 0 \tag{3.83b}$$

whereas in general we write

$$\nabla^2 u = 0 \tag{3.83c}$$

regardless of the coordinate system or number of dimensions.

Laplace's equation arises in steady-state heat conduction problems involving homogeneous solids.* This same equation is satisfied by the

* The heat equation $\nabla^2 u = a^{-2}u_t$ reduces to Laplace's equation when $u_t = 0$, i.e., when u is independent of time t.

gravitational potential in free space, the electrostatic potential in a uniform dielectric, the magnetic potential in free space, the electric potential in the steady flow of currents in solid conductors, and the velocity potential of inviscid, irrotational fluids. The mathematical formulation of all potential problems is essentially the same despite the physical differences of the applications. Because of this, all solutions of the potential equation are collectively called *potential functions*, and the study of the many properties associated with these functions forms that branch of mathematics known as *potential theory*.*

One of the most fundamental properties of the continuous solutions of Laplace's equation is *smoothness*. This property is a consequence of the fact that the equation describes "steady states." However, not all solutions are continuous. For example, the function $u(x,y) = \log(x^2 + y^2)$ satisfies (3.83a), but is discontinuous at (0,0). The continuous solutions of Laplace's equation that also have continuous second partial derivatives in some domain R are commonly called *harmonic functions*.

A properly-posed problem involving Laplace's equation consists of finding a harmonic function in a region R subject to a *single* boundary condition. The most common boundary conditions fall mainly into two categories, giving us two primary types of boundary-value problems. If R denotes a region in the plane and C its boundary curve, then one type of problem is characterized by

$$\begin{aligned} \nabla^2 u &= 0 \text{ in } R \\ u &= f \text{ on } C \end{aligned} \qquad (3.84)$$

which is called a *Dirichlet problem* or *boundary-value problem of the first kind*. In this problem we specify the value of u at each point of the (finite) boundary. An example of a Dirichlet problem is to find the steady-state temperature distribution in a region R given that the temperature is known everywhere on the boundary of R. Another problem is characterized by

$$\begin{aligned} \nabla^2 u &= 0 \text{ in } R \\ \partial u/\partial n &= f \text{ on } C \end{aligned} \qquad (3.85)$$

which is known as a *Neumann problem* or *boundary-value problem of the second kind*. The derivative $\partial u/\partial n$ is called the *normal derivative* of u and is positive in the direction of the outward normal to the boundary curve C.† In steady-state temperature problems, the normal derivative specifies the heat flow across the boundary of R. There is also a third

---

* For example, see O. D. Kellog, *Foundations of Potential Theory*, New York: Dover, 1953.

† Recall that $\partial u/\partial n = \nabla u \cdot \mathbf{n}$, where $\mathbf{n}$ is the outward unit normal to C.

boundary-value problem, called *Robin's problem*, in which the boundary condition is a linear combination of u and its normal derivative. We will, however, not give separate treatment of it.

3.5.1 *Potential Problems in the Half-Plane*

Let us consider the problem of finding a potential function u satisfying the conditions

$$u_{xx} + u_{yy} = 0, \qquad -\infty < x < \infty,\ y > 0$$

$$\text{B.C.:} \quad \begin{cases} u(x,0) = f(x), & -\infty < x < \infty \\ u(x,y) \to 0 \text{ as } \rho \to \infty \end{cases} \tag{3.86}$$

where $\rho = (x^2 + y^2)^{1/2}$ (see Fig. 3.3).

Because we have specified u on the finite boundary, we recognize that (3.86) is a Dirichlet problem. A physical situation leading to a problem like this would be to find the steady-state temperature distribution in a large rectangular plate, the flat surfaces of which are insulated, when the temperature is prescribed by $f(x)$ along one edge of the plate and tends to zero along each of the other edges.

In deciding which integral transform to use, we note that x has an infinite range of values while y has a semiinfinite range. On the variable y we might use the Fourier sine transform but on x we use the Fourier exponential transform. In this case the Fourier exponential transform leads to a somewhat easier problem to solve. We leave the solution by means of the sine transform to the exercises (see Prob. 9 in Exer. 3.5). Therefore, by introducing

$$\mathscr{F}\{u(x,y);x\to s\} = U(s,y) \tag{3.87}$$

$$\mathscr{F}\{f(x);s\} = F(s) \tag{3.88}$$

we are led to the transformed problem

$$U_{yy} - s^2U = 0,\ y > 0$$

$$\text{B.C.:} \quad U(s,0) = F(s),\ U(s,y) \to 0 \text{ as } y \to \infty \tag{3.89}$$

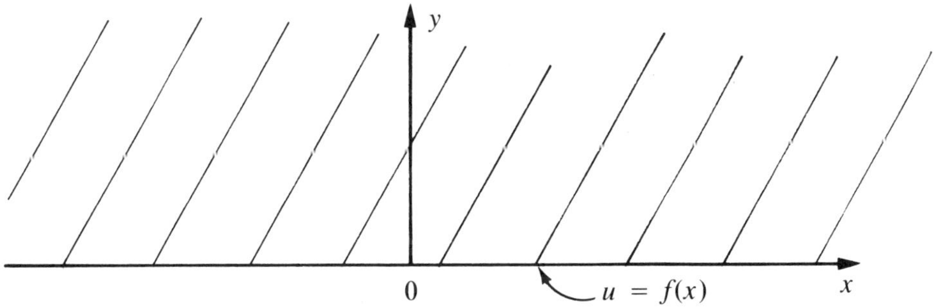

Figure 3.3 Dirichlet problem for half-plane

Remark: Actually, to completely describe the limiting condition in (3.89) we should also state $u(x,y) \to 0$ as $|x| \to \infty$.

The general solution of the differential equation in (3.89) is given by

$$U(s,y) = A(s)e^{-sy} + B(s)e^{sy}$$

where $A(s)$ and $B(s)$ are arbitrary functions of s. A valid solution for all values of s is readily found to be

$$U(s,y) = F(s)e^{-|s|y}, \qquad -\infty < s < \infty \tag{3.90}$$

Thus, recalling that (see Exam. 2.3 in Chap. 2)

$$\mathscr{F}^{-1}\{e^{-|s|y};s\to x\} = \sqrt{\frac{2}{\pi}}\frac{y}{x^2+y^2} \tag{3.91}$$

and applying the convolution theorem, we are led to the solution formula

$$u(x,y) = \frac{y}{\pi}\int_{-\infty}^{\infty}\frac{f(\xi)\,d\xi}{(x-\xi)^2+y^2}, \quad y > 0 \tag{3.92}$$

which is the well-known *Poisson integral formula* for the half-plane.

Example 3.7: Solve the Dirichlet problem (3.86) when

$$f(x) = \begin{cases} T_0, & |x| < b \\ 0, & |x| > b \end{cases}$$

Solution: By substituting f directly into the Poisson integral formula (3.92), we obtain the solution

$$\begin{aligned} u(x,y) &= \frac{yT_0}{\pi}\int_{-b}^{b}\frac{dt}{(t-x)^2+y^2} \\ &= \frac{T_0}{\pi}\left[\tan^{-1}\left(\frac{x+b}{y}\right) - \tan^{-1}\left(\frac{x-b}{y}\right)\right] \end{aligned}$$

but using the trigonometric identity

$$\tan(A-B) = \frac{\tan A - \tan B}{1 + \tan A \tan B}$$

we can express our solution in the more convenient form

$$u(x,y) = \frac{T_0}{\pi}\tan^{-1}\left(\frac{2by}{x^2+y^2-b^2}\right)$$

Curves in the upper half-plane for which the steady-state temperature is constant are called *isotherms*. For our particular problem, these curves are defined by the family of circular arcs (C constant)

$$x^2 + y^2 - Cy = b^2$$

which have centers on the y-axis and endpoints on the x axis at $x = \pm b$ (see Fig. 3.4).

The Neumann problem for the half-plane can be solved by a device using the solution (3.92) of the Dirichlet problem. Let us suppose the boundary condition in the original problem (3.86) is replaced by the Neumann condition

$$\text{B.C.:} \qquad u_y(x,0) = f(x), \quad -\infty < x < \infty \tag{3.93}$$

Physically, specifying the normal derivative on a boundary tells us the heat flux or flow at this boundary. For example, an insulated boundary has zero heat flow.

To solve Laplace's equation subject to the boundary condition (3.93), we define a function $v(x,y) = u_y(x,y)$ which has the properties

$$\begin{aligned} & v_{xx} + v_{yy} = \frac{\partial}{\partial y}(u_{xx} + u_{yy}) = 0 \\ \text{B.C.:} \quad & v(x,0) = u_y(x,0) = f(x) \end{aligned} \tag{3.94}$$

Thus, the function $v(x,y)$ is a solution of the Dirichlet problem described by (3.94), and therefore assumes the form

$$v(x,y) = \frac{y}{\pi}\int_{-\infty}^{\infty} \frac{f(\xi)\, d\xi}{(x - \xi)^2 + y^2}, \qquad y > 0 \tag{3.95}$$

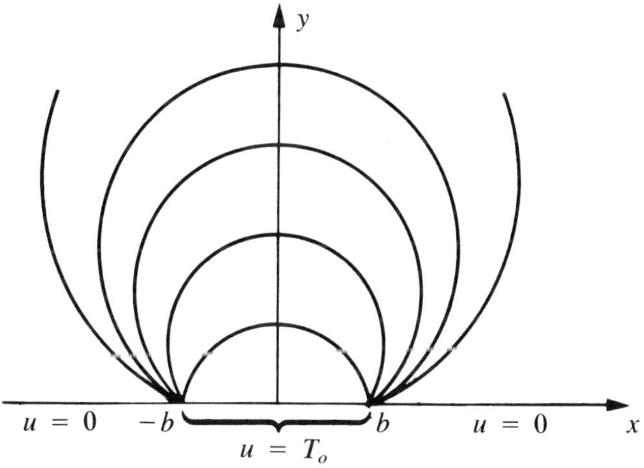

Figure 3.4 Family of isotherms

The solution function $u(x,y)$ that we are seeking can now be obtained by a simple indefinite integration of (3.95), leading to

$$\begin{aligned} u(x,y) &= \int v(x,y)\,dy \\ &= \frac{1}{\pi}\int_{-\infty}^{\infty} f(\xi) \int \frac{y\,dy}{(x-\xi)^2 + y^2}\,d\xi \end{aligned}$$

where we have interchanged the order of integration in this last step. Finally, completing the above indefinite integration, we get

$$u(x,y) = \frac{1}{2\pi}\int_{-\infty}^{\infty} f(\xi)\log[(x-\xi)^2 + y^2]\,d\xi \tag{3.96}$$

as the solution of the Neumann problem on the half-plane $y > 0$. Of course, because we performed an indefinite integration, an arbitrary constant can be added to the solution (3.96). In fact, it can be shown that the solution to any Neumann problem is unique only to within an additive constant.*

3.5.2 *Potential Problems in the Infinite Strip*

Suppose we now consider the Dirichlet problem

$$u_{xx} + u_{yy} = 0, \qquad -\infty < x < \infty,\ 0 < y < a \tag{3.97}$$

$$\text{B.C.:} \quad u(x,0) = f(x),\ u(x,a) = g(x),\ -\infty < x < \infty$$

Physically, this problem might correspond to finding the steady-state temperature in an infinite slab whose faces are maintained at prescribed temperatures (see Fig. 3.5).

Again we use the Fourier transform

$$\mathscr{F}\{u(x,y);x\to s\} = U(s,y) \tag{3.98}$$

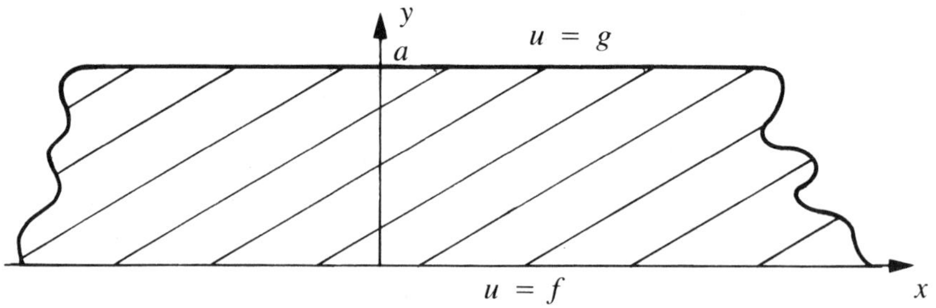

Figure 3.5 Infinite slab

* See L. C. Andrews, *Elementary Partial Differential Equations with Boundary Value Problems*, Orlando: Academic Press, 1986.

which reduces (3.97) to

$$U_{yy} - s^2U = 0, \qquad 0 < y < a \tag{3.99}$$

B.C.: $\quad U(s,0) = F(s), \qquad U(s,a) = G(s)$

where $F(s)$ and $G(s)$ are Fourier transforms, respectively, of $f(x)$ and $g(x)$. The solution of this boundary-value problem is easily shown to be (see Prob. 13 in Exer. 3.5)*

$$U(s,y) = F(s)\frac{\sinh s(a - y)}{\sinh sa} + G(s)\frac{\sinh sy}{\sinh sa} \tag{3.100}$$

Recalling that Fourier transforms, inverse transforms and cosine transforms are all the same for even functions, we deduce that

$$k(x,y) = \mathscr{F}^{-1}\left\{\frac{\sinh sy}{\sinh sa}; s\to x\right\} = \frac{1}{a}\sqrt{\frac{\pi}{2}}\,\frac{\sin(\pi y/a)}{\cosh(\pi x/a) + \cos(\pi y/a)} \tag{3.101}$$

the details of which we leave to the reader.† Hence, using the convolution theorem we can express the inverse transform of (3.100) in the form

$$u(x,y) = \frac{1}{\sqrt{2\pi}}\left[\int_{-\infty}^{\infty} f(\xi)k(x - \xi, a - y)d\xi + \int_{-\infty}^{\infty} g(\xi)k(x - \xi, y)d\xi\right] \tag{3.102}$$

3.5.3 *Potential Problems in the Semiinfinite Strip*

To illustrate the method of integral transforms for a semiinfinite strip, we consider the physical problem of determining the steady-state temperature distribution in a thick slab in which one of the infinite faces of the slab is subjected to a prescribed temperature distribution $f(x)$ and the other two faces are insulated against the flow of heat. The mathematical problem is described by (see Fig. 3.6)

$$u_{xx} + u_{yy} = 0, \qquad 0 < x < \infty, 0 < y < a$$

$$\text{B.C.:}\quad \begin{cases} u(x,0) = f(x), & u_y(x,a) = 0,\ 0 < x < \infty \\ u_x(0,y) = 0, & 0 < y < a \end{cases} \tag{3.103}$$

Remark: The problem described by (3.103) is neither a Dirichlet nor a Neumann problem since the boundary conditions are *mixed*.

---

* As a general rule, we use exponential functions as solutions of $u'' - k^2u = 0$ when the domain is infinite, but find it more convenient to use hyperbolic functions when the domain is finite.

† A special case of the result (3.101) occurs in Prob. 24 in Exer. 2.6.

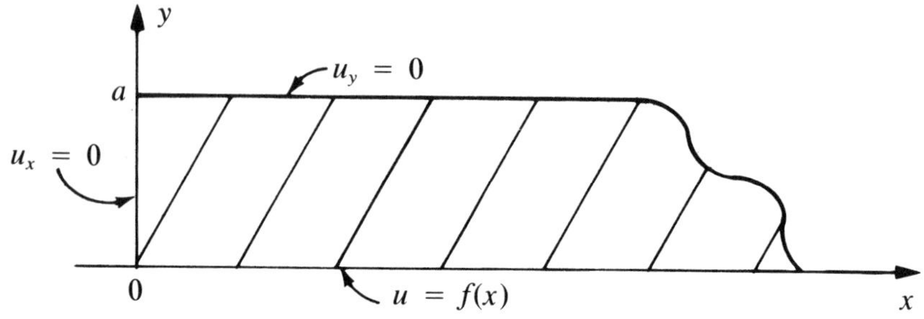

Figure 3.6 Semiinfinite slab

Because the independent variable x is defined on a semiinfinite domain and the boundary condition $u_x(0,y) = 0$ is prescribed, we find the Fourier cosine transform to be the appropriate tool in this case. Therefore, we introduce

$$\mathscr{F}_C\{u(x,y);x\to s\} = U(s,y) \tag{3.104}$$

$$\begin{aligned}\mathscr{F}_C\{u_{xx}(x,y);x\to s\} &= -s^2U(s,y) - \sqrt{2/\pi}u_x(0,y)\\ &= -s^2U(s,y)\end{aligned} \tag{3.105}$$

and

$$\mathscr{F}_C\{f(x);s\} = F(s) \tag{3.106}$$

Using these results, we generate the transformed problem

$$U_{yy} - s^2U = 0,\ 0 < y < a \tag{3.107}$$

B.C.: $\quad U(s,0) = F(s),\ U_y(s,a) = 0$

with solution

$$U(s,y) = F(s)\,\frac{\cosh(a - y)s}{\cosh as} \tag{3.108}$$

If we let (recall Exam. 13 in Sec. 2.6)

$$\begin{aligned}g(x,y) &= \mathscr{F}_C^{-1}\left\{\frac{\cosh(a - y)s}{\cosh as};s\to x\right\}\\ &= \frac{\sin(\pi y/2a)\cosh(\pi x/2a)}{\cosh(\pi x/a) - \cos(\pi y/a)}\end{aligned} \tag{3.109}$$

then through use of the convolution theorem, our solution becomes

$$u(x,y) = \frac{1}{\sqrt{2\pi}}\int_0^\infty f(\xi)\,[g(x + \xi,\ y) + g(|x - \xi|,\ y)]\,d\xi \tag{3.110}$$

EXERCISES 3.5

In Prob. 1–4, use the prescribed function f to find a solution of the Dirichlet problem (3.86).

1. $f(x) = T_0, \quad -\infty < x < \infty$

2. $f(x) = \begin{cases} 0, & x < 0 \\ 1, & x > 0 \end{cases}$

3. $f(x) = \begin{cases} 1, & x < 0 \\ 0, & x > 0 \end{cases}$

4. $f(x) = \begin{cases} T_0, & 0 < x < 1 \\ 0, & \text{otherwise} \end{cases}$

5. By assuming that (3.86) has a solution of the form

$$u(x,y) = \int_0^\infty Y(s,y)\,[A(s)\cos sx + B(s)\sin sx]\,ds$$

(a) show that direct substitution of this expression into Laplace's equation leads to the conclusion $Y(s,y) = e^{-sy}$.

(b) By imposing the boundary condition in (3.86) upon the solution given in (a), deduce that

$$A(s) = \frac{1}{\pi}\int_{-\infty}^\infty f(x)\cos sx\,dx, \qquad B(s) = \frac{1}{\pi}\int_{-\infty}^\infty f(x)\sin sx\,dx$$

In Probs. 6 and 7, use the result of Prob. 5 and the prescribed function f to find an integral representation of the solution of (3.86).

6. $f(x) = \begin{cases} 1 - x^2, & |x| < 1 \\ 0, & |x| > 1 \end{cases}$

7. $f(x) = \begin{cases} 0, & x < 0 \\ e^{-x}, & x > 0 \end{cases}$

8. Use the Fourier sine transform to show that

$$u_{xx} + u_{yy} = 0, \qquad 0 < x < \infty,\ y > 0$$

$$\text{B.C.:} \quad \begin{cases} u(0,y) = 0, & u(x,0) = f(x), \\ u(x,y) \to 0 & \text{as} \quad (x^2 + y^2) \to \infty \end{cases}$$

has the formal solution

(a) $$u(x,y) = \frac{y}{\pi}\int_0^\infty f(t)\left[\frac{1}{(t-x)^2 + y^2} - \frac{1}{(t+x)^2 + y^2}\right]dt$$

(b) Show that when $f(x) = 1$, the solution in (a) reduces to $u(x,y) = (2/\pi)\tan^{-1}(x/y)$.

9. Use the Fourier sine transform to solve the problem described by (3.86).

10. Use the Fourier cosine transform to solve

$$u_{xx} + u_{yy} = 0, \qquad 0 < x < \infty,\ y > 0$$

B.C.: $\begin{cases} u_x(0,y) = 0, \ u(x,0) = f(x) \\ u(x,y) \to 0 \quad \text{as} \quad (x^2 + y^2) \to \infty \end{cases}$

and show that the solution has a form similar to that in Prob. 8.

11. Solve Prob. 10 when

$$f(x) = \begin{cases} 1, & 0 < x < c \\ 0, & c < x < \infty \end{cases}$$

12. Given the nonhomogeneous boundary-value problem

$$u_{xx} + u_{yy} = -xe^{-x^2}, \qquad -\infty < x < \infty,\ y > 0$$

B.C.: $u(x,0) = 0,\ u(x,y)$ ınite as $y \to \infty$,

(a) show that the transformed problem via the Fourier transform has the solution

$$U(s,y) = (1 - e^{-|s|y})\frac{i}{2\sqrt{2}s}e^{-s^2/4}$$

(b) Taking the inverse Fourier transform of (a), deduce that

$$u(x,y) = \frac{1}{2\pi}\int_0^\infty (1 - e^{-sy})\frac{\sin xs}{s}e^{-s^2/4}\,ds$$

13. Verify that (3.100) is the solution of the transformed problem (3.99).

14. Find a solution similar to (3.102) for the potential problem

$$u_{xx} + u_{yy} = 0, \qquad -\infty < x < \infty,\ y > 0$$

B.C.: $u(x,0) = f(x), \qquad u_y(x,a) = 0$

15. Find a solution similar to (3.110) for the potential problem

$$u_{xx} + u_{yy} = 0, \qquad 0 < x < \infty,\ 0 < y < a$$

B.C.: $\begin{cases} u(0,y) = 0 \\ u(x,0) = f(x),\ u(x,a) = g(x) \end{cases}$

16. Use the double Fourier transform to solve the three-dimensional Dirichlet problem

$$u_{xx} + u_{yy} + u_{zz} = 0, \qquad -\infty < x < \infty,\ -\infty < y < \infty,\ z > 0$$

B.C.: $\begin{cases} u(x,y,0) = f(x,y) \\ u(x,y,z) \to 0 \text{ as } (x^2 + y^2 + z^2)^{1/2} \to \infty \end{cases}$

and show that

$$u(x,y,z) = \frac{1}{2\pi}\int_{-\infty}^{\infty}\int_{-\infty}^{\infty} f(\xi,\eta)g(x-\xi,y-\eta,z)\,d\xi\,d\eta$$

where $g(x,y,z) = z(x^2 + y^2 + z^2)^{-3/2}$.

3.6 *Hydrodynamics*

A fluid flow in three-dimensional space is called *two-dimensional* if the velocity vector **V** is always parallel to a fixed plane (xy plane), and if the velocity components parallel to this plane along with the pressure p and fluid density ρ are all constant along any normal to the plane. This permits us to confine our attention to just a single plane which we interpret as a cross section of the three-dimensional region under consideration. Our discussion here will be limited to two-dimensional flow problems.

An *ideal fluid* is one in which the stress on an element of area is wholly normal and independent of the orientation of the area.* In contrast, the stress on a small area is no longer normal to that area for a *viscous fluid* in motion. If the density ρ is constant, we say the flow is *incompressible*. Of course, the notions of an ideal fluid or incompressible fluid are only idealizations that are valid when certain effects can be safely neglected in the analysis of a real fluid.

The velocity, pressure, and fluid density are all interrelated through a set of differential equations consisting of a continuity equation, equation of motion, and an equation of state (such as the density equal to a constant, etc.). The *continuity equation* is an expression of the conservation of mass, which states that the flow of mass into a region equals the flow of mass out of it, and assumes the form

$$\frac{\partial \rho}{\partial t} + \nabla \cdot (\rho \mathbf{V}) = 0 \tag{3.111}$$

For an incompressible fluid, this reduces to

$$\nabla \cdot \mathbf{V} = 0 \tag{3.112}$$

which implies there are no *sources* nor *sinks* within the region of interest (i.e., points at which fluid appears or disappears). The *equation of motion* is a consequence of Newton's second law of motion. By equating the rate of change of momentum of an element of fluid to the total force (i.e., the resultant **F** of the external forces and the net stresses acting on the element), we obtain†

$$\frac{\partial \mathbf{V}}{\partial t} + (\mathbf{V} \cdot \nabla)\mathbf{V} = \mathbf{F} - \frac{1}{\rho}\nabla p + \nu \nabla^2 \mathbf{V} + \frac{1}{3}\nu \nabla(\nabla \cdot \mathbf{V}) \tag{3.113}$$

where ν is the kinematic viscosity of the fluid. If the force **F** is conservative,

---

* An ideal fluid is also called a *perfect fluid* or *inviscid fluid*.

† For details, see B. K. Shivamoggi, *Theoretical Fluid Dynamics*, Dordrecht: Martinus Nijhoff, 1985.

there exists a scalar potential function Λ such that

$$\mathbf{F} = -\nabla\Lambda, \tag{3.114}$$

and further, if the fluid is incompressible, then (3.113) reduces to

$$\frac{\partial \mathbf{V}}{\partial t} - \mathbf{V} \times \Omega + \nabla\left(\Lambda + \frac{1}{2}V^2 + \frac{p}{\rho}\right) + \nu\nabla\times\Omega = \mathbf{0} \tag{3.115}$$

where $\Omega = \nabla\times\mathbf{V}$ is the *vorticity* and $\mathbf{V}$ is the magnitude of the velocity vector. Upon taking the curl of Eq. (3.115), we get the two-dimensional equation of motion

$$\frac{\partial\omega}{\partial t} + u\frac{\partial\omega}{\partial x} + v\frac{\partial\omega}{\partial y} = \nu\nabla^2\omega \tag{3.116}$$

where

$$\Omega = (0,0,\omega), \quad \mathbf{V} = (u,v,0) \tag{3.117}$$

and $\nabla^2 = \partial^2/\partial x^2 + \partial^2/\partial y^2$. Under these conditions, (3.112) now becomes

$$u_x + v_y = 0 \tag{3.118}$$

Equation (3.118) can be satisfied identically by introducing a scalar function ψ, called the *stream function*, according to

$$u = -\psi_y, \quad v = \psi_x \tag{3.119}$$

The stream function can be shown to be constant along any streamline, and therefore the streamlines of a particular flow are the family of curves ψ = constant. Also, it follows that

$$\omega = v_x - u_y = \nabla^2\psi \tag{3.120}$$

so that in terms of the stream function, the equation of motion (3.116) takes the form

$$\frac{\partial}{\partial t}\nabla^2\psi + \left(\frac{\partial\psi}{\partial y}\frac{\partial}{\partial x} - \frac{\partial\psi}{\partial x}\frac{\partial}{\partial y}\right)\nabla^2\psi = \nu\nabla^4\psi \tag{3.121}$$

where $\nabla^4\psi = \nabla^2\nabla^2\psi$.

Finally, a flow is called *steady* if the velocity, pressure, and fluid density are independent of time. If the velocity components are also small and the viscosity large, all terms on the left-hand side of Eq. (3.121), which are due to inertia of the fluid, may be neglected in a first approximation. Hence, in this case we find that (3.121) can be replaced by the *biharmonic equation*

$$\nabla^4\psi = 0 \tag{3.122}$$

which in two dimensions has the explicit form

$$\psi_{xxxx} + 2\psi_{xxyy} + \psi_{yyyy} = 0 \tag{3.123}$$

By solving (3.123) with appropriately prescribed boundary conditions, the velocity components of the flow can ultimately be determined through the use of Eqs. (3.119).

3.6.1 *Irrotational Flow of an Ideal Fluid*

If the vorticity $\mathbf{\Omega}$ is zero at every point within the region of interest, we say the flow is *irrotational*. This means that $\nabla \times \mathbf{V} = \mathbf{0}$, which in two dimensions is described by

$$\omega = v_x - u_y = 0 \tag{3.124}$$

This relation combined with (3.119) leads to Laplace's equation

$$\psi_{xx} + \psi_{yy} = 0 \tag{3.125}$$

Clearly, solutions of Laplace's equation $\nabla^2\psi = 0$ are also solutions of the equation of motion given by (3.121).

The irrotational flow of an ideal fluid can also be described in terms of a velocity potential function ϕ. That is, the condition $\nabla \times \mathbf{V} = \mathbf{0}$ implies the existence of a potential function ϕ such that $\mathbf{V} = -\nabla\phi$, or

$$u = -\phi_x, \qquad v = -\phi_y \tag{3.126}$$

By combining (3.126) with (3.118), we find that the velocity potential ϕ is likewise a solution of Laplace's equation

$$\phi_{xx} + \phi_{yy} = 0 \tag{3.127}$$

Thus, for irrotational flows we have the choice of solving (3.125) for the stream function ψ or solving (3.127) for the potential function ϕ.

Example 3.8: Consider the irrotational flow of an ideal fluid filling the half-space $y \geq 0$ through the strip $|x| \leq a$ (see Fig. 3.7). If the fluid is introduced normal to the region with prescribed velocity $v = f(x)$, find the resulting velocity components $u(x,y)$ and $v(x,y)$ within the half-space.

Solution: Let us formulate and solve the problem in terms of the velocity potential. In this case the problem is characterized by

$$\phi_{xx} + \phi_{yy} = 0, \qquad -\infty < x < \infty,\ y > 0$$

$$\text{B.C.:} \qquad \phi_y(x,0) = -f(x)h(a - |x|), \qquad -\infty < x < \infty$$

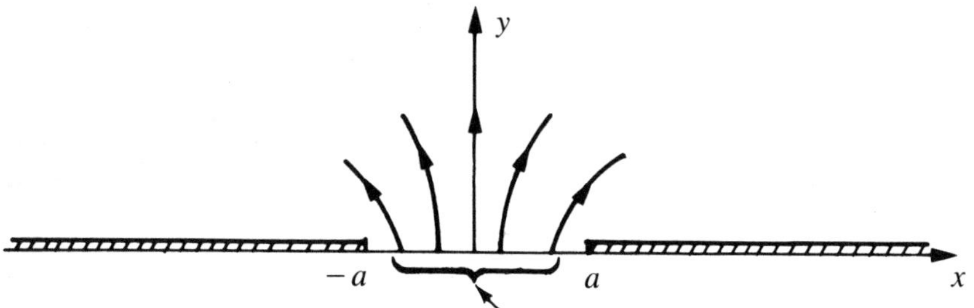

Figure 3.7 Fluid flow into the upper half-plane

In addition, it is customary to assume that the fluid is at rest at large distances from the plane $y = 0$, i.e.,

B.C.: $$\phi_x,\ \phi_y \to 0 \text{ as } y \to \infty$$

This problem is simply a Neumann problem for the half-plane, and thus by use of (3.96) we immediately deduce that

$$\phi(x,y) = -\frac{1}{2\pi}\int_{-a}^{a} f(\xi)\log[(x-\xi)^2 + y^2]\, d\xi$$

Recalling the relations (3.126), it follows that the velocity components are given by

$$u(x,y) = -\phi_x(x,y) = \frac{1}{\pi}\int_{-a}^{a} \frac{(x-\xi)f(\xi)}{(x-\xi)^2 + y^2}\, d\xi$$

and

$$v(x,y) = -\phi_y(x,y) = \frac{y}{\pi}\int_{-a}^{a} \frac{f(\xi)}{(x-\xi)^2 + y^2}\, d\xi$$

The solution of Exam. 3.8 in terms of the stream function ψ is left to the exercises (see Prob. 4 in Exers. 3.6).

3.6.2 *Surface Waves*

We consider here the development of two-dimensional gravity waves on a semiinfinite body of fluid $y \le 0$ from the action of an impulsive pressure on the free surface $y = 0$ of the fluid. We regard the fluid as ideal and initially at rest. It then follows from the laws of hydrodynamics that the motion will be irrotational throughout all time, and hence we can describe such motion in terms of a velocity potential ϕ satisfying Laplace's equation

$$\phi_{xx} + \phi_{yy} = 0 \tag{3.128}$$

Also, in this case the equation of motion (3.115) reduces to

$$\nabla\left(\Lambda + \frac{1}{2}V^2 + \frac{p}{\rho} - \phi_t\right) = 0 \tag{3.129}$$

If $\mathbf{F} = -\nabla\Lambda$ represents the force of gravity and y is the distance measured from the free surface, then $\Lambda = gy$, where g is the acceleration of gravity. Thus, the integrated form of (3.129) becomes*

$$p = -\rho(\phi_t + gy + \tfrac{1}{2}V^2) \tag{3.130}$$

Let us introduce the function $\eta(x,t)$ to denote the elevation of the free surface relative to its equilibrium position $y = 0$. For sufficiently small displacements η, we may linearize the equation of motion (3.130) by neglecting all terms of second order in amplitude. This assumption permits us to omit the term $\frac{1}{2}\rho V^2$ in (3.130) and to evaluate the potential function ϕ and its derivatives at $y = 0$ in the free-surface boundary conditions, rather than at $y = \eta$. The kinematical boundary condition on the vertical velocity component $v = \eta_t$ at this surface is

$$\phi_y(x,0,t) = \eta_t(x,t) \tag{3.131}$$

which expresses the fact that a fluid particle initially on the free surface continues to remain on the free surface during the wave motion. The dynamical boundary condition, corresponding to the requirement $p = 0$ at the free surface, is

$$\phi_t(x,0,t) - g\eta(x,t) = 0 \tag{3.132}$$

Combining (3.131) and (3.132), we obtain the single boundary condition

B.C.: $$\phi_{tt}(x,0,t) + g\phi_y(x,0,t) = 0 \tag{3.133}$$

Finally, we complete the prescription of boundary conditions by invoking the finiteness condition

B.C.: $$\phi_x, \phi_y \to 0 \quad \text{as} \quad y \to -\infty \tag{3.134}$$

Because we have assumed the waves are generated by the action of an impulsive pressure on the free surface $y = 0$ at time $t = 0$, we find from (3.130) that

I.C.: $$\phi(x,0,0) = (1/\rho)\, q(x) \tag{3.135}$$

where $q(x)$ is the impulsive pressure defined by

$$q(x) = \int p(x,0,t)\, dt$$

---

* The constant of integration can be absorbed in the potential function ϕ.

Finally, we assume the initial condition $\eta = 0$ when $t = 0$, which means that [see (3.132)]

$$\text{I.C.:} \qquad \phi_t(x,0,0) = 0 \tag{3.136}$$

To determine the wave system produced by the above conditions, we must solve Laplace's equation (3.128) subject to conditions (3.133)–(3.136). If $\Phi(s,y,t)$ and $Q(s)$ are the Fourier transforms, respectively, of $\phi(x,y,t)$ and $q(x)$, the transformed problem we need to solve is described by

$$\Phi_{yy} - s^2\Phi = 0, \qquad -\infty < y < 0, t > 0$$

$$\text{B.C.:} \qquad \begin{cases} \Phi_{tt}(s,0,t) + g\Phi_y(s,0,t) = 0, & t > 0 \\ \Phi, \Phi_y \to 0 \text{ as } y \to -\infty \end{cases} \tag{3.137}$$

$$\text{I.C.:} \qquad \Phi(s,0,0) = (1/\rho)Q(s), \qquad \Phi_t(s,0,0) = 0$$

The solution of (3.137) is (see Prob. 5 in Exer. 3.6)

$$\Phi(s,y,t) = (1/\rho)Q(s)\cos(\sqrt{g|s|}t)e^{|s|y} \tag{3.138}$$

and consequently, by the inversion formula,

$$\phi(x,y,t) = \frac{1}{\rho\sqrt{2\pi}}\int_{-\infty}^{\infty} Q(s)\cos(\sqrt{g|s|}t)e^{|s|y-isx}\,ds \tag{3.139}$$

If $Q(s)$ is an *even* function of s, this last result becomes

$$\phi(x,y,t) = \frac{1}{\rho\sqrt{2\pi}}\left[\int_0^{\infty} e^{sy}\,Q(s)\cos(sx - \sqrt{gs}t)\,ds + \int_0^{\infty} e^{sy}\,Q(s)\cos(sx + \sqrt{gs}t)\,ds\right] \tag{3.140}$$

Generally speaking, the evaluation of integrals like (3.140) must be accomplished by the use of numerical or approximation methods. However, the asymptotic evaluation of (3.140) as $t \to \infty$ can be achieved by Kelvin's *method of stationary phase* (see Sec. 2.9.1). To recall, if s_0 is a point on the interval $a < s_0 < b$ for which $G'(s_0) = 0$ and $G'(s_0) \neq 0$, then

$$\frac{1}{\sqrt{2\pi}}\int_a^b F(s)e^{itG(s)}\,ds \sim \frac{F(s_0)}{\sqrt{t|G''(s_0)|}}\exp\left[itG(s_0) \pm \frac{i\pi}{4}\right], \quad t \to \infty \tag{3.141}$$

where the plus sign goes with $G''(s_0) > 0$ and the minus sign with $G''(s_0) < 0$.

Now, in (3.140) there are two symmetrical groups of waves moving in the two directions from the origin. If we consider only the right-running waves, then we may drop the second of the two integrals. Writing the first integral as

$$\frac{1}{\rho\sqrt{2\pi}}\int_0^\infty e^{sy}Q(s)\cos(sx - \sqrt{gs}t)\,ds$$

$$= \frac{1}{\rho\sqrt{2\pi}}\operatorname{Re}\left\{\int_0^\infty e^{sy}Q(s)e^{it(\sqrt{gs} - sx/t)}\,ds\right\} \tag{3.142}$$

we can identify $G(s) = \sqrt{gs} - sx/t$. Then setting

$$G'(s) = \frac{\sqrt{g}t - 2x\sqrt{s}}{2t\sqrt{s}} = 0 \tag{3.143}$$

we see that $s_0 = gt^2/4x^2$, and hence by applying the result of (3.141) to the above integral, we deduce that (3.140) reduces to

$$\phi(x,y,t) \sim \frac{1}{\rho}\sqrt{\frac{gt^2}{2x^3}}\,e^{gt^2y/4x^2}Q\left(\frac{gt^2}{4x^2}\right)\cos\left(\frac{gt^2}{4x} - \frac{\pi}{4}\right), \qquad t \to \infty \tag{3.144}$$

3.6.3 *Steady Slow Motion of a Viscous Fluid*

Let us now consider the problem of a viscous, incompressible fluid filling the half-space $y \geq 0$ as discussed in Exam. 3.8 in Sec. 3.6.1. A viscous fluid cannot be irrotational so we must work directly with the stream function ψ in this case. If the flow is steady and the velocity components small, then the governing equation is the biharmonic equation (3.124). Assuming that all other conditions of the problem are the same as described in Exam. 3.8, the present problem is characterized by

$$\begin{aligned} &\psi_{xxxx} + 2\psi_{xxyy} + \psi_{yyyy} = 0, \qquad -\infty < x < \infty,\ y > 0 \\ \text{B.C.:}\quad &\psi_y(x,0) = 0, \qquad \psi_x(x,0) = f(x)h(a - |x|),\ -\infty < x < \infty \end{aligned} \tag{3.145}$$

Because the governing equation in (3.145) is fourth-order, it was necessary to prescribe a boundary condition in addition to that in Exam. 3.8 to maintain a well-posed problem. The above condition $\psi_y(x,0) = 0$ merely states that the horizontal velocity component u is zero along the x axis.* As before, we assume the fluid is at rest at large distances from the plane $y = 0$, i.e.,

$$\text{B.C.:}\qquad \psi_x, \psi_y \to 0 \text{ as } y \to \infty \tag{3.146}$$

To solve (3.145), we will use the Fourier transform with respect to x. Thus, we write

$$\mathscr{F}\{\psi(x,y); x \to s\} = \Psi(s,y) \tag{3.147}$$

---

* A moving viscous fluid tends to adhere to the surface of an obstacle placed in its path.

and the transformed problem takes the form of a fourth-order boundary-value problem

$$\Psi_{yyyy} - 2s^2\Psi_{yy} + s^4\Psi = 0, \; y > 0$$

$$\text{B.C.:} \qquad \Psi_y(s,0) = 0, \qquad -is\Psi(s,0) = F_+(s) \tag{3.148}$$

where

$$F_+(s) = (1/\sqrt{2\pi}) \int_{-a}^{a} e^{isx} f(x)\, dx \tag{3.149}$$

The characteristic polynomial of this differential equation is*

$$m^4 - 2s^2m^2 + s^4 = 0$$

with $m = \pm s$ as double roots leading to the independent solutions

$$e^{-sy},\; ye^{-sy},\; e^{sy},\; ye^{sy}$$

From physical considerations we must select only those solutions which tend to zero as $y \to \infty$. Therefore, the bounded general solution of the problem (3.148) is given by

$$\Psi(s,y) = [A(s) + B(s)y]e^{-|s|y} \tag{3.150}$$

where $A(s)$ and $B(s)$ are arbitrary functions. Application of the boundary conditions in (3.148) yields the relations

$$\Psi_y(s,0) = -|s|A(s) + B(s) = 0$$

$$is\Psi(s,0) = -isA(s) = F_+(s)$$

from which we deduce $A(s) = iF_+(s)/s$ and $B(s) = |s|A(s)$. Hence,

$$\Psi(s,y) = is^{-1}(1 + |s|y)F_+(s)e^{-|s|y} \tag{3.151}$$

In order to invert (3.151), we will use the convolution theorem of Fourier. In this regard it is convenient to define

$$G(s,y) = is^{-1}(1 + |s|y)e^{-|s|y} \tag{3.152}$$

and then determine

$$\mathscr{F}^{-1}\{G(s,y);s\to x\} = \mathscr{F}^{-1}\{is^{-1}e^{-|s|y};s\to x\} + iy\mathscr{F}^{-1}\{e^{-|s|y}\operatorname{sgn}(s);s\to x\}$$

Based on Exam. 2.5 in Chap. 2, we have that

$$\begin{aligned} \mathscr{F}^{-1}\{is^{-1}e^{-|s|y};s\to x\} &= \mathscr{F}_S^{-1}\{s^{-1}e^{-sy};s\to x\} \\ &= \sqrt{2/\pi}\,\tan^{-1}(x/y), \qquad y > 0 \end{aligned} \tag{3.153}$$

---

* We obtain the characteristic polynomial by assuming $\Psi = e^{my}$.

and similarly,

$$\mathscr{F}^{-1}\{ie^{-|s|y}\operatorname{sgn}(s);s\to x\} = \mathscr{F}_S^{-1}\{e^{-sy};s\to x\} \tag{3.154}$$
$$= \sqrt{2/\pi}\, x/(x^2 + y^2), \qquad y > 0$$

Thus,

$$\mathscr{F}^{-1}\{G(s,y);s\to x\} = \sqrt{2/\pi}\,[\tan^{-1}(x/y) + xy/(x^2 + y^2)] \tag{3.155}$$

and through use of the convolution theorem we deduce the solution

$$\psi(x,y) = \frac{1}{\pi}\int_{-a}^{a} f(\xi)\left[\tan^{-1}\left(\frac{x-\xi}{y}\right) + \frac{(x-\xi)y}{(x-\xi)^2 + y^2}\right] d\xi \tag{3.156}$$

The velocity components $u(x,y)$ and $v(x,y)$ can now be determined from Eq. (3.119) (see Prob. 9 in Exers. 3.6).

EXERCISES 3.6

1. Show that the elimination of u and v between (3.119) and (3.126) leads to the *Cauchy–Riemann equations*

$$\phi_x = \psi_y, \qquad \psi_x = -\phi_y$$

and use these relations to calculate the stream function ψ from the expression for ϕ given in Exam. 3.8.

2. Solve Exam. 3.8 for the special case where $f(x) = V_0$, constant.

3. For the special case of Exam. 3.8 where $f(x) = (a^2 - x^2)^{-1/2}$,
(a) show that

$$\phi(x,y) = \frac{1}{2}\int_{-\infty}^{\infty} \frac{J_0(a\xi)}{|\xi|}\, e^{i\xi x - |\xi|y}\, d\xi$$

(b) From (a), determine integral representations for the velocity components and verify that $u(x,0) = 0$ when $|x| < a$.

4. In terms of the stream function, the problem discussed in Exam. 3.8 is characterized by

$$\psi_{xx} + \psi_{yy} = 0, \qquad -\infty < x < \infty,\ y > 0$$

$$\text{B.C.:} \qquad \psi_x(x,0) = f(x)h(a - |x|), \qquad -\infty < x < \infty.$$

Solve this problem for ψ and show that it leads to the same expressions for the velocity components $u(x,y)$ and $v(x,y)$ as derived in Exam. 3.8 from the potential function.

5. Verify that (3.138) is the solution of the transformed problem given by (3.137).

6. Solve the problem described by (3.128), (3.133)–(3.136) for the special case $q(x) = P_0\delta(x)$, where P_0 is a constant, and show that

(a) $\phi(x,0,t) = (P_0/2\pi\rho)\int_0^\infty [\cos(sx - \sqrt{gs}t) + \cos(sx + \sqrt{gs}t)]\,ds$

(b) By making the substitutions

$$\zeta = \sqrt{x/g}\,(\sqrt{gs} \mp gt/2x), \quad \omega = \sqrt{gt^2/4x}$$

show that

$$\phi(x,0,t) = -\frac{P_0}{\pi\rho g}\frac{dJ}{dt}$$

where

$$J = -2\sqrt{g/x}\int_0^\omega \sin(\omega^2 - \zeta^2)\,d\zeta$$

(c) From (b), deduce that

$$\phi(x,0,t) = \frac{P_0 u}{\rho x}[\cos(\tfrac{1}{2}\pi u^2)C(u) + \sin(\tfrac{1}{2}\pi u^2)S(u)]$$

where $C(u)$ and $S(u)$ denote the Fresnel integrals (see Sec. 1.3.2) and where $u^2 = gt^2/2\pi x$.

7. Determine the wave motion produced on the surface of a fluid of infinite depth $y \le 0$ by an initial displacement $\eta = f(x)$ of the surface. Use the method of stationary phase to evaluate this wave motion asymptotically as $t \to \infty$.

8. Determine the wave motion produced by an initial displacement $\eta = f(x)$ on the surface of a fluid of finite depth h. One now has for the potential ϕ the following problem:

$$\phi_{xx} + \phi_{yy} = 0, \qquad -\infty < x < \infty,\ -h < y < 0$$

$$\text{B.C.:}\quad \begin{cases} \phi_{tt}(x,0,t) + g\phi_y(x,0,t) = 0 \\ \phi_y(x,-h,t) = 0 \end{cases}$$

$$\text{I.C.:}\quad \phi_t(x,0,0) = f(x)$$

9. Use the solution (3.156) to determine the velocity components $u(x,y)$ and $v(x,y)$ for the special case $f(x) = V_0$, constant.

10. Consider the steady, slow motion of a viscous, incompressible fluid filling the half-space $y \ge 0$. In terms of the stream function, this boundary value problem is described by

$$\psi_{xxxx} + 2\psi_{xxyy} + \psi_{yyyy} = 0, \qquad -\infty < x < \infty,\ y > 0$$

$$\text{B.C.:}\quad \begin{cases} \psi_y(x,0) = g(x)h(a - |x|), & \psi_x(x,0) = 0 \\ \psi_x, \psi_y \to 0 \text{ as } y \to \infty \end{cases}$$

(a) Determine the solution for arbitrary $g(x)$.
(b) Determine the velocity components $u(x,y)$ and $v(x,y)$ for the special case when $g(x) = V_0$, constant.

3.7 *Elasticity in Two Dimensions*

Many of the great mathematicians since the time of Euler have attacked the problem concerning the state of stress in various elastic bodies under the action of given forces. In this section we will briefly discuss some two-dimensional problems for which the solution may be obtained by Fourier transform methods.

The effect of body or surface forces on a two-dimensional body will be to produce internal forces between various parts of the body. The magnitudes of these internal forces are defined by the ratio of the force to the area over which it acts, called the *average stress*. In the limit as the area shrinks to zero, we obtain the components of stress at a point in the elastic medium. This stress is composed of two *normal components*, σ_{xx} and σ_{yy}, and two *shearing components*, σ_{xy} and σ_{yx}, for which $\sigma_{xy} = \sigma_{yx}$.* We adopt the convention that stresses are positive when a tension is produced and negative when a compression occurs.

The DEs satisfied by the components of stress in an elastic medium under the action of a force per unit mass having components (F_x, F_y) may be obtained by applying Newton's second law of motion to a small rectangular element of the medium. Writing the displacement vector as $\mathbf{u} = (u,v)$, these equations of motion are†

$$\frac{\partial \sigma_{xx}}{\partial x} + \frac{\partial \sigma_{xy}}{\partial y} + \rho F_x = \rho \frac{\partial^2 u}{\partial t^2} \tag{3.157a}$$

$$\frac{\partial \sigma_{xy}}{\partial x} + \frac{\partial \sigma_{yy}}{\partial y} + \rho F_y = \rho \frac{\partial^2 v}{\partial t^2} \tag{3.157b}$$

where ρ is the mass density of the elastic body. For *equilibrium* problems the time derivatives on the right-hand sides can be set to zero. Also, in the absence of body forces we have $F_x = F_y = 0$, and in these cases the equations of equilibrium take the simpler form

$$\frac{\partial \sigma_{xx}}{\partial x} + \frac{\partial \sigma_{xy}}{\partial y} = 0 \tag{3.158a}$$

$$\frac{\partial \sigma_{xy}}{\partial x} + \frac{\partial \sigma_{yy}}{\partial y} = 0 \tag{3.158b}$$

---

* In this section the subscripted variables no longer represent partial derivatives. Instead, we will use the standard Leibniz notation for any partial derivatives.

† For the derivation of these equations and a general discussion of two-dimensional elasticity problems, see R. M. Little, *Elasticity*, Englewood Cliffs: Prentice-Hall, 1972.

The above equations of equilibrium together with appropriate boundary conditions are still not sufficient for the determination of the stresses. That is, the complete solution requires us to take into account a *compatibility condition* for the distribution of stress that arises from the existence of a continuous displacement vector. This condition, specified only in terms of the stresses, is given by the relation*

$$\frac{\partial^2}{\partial y^2}[\sigma_{xx} - \nu(\sigma_{xx} + \sigma_{yy})] + \frac{\partial^2}{\partial x^2}[\sigma_{yy} - \nu(\sigma_{xx} + \sigma_{yy})] = 2\frac{\partial \sigma_{xy}}{\partial x\, \partial y} \tag{3.159}$$

where ν is the Poisson ratio of the material.

To solve this system of equations it is convenient to introduce a scalar function χ, called the *Airy stress function*, by setting

$$\sigma_{xy} = -\frac{\partial^2 \chi}{\partial x \partial y} \tag{3.160}$$

Using this relation, we find the equations of equilibrium (3.158) reduce to

$$\frac{\partial}{\partial x}\left(\sigma_{xx} - \frac{\partial^2 \chi}{\partial y^2}\right) = 0 \tag{3.161a}$$

$$\frac{\partial}{\partial y}\left(\sigma_{yy} - \frac{\partial^2 \chi}{\partial x^2}\right) = 0 \tag{3.161b}$$

Hence, it follows immediately that the equations of equilibrium (3.158) are satisfied by

$$\sigma_{xx} = \frac{\partial^2 \chi}{\partial y^2}, \quad \sigma_{yy} = \frac{\partial^2 \chi}{\partial x^2}, \quad \sigma_{xy} = -\frac{\partial^2 \chi}{\partial x \partial y} \tag{3.162}$$

Lastly, the substitution of these expressions into the compatibility condition (3.159) yields the *biharmonic equation*

$$\frac{\partial^4 \chi}{\partial x^4} + 2\frac{\partial^4 \chi}{\partial x^2 \partial y^2} + \frac{\partial^4 \chi}{\partial y^4} = 0 \tag{3.163}$$

Solving (3.163) for the Airy stress function, subject to appropriately prescribed boundary conditions, leads to the stress components through use of Eq. (3.162).

3.7.1 *Elastic Equilibrium under Surface Forces*

Let us consider the equilibrium of a solid body deformed by the application of pressure to its bounding surfaces. We will assume the body forces are zero throughout the body so that (3.163) is the governing equation.

---

* For example, see S. P. Timoshenko and J. N. Goodier, *Theory of Elasticity*, New York: McGraw-Hill, 1965.

The elastic body that we consider is bounded by a plane of infinite extent defined by $x = 0$. The x-axis is then normal to this plane and taken positive in the direction into the medium. If the domain is compressed by a surface pressure p, which varies along the surface, the problem is mathematically described by

$$\frac{\partial^4 \chi}{\partial x^4} + 2\frac{\partial^4 \chi}{\partial x^2 \partial y^2} + \frac{\partial^4 y}{\partial y^4} = 0, \qquad 0 < x < \infty,\ -\infty < y < \infty \tag{3.164}$$

B.C.: $\sigma_{xx}(0,y) = -\,p(y), \qquad \sigma_{xy}(0,y) = 0,\ -\infty < y < \infty$

Of course, all stress components must satisfy the limiting condition

$$\sigma_{xx}, \sigma_{yy}, \sigma_{xy} \to 0 \text{ as } x \to \infty \tag{3.165}$$

Formulated in terms of the Airy stress function by using (3.162), the boundary conditions in (3.164) are equivalent to

$$\text{B.C.:} \qquad \frac{\partial^2 \chi}{\partial y^2}(0,y) = -\,p(y),\ \frac{\partial^2 \chi}{\partial x \partial y}(0,y) = 0,\ -\infty < y < \infty \tag{3.166}$$

By applying the Fourier transform

$$\mathscr{F}\{\chi(x,y); y \to s\} = G(x,s) \tag{3.167}$$

to the governing equation in (3.164) and boundary conditions (3.166), we obtain the transformed problem

$$\frac{d^4 G}{dx^4} - 2s^2\frac{d^2 G}{dx^2} + s^4 G = 0, \qquad x > 0 \tag{3.168}$$

$$\text{B.C.:} \qquad s^2 G(0,s) = P(s),\ -is\frac{dG}{dx}(0,s) = 0$$

where $P(s)$ is the Fourier transform of $p(y)$. The bounded solutions of this ODE are given by

$$G(x,s) = [A(s) + B(s)x]e^{-|s|x} \tag{3.169}$$

where $A(s)$ and $B(s)$ are unknown "constants." By application of the boundary conditions in (3.168), we see that $A(s) = P(s)/s^2$ and $B(s) = P(s)/|s|$; hence,

$$G(x,s) = \frac{P(s)}{s^2}(1 + |s|x)e^{-|s|x} \tag{3.170}$$

from which we deduce

$$\chi(x,y) = \frac{1}{\sqrt{2\pi}}\int_{-\infty}^{\infty} \frac{P(s)}{s^2}(1 + |s|x)e^{-|s|x - isy}\, ds \tag{3.171}$$

The stress components that we desire are now calculated using (3.162), and are given by

$$\sigma_{xx}(x,y) = -\frac{1}{\sqrt{2\pi}} \int_{-\infty}^{\infty} P(s)(1 + |s|x)e^{-|s|x - isy}\, ds \tag{3.172}$$

$$\sigma_{yy}(x,y) = -\frac{1}{\sqrt{2\pi}} \int_{-\infty}^{\infty} P(s)(1 - |s|x)e^{-|s|x - isy}\, ds \tag{3.173}$$

$$\sigma_{xy}(x,y) = -\frac{ix}{\sqrt{2\pi}} \int_{-\infty}^{\infty} sP(s)e^{-|s|x - isy}\, ds \tag{3.174}$$

For the special cases where $P(s)$ is either an even function or an odd function, these equations take on a simpler form. Specifically, if $P(s)$ is an even function, then

$$\sigma_{xx}(x,y) = -\sqrt{2/\pi} \int_{0}^{\infty} P(s)(1 + sx)e^{-sx}\cos sy\, ds \tag{3.175}$$

$$\sigma_{yy}(x,y) = -\sqrt{2/\pi} \int_{0}^{\infty} P(s)(1 - sx)e^{-sx}\cos sy\, ds \tag{3.176}$$

$$\sigma_{xy}(x,y) = -\sqrt{2/\pi}\, x \int_{0}^{\infty} sP(s)e^{-sx}\sin sy\, ds \tag{3.177}$$

whereas when $P(s)$ is an odd function, we find

$$\sigma_{xx}(x,y) = -\sqrt{2/\pi} \int_{0}^{\infty} P(s)(1 + sx)e^{-sx}\sin sy\, ds \tag{3.178}$$

$$\sigma_{yy}(x,y) = -\sqrt{2/\pi} \int_{0}^{\infty} P(s)(1 - sx)e^{-sx}\sin sy\, ds \tag{3.179}$$

$$\sigma_{xy}(x,y) = \sqrt{2/\pi}\, x \int_{0}^{\infty} sP(s)e^{-sx}\cos sy\, ds \tag{3.180}$$

By invoking the convolution theorems of Fourier, another representation of these results is also possible (see Probs. 1 and 2 in Exer. 3.7).

EXERCISES 3.7

1. Use the convolution theorem to show that Eqs. (3.172)–(3.174) can be expressed in the alternate form

$$\sigma_{xx} = -\frac{2x^3}{\pi} \int_{-\infty}^{\infty} \frac{p(y - \eta)}{(x^2 + \eta^2)^2}\, d\eta$$

$$\sigma_{yy} = -\frac{2x}{\pi} \int_{-\infty}^{\infty} \frac{\eta^2 p(y - \eta)}{(x^2 + \eta^2)^2}\, d\eta$$

$$\sigma_{xy} = -\frac{2x^2}{\pi} \int_{-\infty}^{\infty} \frac{\eta p(y - \eta)}{(x^2 + \eta^2)^2}\, d\eta$$

2. Find a result similar to that in Prob. 1 for the special case when $p(y)$ is an *even* function.

3. Use the result of Prob. 1 to show that, when the prescribed pressure is $p(y) = p_0 h(y)$, p_0 constant,

(a) the stress components of the problem described by (3.164) become

$$\sigma_{xx} = -p_0\left(\frac{1}{2} + \frac{\theta}{\pi} + \frac{1}{2\pi}\sin 2\theta\right)$$

$$\sigma_{yy} = -p_0\left(\frac{1}{2} + \frac{\theta}{\pi} - \frac{1}{2\pi}\sin 2\theta\right)$$

$$\sigma_{xy} = \frac{p_0}{\pi}\cos^2\theta$$

where $x = r\cos\theta$ and $y = r\sin\theta$

(b) Find the *maximum shear stress* defined by

$$\tau = \tfrac{1}{4}(\sigma_{xx} - \sigma_{yy})^2 + \sigma_{xy}^2]^{1/2}$$

4. Repeat Prob. 3 for the case $p(y) = p_0 h(a - |y|)$, p_0 constant.

5. In the problem given by (3.164) let the pressure be prescribed by

$$p(y) = \frac{p_0}{\pi}(a^2 - y^2)^{-1/2}\, h(a - |y|),\ p_0 \text{ constant}$$

Show that

(a) $\sigma_{xx} + \sigma_{yy} = -\sqrt{2/\pi}\, p_0 \int_0^\infty J_0(as)e^{-sx}\cos sy\, ds$

(b) $\sigma_{yy} - \sigma_{xx} + 2i\sigma_{xy} = \sqrt{\dfrac{2}{\pi}}\, p_0 x \int_0^\infty sJ_0(as)e^{-s(x+iy)}\, ds$

(c) At $y = 0$, show that (a) and (b) reduce to

$$\sigma_{xx} + \sigma_{yy} = -\sqrt{2/\pi}\, p_0(x^2 + a^2)^{-1/2}$$

$$\sigma_{yy} - \sigma_{xx} + 2i\sigma_{xy} = \sqrt{2/\pi}\, p_0 x^2(x^2 + a^2)^{-3/2}$$

6. Find the stress components inside the semiinfinite medium $y \geq 0$ such that

$$\frac{\partial^4\chi}{\partial x^4} + 2\frac{\partial^4\chi}{\partial x^2\partial y^2} + \frac{\partial^4\chi}{\partial y^4} = 0, \qquad -\infty < x < \infty,\ y > 0$$

B.C.: $\sigma_{yy}(x,0) = 0, \qquad \sigma_{xy}(x,0) = q(x)$

3.8 *Probability and Statistics*

Suppose that X is a random variable. The function $P(x)$, called the *distribution function*, represents the probability that $X < x$, where x is a real number. The distribution function has the following properties:

1. $\lim_{x \to -\infty} P(x) = 0$
2. $\lim_{x \to \infty} P(x) = 1$
3. $P(x_1) \leq P(x_2)$ when $x_1 \leq x_2$

If we think of X as a continuous variable, then there usually exists a related function $p(x)$ such that

$$P(x) = \int_{-\infty}^{x} p(u)\,du \tag{3.181}$$

The function $p(x)$ is called the *probability density function* (PDF) of the random variable X. Once $p(x)$ has been determined, various properties of the random variable X can be calculated, such as the *statistical moments* of X.

In statistics, the moments $m_1, m_2, \ldots,$ of the random variable X are defined in terms of the expectation operator E. For example,

$$m_1 = E[X] = \int_{-\infty}^{\infty} xp(x)dx \tag{3.182a}$$

$$m_2 = E[X^2] = \int_{-\infty}^{\infty} x^2p(x)dx \tag{3.182b}$$

whereas in general,

$$m_k = E[X^k] = \int_{-\infty}^{\infty} x^kp(x)dx, \qquad k = 1,2,3,\ldots \tag{3.183}$$

The first moment gives the average value of a random variable, and the higher-order moments give additional information about the spread of the distribution defining the random variable. The *variance* of the distribution is defined by

$$\sigma^2 = \int_{-\infty}^{\infty} (x - m_1)^2p(x)\,dx = m_2 - m_1^2 \tag{3.184}$$

which follows by expanding the square and using relations (3.183).

3.8.1 *Characteristic Functions*

In many situations it is convenient to introduce the notion of a *characteristic function* $C(t)$ from which the statistical moments of X also can be found.

In certain applications the characteristic function of a random variable is easier to calculate than its PDF, and thus this function can be very useful in such cases. We define this new function by the expectation

$$C(t) = E[e^{itx}] = \int_{-\infty}^{\infty} e^{itx} p(x) dx \tag{3.185}$$

which we recognize as a Fourier transform relation given by

$$C(t) = \sqrt{2\pi}\, \mathscr{F}\{p(x);t\} \tag{3.186}$$

Hence, it follows that

$$C(0) = \int_{-\infty}^{\infty} p(x)\, dx = 1 \tag{3.186a}$$

$$C'(0) = i\int_{-\infty}^{\infty} xp(x)\, dx = im_1 \tag{3.186b}$$

$$C''(0) = -\int_{-\infty}^{\infty} x^2 p(x)\, dx = -m_2 \tag{3.186c}$$

while in general, we deduce

$$m_k = (-i)^k C^{(k)}(0),\ k = 1,2,3,\ldots \tag{3.187}$$

This says that m_k is the coefficient of $(it)^k/k!$ in the Maclaurin series expansion of the characteristic function.

Finally, it also follows from properties of inverse Fourier tranforms that if the characteristic function of a random variable is known, its probability density function is related by*

$$p(x) = \frac{1}{2\pi}\int_{-\infty}^{\infty} e^{-itx}\, C(t)\, dt \tag{3.188}$$

Example 3.9: Find the characteristic function associated with the *normal* or *Gaussian density function*

$$p(x) = \frac{1}{\sqrt{2\pi}\sigma} e^{-(x-m)^2/2\sigma^2}$$

where m denotes the mean (first moment) and σ^2 the variance.

Solution: From (3.186), we have

* For a more thorough discussion of random variables and characteristic functions, see A. Papoulis, *Probability, Random Variables, and Stochastic Processes*, New York: McGraw-Hill, 1965, or C. W. Helstrom, *Probability and Stochastic Processes for Engineers*, New York: Macmillan, 1984.

$$\begin{aligned} C(t) &= \sqrt{2\pi}\, \mathscr{F}\{p(x);t\} \\ &= (1/\sigma)\mathscr{F}\{e^{-(x-m)^2/2\sigma^2};t\} \\ &= e^{itm}\mathscr{F}\{e^{-u^2/2};u\to\sigma t\} \end{aligned}$$

Recalling Exam. 2.6 in Chap. 2, we deduce

$$C(t) = e^{itm-\sigma^2t^2/2}$$

Our next example illustrates the way in which the characteristic function is used to provide the desired PDF under a transformation of random variables.

Example 3.10: Given that X and Y are independent normal random variables with means zero and unit variances, find the PDF of the random variable $Z = XY$.

Solution: To find $p_Z(z)$,* we will first determine the characteristic function for Z and then invert it according to (3.188).

Since $Z = XY$, the characteristic function of Z can be determined by calculating

$$C_Z(t) = E[e^{itz}] = \int_{-\infty}^{\infty}\int_{-\infty}^{\infty} e^{itxy}\, p_{X,Y}(x,y)dx\,dy$$

where $p_{X,Y}(x,y)$ is the joint PDF of X and Y. Because X and Y are assumed to be independent, it follows that the joint PDF is simply the product of their individual PDFs. Hence, we have

$$\begin{aligned} P_{X,Y}(x,y) &= p_X(x)p_Y(y) \\ &= \frac{1}{\sqrt{2\pi}}e^{-x^2/2}\cdot\frac{1}{\sqrt{2\pi}}e^{-y^2/2} \\ &= \frac{1}{2\pi}e^{-(x^2+y^2)/2} \end{aligned}$$

which leads to the double integral

$$C_Z(t) = \frac{1}{2\pi}\int_{-\infty}^{\infty} e^{-y^2/2}\int_{-\infty}^{\infty} e^{itxy}\, e^{-x^2/2}\,dx\,dy$$

Working with the innermost integral, we find

$$\begin{aligned} \int_{-\infty}^{\infty} e^{itxy}\, e^{-x^2/2}\,dx &= \sqrt{2\pi}\,\mathscr{F}\{e^{-x^2/2};ty\} \\ &= e^{-t^2y^2/2} \end{aligned}$$

* When working with more than one PDF, we find it convenient to use the more distinguishing notation $p_X(x)$ instead of just $p(x)$.

Using this result, the remaining integral yields

$$C_Z(t) = \frac{1}{\sqrt{2\pi}} \int_{-\infty}^{\infty} e^{-(1 + t^2)y^2/2}$$
$$= \frac{1}{\sqrt{1 + t^2}}$$

Substituting this expression for $C_Z(t)$ into (3.188), we obtain

$$p_Z(z) = \frac{1}{2\pi} \int_{-\infty}^{\infty} e^{-itz}\, C_Z(t)\, dt$$
$$= \frac{1}{2\pi} \int_{-\infty}^{\infty} \frac{e^{-itz}}{\sqrt{1 + t^2}}\, dt$$
$$= \frac{1}{\pi} \int_{0}^{\infty} \frac{\cos zt}{\sqrt{1 + t^2}}\, dt$$

where we have used the fact that $C_Z(t)$ is an even function. This last integral is not an elementary integral nor does it lend itself to evaluation by conventional means using basic complex variable theory. Nonetheless, it is a well-known integral which leads to the final result*

$$p_Z(z) = \frac{1}{\pi} K_0(|z|),$$

where $K_0(x)$ is a *modified Bessel function of the second kind* and order zero.

EXERCISES 3.8

1. A certain random variable X is known to have the characteristic function given by $C(t) = (1 - 2it)^{-1}$.

(a) Determine the first moment m_1 and variance σ^2 of X.
(b) Find the PDF of X.

2. The *uniform* or *rectangular distribution* of a random variable X is defined by

$$p(x) = \begin{cases} 1/2\sigma, & |x - m| \le \sigma \\ 0, & |x - m| > \sigma \end{cases}$$

(a) Find the characteristic function of X.
(b) Show that the kth moment of X is given by

* For example, see I. S. Gradshteyn and I. M. Ryzhik, *Table of Integrals, Series, and Products*, New York: Academic Press, 1980.

$$m_k = k! \sum_{j=1}^{[k/2]} \frac{\sigma^{2j} m_{k-2j}}{(2j + 1)!(k - 2j)!}$$

where

$$[k/2] = \begin{cases} k/2, & k \text{ even} \\ (k-1)/2, & k \text{ odd} \end{cases}$$

3. The *Laplace distribution* of a random variable X is defined by

$$p(x) = \frac{1}{2\sigma} e^{-|x-m|/\sigma}$$

(a) Show that the characteristic function of X is

$$C(t) = e^{imt}/(1 + \sigma^2 t^2)$$

(b) Use (3.188) to invert the characteristic function in (a) and show that it leads to the Laplace PDF.

4. The *negative exponential distribution* of a random variable X is defined by

$$p(x) = \frac{1}{2\sigma^2} e^{-x/2\sigma^2} h(x)$$

where $h(x)$ is the Heaviside unit function.

(a) Find the characteristic function of X.
(b) Find the kth moment of X.

5. The *Rayleigh distribution* of a random variable R is defined by

$$p(r) = \frac{r}{\sigma^2} e^{-r^2/2\sigma^2} h(r)$$

(a) Find the characteristic function of R.
(b) Find the kth moment of R.

6. Given that $Z = X + Y$, where X and Y are independent random variables,

(a) show that the characteristic functions of X, Y, and Z are related by $C_Z(t) = C_X(t)C_Y(t)$.
(b) If Z is a sum of N identically distributed but independent random variables, i.e.,

$$Z = X_1 + X_2 + \dots + X_N$$

show that

$$C_Z(t) = [C_X(t)]^N$$

where $C_X(t)$ denotes the characteristic function of any of the N random variables $X_1, X_2, \ldots, X_N$.

7. Let the random variable Z be defined by the sum

$$Z = X_1Y_1 + X_2Y_2 + \ldots + X_NY_N$$

Assuming that each X_j and Y_j are independent normal random variables with means zero and equal variances σ^2, use Prob. 6(b) to determine $p_Z(z)$

(a) when $N = 2$.
(b) when $N = 4$.
(c) Compute the variance of Z for arbitrary N.

8. Objects illuminated by laser light exhibit what is called a speckle pattern. The normalized intensity E of the speckle pattern is usually governed by the negative exponential PDF, $p_E(x) = e^{-x}h(x)$. The addition of N independent speckle patterns on an intensity basis leads to a total intensity pattern described by

$$I = E_1 + E_2 + \ldots + E_N$$

Assuming each individual speckle pattern satisfies negative exponential statistics, find the PDF $p_I(x)$ for the total intensity.

Hint: Use Prob. 6(b).

9. A square-law device consists of a squaring device followed by a lowpass filter (see accompanying figure). If the input X to the square-law device is the sum of a deterministic signal and zero-mean Gaussian noise, the output Y is known to have the PDF

$$p_Y(y) = \tfrac{1}{2}e^{-(A^2+y)/2}I_0(A\sqrt{y})h(y)$$

If the output Y is then integrated to produce the output Z, which is assumed to be a sum of N independent samples of Y, what is the PDF of Z?

Hint: Use Prob. 6(b).

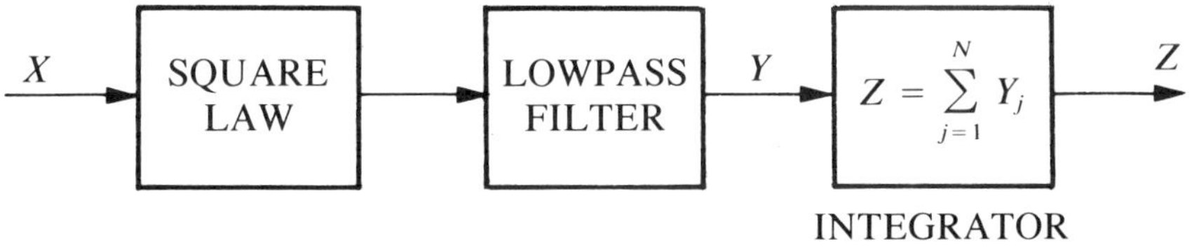

4

The Laplace Transform

4.1 *Introduction*

Although the Fourier transform has been shown to be a useful tool in a variety of applications, there are others for which it is not particularly well suited. Generally these other applications involve initial value problems for which the auxiliary data are prescribed at $t = 0$. Also, many of the functions which commonly arise in engineering and science applications — like sinusoidal functions and polynomials — do not have Fourier transforms in the usual sense without the introduction of generalized functions. For these reasons, among others, it is useful to develop other integral transforms.

There are numerous integral transforms that have been developed over the years, many of which are highly specialized. The most versatile of all integral transforms, including the Fourier transform, is the Laplace transform. Laplace transforms date back to the French mathematician Laplace who made use of the transform integral in his work on probability theory in the 1780s. S. D. Poisson (1781–1840) also knew of the Laplace transform integral in the 1820s and it occurred in Fourier's famous 1811 paper on heat conduction. Nonetheless, it was Oliver Heaviside* who

* Oliver Heaviside (1850–1925) was an English electrical engineer who set himself apart from the established scientists of the day by studying alone primarily. He was a self-made man, without academic credentials, but made significant contributions to the development and application of electromagnetic theory. He called mathematicians "woodenheaded"

popularized the use of the Laplace transform as a computational tool in elementary differential equations and electrical engineering.

We can define the Laplace transform outright, but it is instructive to formally derive it and its inversion formula directly from the Fourier integral theorem. In this way we will have a better perspective on the relation between the Fourier and Laplace transforms. To begin, let us suppose that f and its derivative f' are both piecewise continuous functions for all $t \geq 0$. In this case the Fourier transform of $f(t)$ may not exist. Regardless, it often turns out that we are interested in only the response of a system due to the action of an agent f, which we generally assume is zero for $t < 0$. Such functions are referred to as *causal functions* in the engineering literature and are usually best handled by the Laplace transform.

While we don't require f to be absolutely integrable, let us assume the related function

$$g(t) = e^{-ct} f(t)h(t) \tag{4.1}$$

does have this property, where c is a positive real constant and $h(t)$ is the Heaviside unit function. It follows therefore that f must satisfy the condition

$$\int_0^\infty e^{-ct} |f(t)| \, dt < \infty$$

Because the function g satisfies the conditions of the Fourier integral theorem (Theor. 2.1), we can write [see Eq. (2.26) in Sec. 2.4]

$$g(t) = \frac{1}{2\pi} \int_{-\infty}^\infty e^{-ist} \int_{-\infty}^\infty g(x) e^{isx} dx \, ds$$

or equivalently,

$$f(t)h(t) = \frac{e^{ct}}{2\pi} \int_{-\infty}^\infty e^{-ist} \int_0^\infty f(x) e^{-(c-is)x} dx \, ds \tag{4.2}$$

If we now introduce the change of variable $p = c - is$, then (4.2) takes the form

$$f(t)h(t) = \frac{1}{2\pi} \int_{c-i\infty}^{c+i\infty} e^{pt} \int_0^\infty e^{-px} f(x) dx \, dp \tag{4.3}$$

---

when his powerful new mathematical methods perplexed them, and he also alienated his fellow electrical engineers who didn't know what to make of him. Because of poor relations with his fellow scientists, much of his work did not receive the credit he deserved. Even today his name is often not mentioned in connection with some of his most important contributions to science, such as his pioneering work on discontinuous functions using operational methods and his mathematical solution for the distortionless transmission line.

Thus, in a purely heuristic manner we have derived the pair of transform formulas

$$F(p) = \int_0^\infty e^{-pt} f(t)\, dt = \mathscr{L}\{f(t);p\} \tag{4.4}$$

called the *Laplace transform* of $f(t)$, and

$$f(t)h(t) = \frac{1}{2\pi i}\int_{c-i\infty}^{c+i\infty} e^{pt} F(p)\, dp = \mathscr{L}^{-1}\{F(p);t\} \tag{4.5}$$

denoting the *inverse Laplace transform*. The path of integration in (4.5) appears as illustrated in Fig. 4.1. It will be shown in Sec. 4.6 that this integral converges if the real part of p exceeds some minimum value, say c_0, such that all singularities of $F(p)$ lie to the left of the line $\mathrm{Re}(p) = c_0$. The inverse Laplace transform given in (4.5) then exists for all $c > c_0$.

In our subsequent discussion of Laplace transforms it is to be understood that $f(t)$ is defined only for $t \geq 0$. Hence, we will no longer write $f(t)h(t)$ in the inverse transform relation.

4.2 *The Transforms of Some Typical Functions*

The Laplace transform of many functions can be obtained through routine formal integrations of the defining integral (4.4) by treating p as if it were

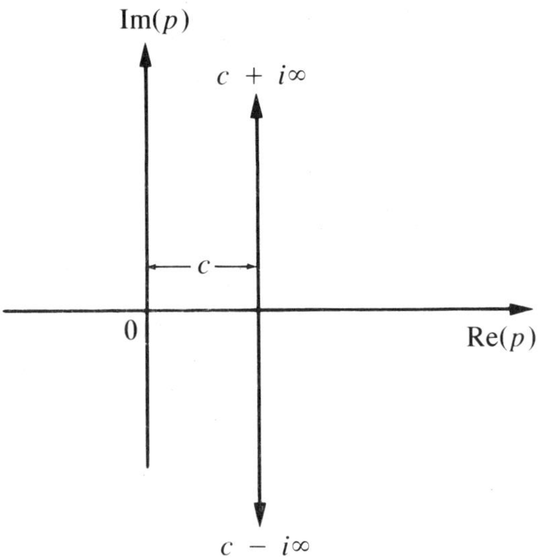

Figure 4.1 Integration path of inverse Laplace transform

a real variable. For now we will proceed with such integrations and later examine the conditions under which this formalism is valid.

Example 4.1: Find the Laplace transform of e^{at}.

Solution: By definition,

$$\begin{aligned}\mathscr{L}\{e^{at};p\} &= \int_0^\infty e^{-pt}\,e^{at}dt\\ &= \int_0^\infty e^{-(p-a)t}dt\\ &= \left.\frac{e^{-(p-a)t}}{-(p-a)}\right|_0^\infty\end{aligned}$$

Thus, the integral diverges for $\text{Re}(p) \le a$, while for $\text{Re}(p) > a$, we get

$$\mathscr{L}\{e^{at};p\} = 1/(p-a), \qquad \text{Re}(p) > a$$

The restriction $\text{Re}(p) > a$ that is required in Exam. 4.1 is typical in the evaluation of Laplace transforms. That is, the transform integral (4.4) will be meaningful only for complex values of p in some half-plane and not for others, if indeed the integral converges at all. Such restrictions are important in developing the general theory of Laplace transforms but are of little consequence in many of the applications of the transform. That is, in practice we are generally satisfied if we know that the transform exists for some values of p without knowing specifically for which values.

Notice that by allowing $a \to 0^+$ in the result of Exam. 4.1, we get the limiting case

$$\mathscr{L}\{1;p\} = 1/p, \qquad \text{Re}(p) > 0 \tag{4.6}$$

Example 4.2: Find the Laplace transforms of cos at and sin at.

Solution: From definition we have

$$\mathscr{L}\{\cos at;p\} = \int_0^\infty e^{-pt}\cos at\,dt = p/(p^2+a^2), \qquad \text{Re}(p) > 0$$

$$\mathscr{L}\{\sin at;p\} = \int_0^\infty e^{-pt}\sin at\,dt = a/(p^2+a^2), \qquad \text{Re}(p) > 0$$

which follow directly from the results of Eqs. (2.44) and (2.45) in Chap. 2. Also, we note the relations

$$\mathscr{L}\{\cos at;p\} = \sqrt{\pi/2}\,\mathscr{F}_C\{e^{-pt};a\}$$
$$\mathscr{L}\{\sin at;p\} = \sqrt{\pi/2}\,\mathscr{F}_S\{e^{-pt};a\}$$

where $\mathscr{F}_C$ and $\mathscr{F}_S$ denote the Fourier cosine and Fourier sine transforms, respectively.

Example 4.3: Find the Laplace transform of t^x, where $x > -1$.

Solution: From the defining integral,

$$\mathscr{L}\{t^x;p\} = \int_0^\infty e^{-pt}\, t^x\, dt = \frac{1}{p^{x+1}} \int_0^\infty e^{-u}\, u^x\, du$$

where we have made the change of variable $u = pt$. Using properties of the gamma function (see Sec. 1.2), we now deduce that

$$\mathscr{L}\{t^x;p\} = \Gamma(x + 1)/p^{x+1},\, x > -1,\, \mathrm{Re}(p) > 0$$

We will consider some additional transforms in Sec. 4.4. For now, let us examine conditions under which the Laplace transform exists.

4.2.1 *Existence Theorem*

Our evaluation of Laplace transforms thus far has been purely formal, using elementary integration techniques of real variables. We have not addressed the question as to which class of functions actually have Laplace transforms. That is, like the Fourier transform, not all functions (even continuous functions) have a Laplace transform.

Suppose that f is a piecewise continuous function with the further property that there exists a real number c_0 such that

$$\lim_{t\to\infty} |f(t)|e^{-ct} = \begin{cases} 0, & c > c_0 \\ \text{no limit}, & c < c_0 \end{cases} \tag{4.7}$$

A function f satisfying this condition is said to be of *exponential order* c_0, also written $0(e^{c_0 t})$. Equation (4.7) may or may not be satisfied if $c = c_0$.

Bounded functions $f(t)$, such as current and velocity, occurring in the solution for time response of stable linear systems are of exponential order zero. That is to say, the product $|f(t)|e^{-ct}$ in such cases approaches zero as $t \to \infty$ for all $c > 0$. Even some unbounded functions, like electrical charge or mechanical displacement in systems displaying resonance, may have exponential order zero if they increase like t^n, $n > 0$. In some unstable systems the response function may increase exponentially like e^{at}. Here we see that

$$\lim_{t\to\infty} e^{at}\, e^{-ct} = 0, \qquad c > a$$

and thus deduce that e^{at} is of exponential order a. An example of a function not of exponential order is $f(t) = e^{t^2}$, since

$$\lim_{t\to\infty} e^{t^2} e^{-ct} = \infty$$

for any constant $c > -\infty$.

Remark: Functions that are identically zero for $t \geq t_0 > 0$ are said to be of exponential order $-\infty$.

To establish that a given function $f(t)$ has a Laplace transform $F(p)$, we must show that the Laplace transform integral

$$F(p) = \int_0^\infty e^{-pt} f(t)\, dt$$

converges. This will be the case provided

$$|F(p)| \leq \int_0^\infty |e^{-pt} f(t)|\, dt = \int_0^\infty e^{-ct}\, |f(t)|\, dt < \infty$$

where $c = \text{Re}(p)$. Let f be piecewise continuous on $t \geq 0$ and of $0(e^{c_0 t})$, and let c_1 be a number such that $c_0 < c_1 < c$. Because $f(t) = 0(e^{c_0 t})$, it follows that for any given small positive constant ε, there exists some t_0 such that

$$|f(t)|e^{-c_1 t} < \varepsilon \text{ when } t > t_0$$

We now write

$$\int_0^\infty e^{-ct}|f(t)|\, dt = \int_0^{t_0} e^{-ct}|f(t)|\, dt + \int_{t_0}^\infty e^{-ct}|f(t)|\, dt$$

where the first integral with finite limits exists because f is piecewise continuous. Furthermore, the second integral satisfies

$$\int_{t_0}^\infty e^{-ct}|f(t)|\, dt = \int_{t_0}^\infty e^{-(c-c_1)t}|f(t)|e^{-c_1 t}\, dt$$
$$< \varepsilon \int_{t_0}^\infty e^{-(c-c_1)t}\, dt$$

But this last integral exists for $c > c_1$, and thus we have established conditions under which the Laplace transform integral *converges absolutely* in the half-plane $\text{Re}(p) > c_0$. It can also be shown that the Laplace transform integral *converges uniformly* for $\text{Re}(p) \geq c_2 > c_0$, where c_2 is any real number satisfying $c_0 < c_2 \leq c$.

In summary, we have the following existence theorem.

Theorem 4.1 (*Existence theorem*). If f is piecewise continuous on $t \geq 0$ and is $0(e^{c_0 t})$, then $f(t)$ has a Laplace transform $F(p)$ in the half-plane

$\text{Re}(p) > c_0$. Moreover, the Laplace transform integral converges both absolutely and uniformly for $\text{Re}(p) \geq c_2 > c_0$.

Most functions met in practice satisfy the conditions of Theor. 4.1. However, these conditions are *sufficient* rather than *necessary* to ensure that a given function has a Laplace transform. For example, both t^{-1} and $t^{-1/2}$ have infinite discontinuities at $t = 0$ and are therefore not piecewise continuous on $t \geq 0$. Yet, while the integral

$$\mathscr{L}\{t^{-1};p\} = \int_0^\infty e^{-pt}\, t^{-1}\, dt$$

does not exist, we have shown in Exam. 4.3 that*

$$\mathscr{L}\{t^{-1/2};p\} = \int_0^\infty e^{-pt}\, t^{-1/2}\, dt = \sqrt{\pi/p}$$

Also, the Laplace transform may exist in certain instances when f is not of exponential order, although we will not provide any general discussion of this case.†

4.2.2 *Analytic Continuation*

According to Theor. 4.1, every function $f(t)$ that is of exponential order and belongs to the class of piecewise continuous functions has a Laplace transform $F(p)$. Due to properties of definite integrals, we also know that $F(p)$ is unique. Lastly, the transform function $F(p)$ has the following important property which we do not prove.†

Theorem 4.2. If $f(t)$ is piecewise continuous on $t \geq 0$ and is $0(e^{c_0 t})$, its Laplace transform $F(p)$ is an analytic function of the complex variable p in the half-plane $\text{Re}(p) > c_0$.

In some cases the function $F(p)$ may be analytic to the left of the line $\text{Re}(p) = c_0$, although for our purposes it suffices to know that there exists a half-plane where $F(p)$ is indeed analytic.

Up to this point we have produced Laplace transforms

$$F(p) = \int_0^\infty e^{-pt} f(t)\, dt$$

---

* Since p is complex, one really needs also to specify the branch of the multivalued function $p^{1/2}$, which may be either $\sqrt{p}$ or $-\sqrt{p}$ in general.

† For a discussion of the convergence of the Laplace transform integral in the general case, see W. R. LePage, *Complex Variables and the Laplace Transform for Engineers*, New York: Dover, 1980.

† For a proof of Theor. 4.2, see R. V. Churchill, *Operational Mathematics*, New York: McGraw-Hill, 1972.

by formal integration methods applicable to real integrals. In other words, we have treated the complex variable p as if it were a *real* variable x, and then once $F(x)$ was determined we obtained $F(p)$ by simply replacing x with p. It is reassuring to know that this formal procedure is actually valid! What permits this type of formalism is the fact that $F(p)$ is an *analytic function*, and thus we can use *analytic continuation* in obtaining $F(p)$ from $F(x)$, where x is real.

EXERCISES 4.2

In Probs. 1–10, evaluate the Laplace transform of each function directly from the defining integral.

1. $f(t) = t^2$

2. $f(t) = e^{-at} - e^{-bt}$

3. $f(t) = \cosh at$

4. $f(t) = \sinh at$

5. $f(t) = h(t - a), \quad a > 0$

6. $f(t) = te^{2t}$

7. $f(t) = t \sin kt$

8. $f(t) = \cos^2 kt$

9. $f(t) = \sin at \sin bt$

10. $f(t) = e^{at}\cosh kt$

In Probs. 11–20, determine which functions are of exponential order, and for those which are, determine c_0.

11. $f(t) = t^{100}$

12. $f(t) = t^{-1/2}$

13. $f(t) = te^{2t}$

14. $f(t) = \sin t^2$

15. $f(t) = (\sin t)/t$

16. $f(t) = \log t$

17. $f(t) = t^{-3/2}\log t$

18. $f(t) = \sin(e^{t^2})$

19. $f(t) = \sqrt{|\tan t|}$

20. $f(t) = e^{t \log t}$

In Probs. 21–26, verify the Laplace transform relations.

21. $\mathscr{L}\{t^{1/2};p\} = \dfrac{1}{2p}\sqrt{\dfrac{\pi}{p}}$

22. $\mathscr{L}\{t^{5/2};p\} = \dfrac{15}{8p^3}\sqrt{\dfrac{\pi}{p}}$

23. $\mathscr{L}\{e^{-t^2/4};p\} = \sqrt{\pi}\, e^{p^2}\text{erfc}(p)$

24. $\mathscr{L}\{(t + a)^{-1/2};p\} = \sqrt{\pi/p}\, e^{ap}\, \text{erfc}(\sqrt{ap}), \qquad a > 0$

25. $\mathscr{L}\{(t + a)^{-3/2};p\} = \dfrac{2}{\sqrt{a}} - 2\sqrt{\pi p}\, e^{ap}\, \text{erfc}(\sqrt{ap}), \qquad a > 0$

26. $\mathscr{L}\{\log t; p\} = \frac{1}{p}\left[\Gamma'(1) - \log p\right]$

Hint: First determine $\Gamma'(z)$.

In Probs. 27–30, the given functions are various types of pulses of unit height and duration T. Determine their Laplace transforms.

27. $f(t) = \begin{cases} 1, & 0 < t < T \\ 0, & t > T \end{cases}$

28. $f(t) = \begin{cases} 1, & a < t < a + T,\ a > 0 \\ 0, & \text{otherwise} \end{cases}$

29. $f(t) = \begin{cases} 2t/T, & 0 < t < T/2 \\ 2 - 2t/T, & T/2 < t < T \\ 0, & t > T \end{cases}$

30. $f(t) = \begin{cases} \sin at, & 0 < t < \pi/a \\ 0, & t > \pi/a \end{cases}$

31. Prove that $F(p)$ has no finite singularities (i.e., it is an *entire function*), where $F(p)$ is the Laplace transform associated with
(a) Prob. 27.
(b) Prob. 28.
(c) Prob. 29.
(d) Prob. 30.

32. Prove that if $f(t)$ is piecewise continuous and $f(t)$ is identically zero for $t > T$, where T is any positive number, then the transform function $F(p)$ is an entire function.

33. If $f(t)$ is a polynomial, prove that its transform $F(p)$ is a rational function.

4.3 *Basic Operational Properties*

The use of the integral definition to compute Laplace transforms is a tedious task and not required at all times. That is, once we have found several transforms directly from the defining integral it may be possible to find additional ones by using various properties of the integral transform known as *operational properties*. Perhaps the most fundamental of these is the *linearity property*, which we state first.

Theorem 4.3 (*Linearity property*). If $F(p)$ and $G(p)$ are the Laplace transforms, respectively, of $f(t)$ and $g(t)$, then for any constants C_1 and C_2,

$$\mathscr{L}\{C_1 f(t) + C_2 g(t); p\} = C_1 F(p) + C_2 G(p)$$

The proof of Theor. 4.3 is a simple consequence of the linearity property of integrals and is left to the exercises (see Prob. 1 in Exer. 4.3).

For $a > 0$, we find that

$$\mathscr{L}\{f(at);p\} = \int_0^\infty e^{-pt} f(at)\,dt$$
$$= \frac{1}{a}\int_0^\infty e^{-up/a} f(u)\,du$$

the last step of which is the result of making the variable change $u = at$. Thus, if $F(p)$ denotes the Laplace transform of $f(t)$, the above relation suggests that

$$\mathscr{L}\{f(at);p\} = \frac{1}{a}F(p/a), \qquad a > 0 \tag{4.8}$$

which is called the *scaling property*. We illustrate this property in the following example.

Example 4.4: Given that $\mathscr{L}\{e^{-t^2/4};p\} = \sqrt{\pi}\, e^{p^2}\operatorname{erfc}(p)$, find the Laplace transform of e^{-t^2}.

Solution: By writing $e^{-t^2} = e^{-a^2t^2/4}$, we identify $a = 2$ in the scaling property (4.8). Thus, it immediately follows that

$$\mathscr{L}\{e^{-t^2};p\} = \frac{\sqrt{\pi}}{2}\, e^{p^2/4}\operatorname{erfc}(p/2)$$

4.3.1 *The Shifting Theorems*

It sometimes happens that we need to calculate the transform of $e^{at}f(t)$, where the transform of $f(t)$ is known or is readily computed. Transforms of this nature are easily handled because of the exponential function occurring in the defining integral, and leads us to the *first shift property*.

Theorem 4.4 (*Shifting property*). If $F(p)$ is the Laplace transform of $f(t)$, then

$$\mathscr{L}\{e^{at}f(t);p\} = F(p - a)$$

Proof: From definition,

$$\mathscr{L}\{e^{at}f(t);p\} = \int_0^\infty e^{-pt}\, e^{at}f(t)\,dt$$
$$= \int_0^\infty e^{-(p-a)t} f(t)\,dt$$
$$= F(p - a)$$

where $F(p)$ is the Laplace transform of $f(t)$. ■

Example 4.5: Find the Laplace transform of $e^{-2t}\cos 3t$.

Solution: Recalling Example 4.2, we have

$$\mathscr{L}\{\cos 3t;p\} = p/(p^2 + 9)$$

and hence, through the shifting property it follows that

$$\mathscr{L}\{e^{-2t}\cos 3t;p\} = \frac{p+2}{(p+2)^2+9} = \frac{p+2}{p^2+4p+13}$$

The Laplace transform has the effect of taking discontinuous functions in the t-domain and "smoothing" them in the p-domain. Thus, one of the most interesting and useful applications of the Laplace transform is solving linear differential equations with discontinuous or impulsive forcing functions, which are commonplace in circuit analysis problems and some mechanical systems. The Heaviside unit function $h(t - a)$ discussed in Sec. 1.5 is used extensively in dealing with discontinuous and impulsive functions. The property we need now involves the Heaviside unit function and is called *Heaviside's second-shift property*, or the *translation property*.

Theorem 4.5 (*Translation property*). If $F(p)$ is the Laplace transform of $f(t)$, then

$$\mathscr{L}\{f(t-a)h(t-a);p\} = e^{-ap}F(p), \qquad a > 0$$

Proof: Here we see that

$$\mathscr{L}\{f(t-a)h(t-a);p\} = \int_0^\infty e^{-pt} f(t-a)h(t-a)\,dt$$

$$= \int_a^\infty e^{-pt} f(t-a)\,dt$$

Introducing the new variable $u = t - a$, we get

$$\mathscr{L}\{f(t-a)h(t-a);p\} = \int_0^\infty e^{-p(u+a)} f(u)\,du$$

$$= e^{-ap}\int_0^\infty e^{-pu} f(u)\,du$$

$$= e^{-ap} F(p) \qquad \blacksquare$$

The translation property (Theor. 4.5) can also be expressed as

$$\mathscr{L}\{f(t)h(t-a);p\} = e^{-ap}\,\mathscr{L}\{f(t+a);p\} \tag{4.9}$$

which may be a more useful form in certain applications. Also, as a consequence of the translation property, we have that

$$\mathscr{L}\{h(t-a);p\} = \frac{1}{p}e^{-ap}, \qquad a > 0 \tag{4.10}$$

Example 4.6: Find the Laplace transform of the discontinuous function

$$f(t) = \begin{cases} t^2, & 0 < t < 2 \\ 6, & t > 2 \end{cases}$$

Solution: We first express $f(t)$ in terms of the Heaviside unit function, i.e.,

$$\begin{aligned} f(t) &= t^2\,[h(t) - h(t-2)] + 6h(t-2) \\ &= t^2 + (6 - t^2)h(t-2) \end{aligned}$$

Then, using the linearity property and Eq. (4.9), we find

$$\begin{aligned} \mathscr{L}\{f(t);p\} &= \mathscr{L}\{t^2;p\} + \mathscr{L}\{(6-t^2)h(t-2);p\} \\ &= \mathscr{L}\{t^2;p\} + e^{-2p}\,\mathscr{L}\{6-(t+2)^2;p\} \\ &= \mathscr{L}\{t^2;p\} + e^{-2p}\,\mathscr{L}\{2-4t-t^2;p\} \end{aligned}$$

and thus deduce that

$$\mathscr{L}\{f(t);p\} = \frac{2}{p^3} + e^{-2p}\left(\frac{2}{p} - \frac{4}{p^2} - \frac{2}{p^3}\right)$$

4.3.2 *Transforms of Derivatives and Integrals*

The real merit of the Laplace transform is revealed by its effect on derivatives. Here we will derive a relation between the Laplace transform of the derivative of a function and the Laplace transform of the function itself.

Suppose that f is a continuous function with a piecewise continuous derivative f' on the interval $t \geq 0$. We further suppose that both f and f' are of exponential order c_0. Using integration by parts, we obtain

$$\begin{aligned} \mathscr{L}\{f'(t);p\} &= \int_0^\infty e^{-pt} f'(t)\,dt \\ &= e^{-pt} f(t)\Big|_0^\infty + p\int_0^\infty e^{-pt} f(t)\,dt \end{aligned}$$

Because f is $0(e^{c_0 t})$, it follows that $e^{-pt}f(t) \to 0$ as $t \to \infty$, and consequently,

$$\mathscr{L}\{f'(t);p\} = pF(p) - f(0) \tag{4.11}$$

where $F(p)$ is the Laplace transform of $f(t)$.

Remark: If f has a finite discontinuity at $t = 0$, then we replace $f(0)$ in (4.11) with $f(0^+)$.

Similarly, if f and f' are continuous and f'' is piecewise continuous on $t \geq 0$, and if all three functions are of exponential order c_0, we can use (4.11) to obtain

$$\mathscr{L}\{f''(t);p\} = p\mathscr{L}\{f'(t);p\} - f'(0)$$

which simplifies to

$$\mathscr{L}\{f''(t);p\} = p^2F(p) - pf(0) - f'(0) \tag{4.12}$$

By repeated application of (4.11) and (4.12), we arrive at the following general result.

Theorem 4.6 (*Differentiation property*). If $f, f', \ldots, f^{(n-1)}$ are all continuous functions on $t \geq 0$, $f^{(n)}$ is piecewise continuous on $t \geq 0$, and if all are of exponential order c_0, then for $n = 1,2,3,\ldots,$

$$\mathscr{L}\{f^{(n)}(t);p\} = p^nF(p) - p^{n-1}f(0) - p^{n-2}f'(0) - \ldots - f^{(n-1)}(0)$$

where $F(p)$ is the Laplace transform of $f(t)$.

Although the real utility of Theor. 4.6 will not be observed until we use it in the solution of differential equations, the following example provides a novel way in which the differentiation property may be used.

Example 4.7: Find the Laplace transform of t^n, $n = 1,2,3,\ldots$, by using Theor. 4.6.

Solution: The function $f(t) = t^n$ and all its derivatives are continuous functions of exponential order. Also, we see that

$$f(0) = f'(0) = \ldots = f^{(n-1)}(0) = 0$$
$$f^{(n)}(t) = n!$$

Substituting these results into Theor. 4.6 leads to

$$\mathscr{L}\{f^{(n)}(t);p\} = n!\,\mathscr{L}\{1;p\} = p^n\,\mathscr{L}\{t^n;p\} - 0 - 0 - \ldots - 0$$

and therefore

$$\mathscr{L}\{t^n;p\} = n!/p^{n+1},\ n = 1,2,3,\ldots^*$$

where we are using the previous result $\mathscr{L}\{1;p\} = 1/p$.

A different application of Theor. 4.6 involves the Laplace transform of the integral of f.

* Notice that this result is simply a special case of that in Exam. 4.3.

Theorem 4.7 (*Integration property*). If f is piecewise continuous on $t \geq 0$ and is $0(e^{c_0 t})$, then

$$\mathcal{L}\left\{\int_0^t f(u)du;p\right\} = F(p)/p$$

where $F(p)$ is the Laplace transform of $f(t)$.

Proof: Let us define

$$g(t) = \int_0^t f(u)du$$

which is a continuous function since f is piecewise continuous; furthermore, $g(0) = 0$ and $g'(t) = f(t)$. Hence, g satisfies the conditions of Eq. (4.11), which leads to

$$\mathcal{L}\{f(t);p\} = p\,\mathcal{L}\left\{\int_0^t f(u)du;p\right\} - 0$$

or

$$\mathcal{L}\left\{\int_0^t f(u)du;p\right\} = F(p)/p$$

■

Remark: Theorem 4.7 is actually a special case of the *convolution theorem* presented in Sec. 4.5.

In these last properties we have found that differentiation and integration of a given function in the t-domain correspond roughly to multiplication and division, respectively, in the p-domain. In this fashion, the Laplace transform has the effect of replacing the operations of calculus in the t-domain with algebraic operations in the p-domain. It is primarily for this reason that the Laplace transform is so useful in the solution of differential and integral equations.

4.3.3 *Derivatives and Integrals of the Transform*

Sometimes we need to evaluate the transform of a function that is expressed as $t^n f(t)$, where the transform of $f(t)$ is easily obtained. To develop the property we need in this regard, let us start with the transform relation

$$F(p) = \int_0^\infty e^{-pt} f(t)\,dt \tag{4.13}$$

If $f(t)$ is piecewise continuous on $t \geq 0$ and is $0(e^{c_0 t})$, then $F(p)$ is an analytic function in the half-plane $\text{Re}(p) > c_0$ (Theor. 4.2). As a consequence, $F(p)$ has derivatives of all orders which we assume can be formally obtained by differentiating (4.13) under the integral.* Thus, we have

$$F'(p) = \int_0^\infty (-t)e^{-pt} f(t)\,dt = -\int_0^\infty e^{-pt}\,[tf(t)]\,dt$$

from which we deduce

$$\mathscr{L}\{tf(t);p\} = -F'(p) \tag{4.14}$$

Continued differentiation of (4.13) leads to

$$F^{(n)}(p) = \int_0^\infty (-t)^n e^{-pt} f(t)\,dt, \qquad n = 1,2,3,\ldots$$

and thus we have the following theorem.

Theorem 4.8. If f is piecewise continuous on $t \geq 0$ and is $0(e^{C_0 t})$, then

$$\mathscr{L}\{t^n f(t);p\} = (-1)^n F^{(n)}(p), \qquad n = 1,2,3,\ldots$$

where $F(p)$ is the Laplace transform of $f(t)$.

Example 4.8: Evaluate the Laplace transform of $te^{-2t}\cos t$.

Solution: Starting with the transform relation (see Exam. 4.2)

$$\mathscr{L}\{\cos t;p\} = p/(p^2 + 1)$$

we use (4.14) to determine

$$\mathscr{L}\{t\cos t;p\} = -\frac{d}{dp}\left(\frac{p}{p^2+1}\right) = \frac{p^2-1}{(p^2+1)^2}$$

Finally, we apply the shifting property (Theor. 4.4) to obtain

$$\mathscr{L}\{te^{-2t}\cos t;p\} = \frac{(p+2)^2-1}{[(p+2)^2+1]^2} = \frac{p^2+4p+3}{(p^2+4p+5)^2}$$

By replacing p with u and integrating both sides of Eq. (4.13), we find that

$$\begin{aligned}\int_p^\infty F(u)\,du &= \int_p^\infty \int_0^\infty e^{-ut} f(t)\,dt\,du \\ &= \int_0^\infty \left(\int_p^\infty e^{-ut}\,du\right) f(t)\,dt \\ &= \int_0^\infty e^{-pt}\,[f(t)/t]\,dt\end{aligned}$$

---

* Recall from Theor. 4.1 that the integral in (4.13) is uniformly convergent in a half-plane. Thus, differentiation under the integral sign is permitted as long as each new integral produced in this fashion also converges uniformly.

where we have reversed the order of integration. Hence, we have derived the following property.

Theorem 4.9. If f is piecewise continuous on $t \geq 0$ and is $0(e^{c_0 t})$, and $f(t)/t$ has a Laplace transform, then

$$\mathscr{L}\{f(t)/t;p\} = \int_p^\infty F(u)\,du$$

where $F(p)$ is the Laplace transform of $f(t)$.

Example 4.9. Find the Laplace transform of the *sine integral*

$$\mathrm{Si}(t) = \int_0^t \frac{\sin u}{u}\,du$$

Solution: Starting with the transform relation (see Exam. 4.2)

$$\mathscr{L}\{\sin t;p\} = 1/(p^2 + 1)$$

and using Theor. 4.9, we obtain

$$\mathscr{L}\left\{\frac{\sin t}{t};p\right\} = \int_p^\infty \frac{du}{u^2 + 1} = \tan^{-1}\frac{1}{p}$$

Then, applying Theor. 4.7, we deduce that

$$\mathscr{L}\,\{\mathrm{Si}(t);p\} = \mathscr{L}\left\{\int_0^t \frac{\sin u}{u}\,du;p\right\} = \frac{1}{p}\tan^{-1}\frac{1}{p}$$

Some additional properties of the Laplace transform involving integrals are

$$\mathscr{L}\left\{\int_0^p \frac{f(u)}{u}du;p\right\} = \frac{1}{p}\int_p^\infty F(s)\,ds \tag{4.15}$$

$$\mathscr{L}\left\{\int_p^\infty \frac{f(u)}{u}du;p\right\} = \frac{1}{p}\int_0^p F(s)\,ds \tag{4.16}$$

and

$$\mathscr{L}\left\{\int_0^\infty \frac{f(u)}{u}du;p\right\} = \frac{1}{p}\int_0^\infty F(s)\,ds \tag{4.17}$$

the last of which is a simple consequence of the first two. The verification of these properties is left to the exercises.

4.3.4 *Periodic Functions*

A function f is called *periodic* if there exists a constant $T > 0$ for which $f(t + T) = f(t)$ for all $t \geq 0$. The smallest value of T for which the property holds is called the *fundamental period*, or simply, the *period*. Familiar examples of periodic functions are cos t and sin t which have period 2π, but there are many others whose definition is not so easily given. Periodic functions appear in a wide variety of engineering applications where the Laplace transform may be used. Such applications rely heavily on the following theorem concerning periodic functions.

Theorem 4.10. Let f be piecewise continuous on $t \geq 0$ and of $0(e^{c_0 t})$. If f is also periodic with period T, then

$$\mathscr{L}\{f(t);p\} = [1/(1 - e^{-pT})] \int_0^T e^{-pt} f(t)\, dt$$

Proof: Let us write the Laplace transform as

$$\mathscr{L}\{f(t);p\} = \int_0^T e^{-pt} f(t)\, dt + \int_T^\infty e^{-pt} f(t)\, dt$$

By making the change of variable $t = u + T$ in the last integral, we obtain

$$\begin{aligned}\mathscr{L}\{f(t);p\} &= \int_0^T e^{-pt} f(t)\, dt + \int_0^\infty e^{-p(u+T)} f(u + T)\, du \\ &= \int_0^T e^{-pt} f(t)\, dt + e^{-pT} \int_0^\infty e^{-pu} f(u)\, du \\ &= \int_0^T e^{-pt} f(t)\, dt + e^{-pT} \mathscr{L}\{f(t);p\}\end{aligned}$$

Solving for $\mathscr{L}\{f(t);p\}$ yields

$$\mathscr{L}\{f(t);p\} = [1/(1 - e^{-pT})] \int_0^T e^{-pt} f(t)\, dt.$$

■

Example 4.10: Find the Laplace transform of the half-wave rectified sinusoidal (see Fig. 4.2)

$$f(t) = \begin{cases} \sin t, & 0 < t < \pi \\ 0, & \pi < t < 2\pi \end{cases}$$

where $f(t + 2\pi) = f(t)$.

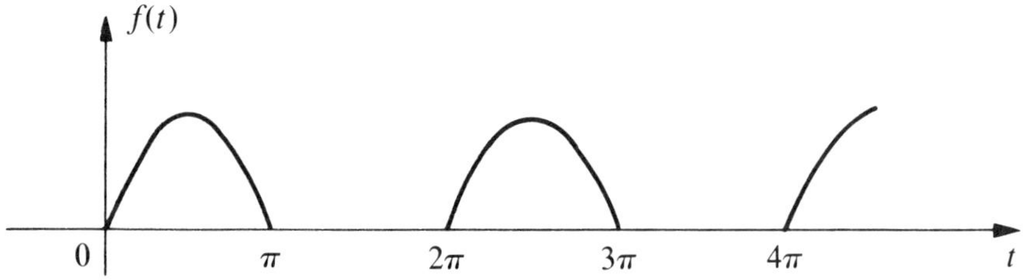

Figure 4.2 Graph of a half wave rectified sinusoidal

Solution: Since $T = 2\pi$, we compute

$$\int_0^{2\pi} e^{-pt} f(t)\,dt = \int_0^{\pi} e^{-pt}\sin t\,dt = \frac{1 + e^{-\pi p}}{p^2 + 1}$$

and therefore

$$\mathscr{L}\{f(t);p\} = \frac{1}{1 - e^{-2\pi p}}\left(\frac{1 + e^{-\pi p}}{p^2 + 1}\right) = \frac{1}{(1 - e^{-\pi p})(p^2 + 1)}$$

EXERCISES 4.3

1. Prove the *linearity property* (Theor. 4.3)

$$\mathscr{L}\{C_1 f(t) + C_2 g(t);p\} = C_1 F(p) + C_2 G(p)$$

2. If $F(p)$ denotes the Laplace transform of $f(t)$, show that

$$\mathscr{L}\{e^{-bt/a} f(t/a);p\} = aF(ap + b), \qquad a > 0$$

3. Given that $\mathscr{L}\{(\sin t)/t;p\} = \tan^{-1}(1/p)$, find the Laplace transform of $(\sin at)/t$ where $a > 0$.

4. Use Probs. 2 and 3 to find the Laplace transform of $e^{3t}(\sin 5t)/t$.

In Probs. 5–20, evaluate the Laplace transform of the given function using known transforms and operational properties.

5. $f(t) = 3t^4 e^{5t}$

6. $f(t) = 2e^{-2t}\cos^2 3t$

7. $f(t) = e^{4t}\cosh 5t$

8. $f(t) = e^{-t}(3\cos 6t - 5\sin 6t)$

9. $f(t) = \begin{cases} 2, & 0 < t < 1 \\ t, & t > 1 \end{cases}$

10. $f(t) = \begin{cases} t^2, & 0 < t < 3 \\ e^{-t}, & t > 3 \end{cases}$

11. $f(t) = \begin{cases} \sin t, & 0 < t < \pi \\ 0, & t > \pi \end{cases}$

12. $f(t) = \begin{cases} \cos t, & 0 < t < \pi \\ \sin t, & t > \pi \end{cases}$

13. $f(t) = \begin{cases} t, & 0 < t < 1 \\ 1, & 1 < t < 2 \\ 3 - t, & 2 < t < 3 \\ 0, & t > 3 \end{cases}$

14. $f(t) = \int_0^t (u^2 - u + e^{-u})\,du$

15. $f(t) = \int_0^t e^{-u^2/4}\,du$

16. $f(t) = \int_0^t e^{4u}\cosh 5u\,du$

17. $f(t) = t^2 \sin kt$

18. $f(t) = 5te^{3t}\sin^2 t$

19. $f(t) = (\sinh t)/t$

20. $f(t) = (e^{-at} - e^{-bt})/t$

21. If $\mathscr{L}\{f(t);p\} = (1/p)e^{-1/p}$, find the Laplace transform of $e^{-2t}f(3t)$.

22. Given that $\mathscr{L}\{\sin\sqrt{t};p\} = (1/2p)\sqrt{\pi/p}\, e^{-1/4p}$,

(a) use Eq. (4.11) to determine the transform of $(1/\sqrt{t})\cos\sqrt{t}$.

(b) use Theor. 4.9 to determine the transform of $(1/t)\sin\sqrt{t}$.

23. Given that $\mathscr{L}\{t^{-1/2};p\} = \sqrt{\pi/p}$, show that

$$\mathscr{L}\left\{\left(\frac{1+2at}{\sqrt{t}}\right)e^{at};p\right\} = \sqrt{\frac{\pi}{p+a}}\left(\frac{p}{p+a}\right)$$

In Probs. 24–30, establish the given transform relation.

24. $\mathscr{L}\left\{\dfrac{\cos at - \cos bt}{t};p\right\} = \dfrac{1}{2}\log\dfrac{p^2+a^2}{p^2+b^2}$

25. $\mathscr{L}\{t\mathrm{Si}(t);p\} = \dfrac{\tan^{-1}p}{p^2} - \dfrac{1}{p(p^2+1)}$

26. $\mathscr{L}\left\{\dfrac{e^{-a\sqrt{t}}}{\sqrt{t}};p\right\} = \sqrt{\dfrac{\pi}{p}}\, e^{a^2/4p}\operatorname{erfc}\left(\dfrac{a}{2\sqrt{p}}\right), \qquad a > 0$

Hint: Use Exam. 4.4.

27. $\mathscr{L}\{e^{-a\sqrt{t}};p\} = \dfrac{1}{p} - \dfrac{a}{2p}\sqrt{\dfrac{\pi}{p}}\, e^{a^2/4p}\operatorname{erfc}\left(\dfrac{a}{2\sqrt{p}}\right), \qquad a > 0$

Hint: Use Exam. 4.4.

28. $\mathscr{L}\left\{\displaystyle\int_0^t \frac{f(u)}{u}\,du;p\right\} = \dfrac{1}{p}\displaystyle\int_p^\infty F(s)\,ds$

29. $\mathscr{L}\left\{\displaystyle\int_t^\infty \frac{f(u)}{u}\,du;p\right\} = \dfrac{1}{p}\displaystyle\int_0^p F(s)\,ds$

30. $\mathscr{L}\left\{\displaystyle\int_0^\infty \frac{f(u)}{u}\,du;p\right\} = \dfrac{1}{p}\displaystyle\int_0^\infty F(s)\,ds$

31. Use the result of Prob. 30 to deduce that

$$\int_0^\infty \frac{f(u)}{u}\,du = \int_0^\infty F(s)\,ds$$

32. Use Prob. 29 to evaluate the Laplace transform of

(a) the *cosine integral* defined by

$$\mathrm{Ci}(t) = \int_\infty^t \frac{\cos u}{u}\,du, \qquad t > 0$$

(b) the *exponential integral* defined by

$$E_1(t) = \int_t^\infty \frac{e^{-u}}{u}\,du, \qquad t > 0.$$

33. The *Laguerre polynomials* are defined by

$$L_n(t) = \frac{e^t}{n!}\frac{d^n}{dt^n}(t^n e^{-t}),\ n = 0,1,2,\ldots$$

Show that

$$\mathscr{L}\{L_n(t);p\} = \frac{1}{p}\left(\frac{p-1}{p}\right)^n$$

34. If $f(t) = (1/t)g'(t)$, show that

$$F(p) = \int_p^\infty sG(s)\,ds$$

35. Considering the integral

$$I(b) = \int_0^\infty e^{-pu^2}\cos bu\,du, \qquad b \geq 0,\ p > 0$$

as a function of the parameter b,
(a) show that I satisfies the first-order linear DE

$$\frac{dI}{db} - \frac{b}{2p}I = 0$$

(b) Evaluate $I(0)$ directly from the integral.
(c) Solve the DE in (a) subject to the initial condition in (b) to deduce the result

$$I(b) = \frac{1}{2}\sqrt{\frac{\pi}{p}}\,e^{-b^2/4p}$$

36. Use the result of Prob. 35 to deduce the Laplace transform of
(a) $(1/\sqrt{t})\cos a\sqrt{t}$.
(b) Differentiate the transform relation in part (a) with respect to a and deduce the Laplace transform of $\sin a\sqrt{t}$.
(c) By integrating the transform relation in part (a) with respect to a from 0 to 1, deduce the Laplace transform of $(1/t)\sin\sqrt{t}$.

In Probs. 37–40, use Theor. 4.10 to find the Laplace transform of the given periodic function.

37. $f(t) = |\sin t|$

38. $f(t) = \begin{cases} 1, & 0 < t < c \\ -1, & c < t < 2c \end{cases} \qquad f(t + 2c) = f(t)$

39. $f(t) = \begin{cases} t, & 0 < t < c \\ 2c - t, & c < t < 2c \end{cases} \qquad f(2t + 2c) = f(t)$

40. $f(t) = \begin{cases} 3t, & 0 < t < 2 \\ 6, & 2 < t < 4 \end{cases} \qquad f(t + 4) = f(t)$

4.4 *Transforms of More Complicated Functions*

When the function whose transform is needed leads to a nonelementary integral, we must usually resort to advanced integration techniques or various tricks and manipulations to accomplish our task. Let us illustrate with some examples.

Example 4.11: Find the Laplace transform of erf(t), which is the *error function* discussed in Sec. 1.3.

Solution: From the defining integral of the Laplace transform and that of the error function, we have

$$\begin{aligned}\mathscr{L}\{\text{erf}(t);p\} &= \int_0^\infty e^{-pt}\,\text{erf}(t)\,dt \\ &= \int_0^\infty e^{-pt}\frac{2}{\sqrt{\pi}}\int_0^t e^{-x^2}\,dx\,dt\end{aligned}$$

Recharacterizing the region of integration $0 \le x \le t$, $0 \le t < \infty$, by $x \le t < \infty$, $0 \le x < \infty$, we can interchange the order of integration (see Fig. 4.3) to get

$$\begin{aligned}\mathscr{L}\{\text{erf}(t);p\} &= \frac{2}{\sqrt{\pi}}\int_0^\infty e^{-x^2}\int_x^\infty e^{-pt}\,dt\,dx \\ &= \frac{2}{p\sqrt{\pi}}\int_0^\infty e^{-(x^2+px)}\,dx \\ &= \frac{2}{p\sqrt{\pi}}e^{p^2/4}\int_0^\infty e^{-(x+p/2)^2}\,dx\end{aligned}$$

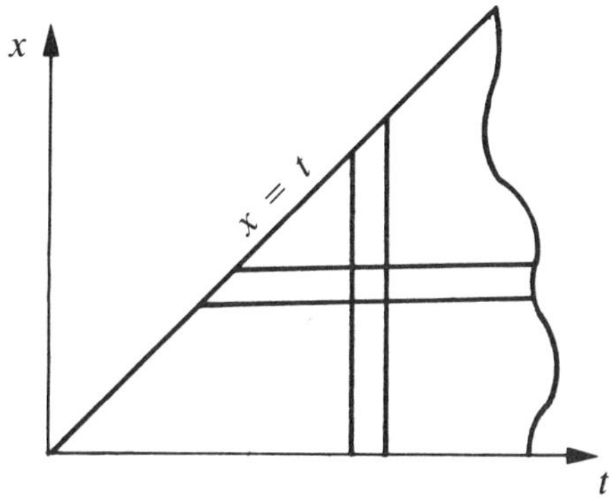

Figure 4.3 Region of integration

where we have written

$$x^2 + px = (x + p/2)^2 - p^2/4$$

Finally, making the change of variable $u = x + p/2$ leads to

$$\mathcal{L}\{\text{erf}(t);p\} = \frac{2}{p\sqrt{\pi}} e^{p^2/4} \int_{p/2}^{\infty} e^{-u^2}\, du$$

and we deduce that

$$\mathcal{L}\{\text{erf}(t);p\} = (1/p)e^{p^2/4}\,\text{erfc}(p/2), \qquad \text{Re}(p) > 0$$

Several Laplace transform relations involve the integral formula

$$\int_0^{\infty} e^{-a^2x^2 - b^2/x^2}\, dx = \frac{\sqrt{\pi}}{2a} e^{-2ab}, \qquad a > 0,\, b \geq 0, \tag{4.18}$$

the derivation of which is left to the exercises (see Prob. 32 in Exer. 4.4).

Example 4.12: Find the Laplace transform of $(1/\sqrt{t})e^{-k/t}$, $k \geq 0$.

Solution: From definition,

$$\mathcal{L}\left\{\frac{1}{\sqrt{t}} e^{-k/t};p\right\} = \int_0^{\infty} e^{-pt} \frac{1}{\sqrt{t}} e^{-k/t}\, dt$$

$$= 2\int_0^{\infty} e^{-px^2 - k/x^2}\, dx$$

where we have set $t = x^2$ in the last step. Referring now to Eq. (4.18), we see that

$$\mathcal{L}\left\{\frac{1}{\sqrt{t}} e^{-k/t};p\right\} = \sqrt{\frac{\pi}{p}}\, e^{-2\sqrt{kp}}, \qquad k \geq 0,\, \text{Re}(p) > 0$$

By formally differentiating both sides of the result of Exam. 4.12 with respect to the parameter k, we obtain the new transform relation

$$\mathcal{L}\left\{\frac{1}{t\sqrt{t}} e^{-k/t};p\right\} = \sqrt{\frac{\pi}{k}}\, e^{-2\sqrt{kp}}, \qquad k > 0,\, \text{Re}(p) > 0. \tag{4.19}$$

Another transform whose calculation involves the integral formula (4.18) is given in the next example.

Example 4.13: Find the Laplace transform of $\text{erfc}(1/\sqrt{t})$.

Solution: Here we see that

$$\mathscr{L}\{\text{erfc}(1/\sqrt{t});p\} = \int_0^\infty e^{-pt}\,\text{erfc}(1/\sqrt{t})\,dt$$
$$= \int_0^\infty e^{-pt}\frac{2}{\sqrt{\pi}}\int_{1/\sqrt{t}}^\infty e^{-x^2}\,dx\,dt$$
$$= \frac{2}{\sqrt{\pi}}\int_0^\infty \int_{1/\sqrt{t}}^\infty e^{-pt-x^2}\,dx\,dt$$

Changing the order of integration (see Fig. 4.4) leads to

$$\mathscr{L}\{\text{erfc}(1/\sqrt{t});p\} = \frac{2}{\sqrt{\pi}}\int_0^\infty e^{-x^2}\int_{1/x^2}^\infty e^{-pt}\,dt\,dx$$
$$= \frac{2}{p\sqrt{\pi}}\int_0^\infty e^{-x^2-p/x^2}\,dx$$

Based on Eq. (4.18), it is clear that

$$\mathscr{L}\{\text{erfc}(1/\sqrt{t});p\} = (1/p)e^{-2\sqrt{p}}, \qquad \text{Re}(p) > 0$$

If we use the scaling property (4.8), we can readily generalize the result of Exam. 4.13 to

$$\mathscr{L}\{\text{erfc}(a/\sqrt{t});p\} = (1/p)e^{-2a\sqrt{p}}, \qquad \text{Re}(p) > 0 \tag{4.20}$$

When the function $f(t)$ that we wish to transform is continuous and has a power series expansion about $t = 0$ that converges for all values $t \geq 0$, it may be useful to first express $f(t)$ in this power series representation and then take the Laplace transform of the series termwise. In this case the transformed series is a Laurent series of the function $F(p)$, and in many cases of interest it is possible to sum this transformed series and obtain an explicit representation of $F(p)$.*

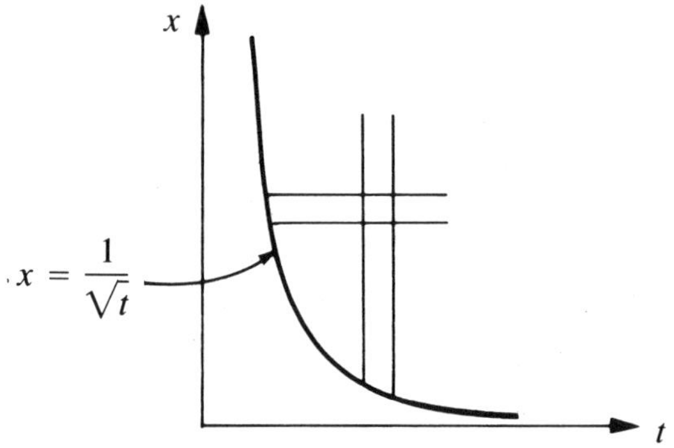

Figure 4.4 Region of integration

* See also Watson's lemma (Theor. 4.16) in Sec. 4.7.

Example 4.14: Find the Laplace transform of $\sin\sqrt{t}$.

Solution: Using the well-known series for $\sin x$, we have

$$\sin\sqrt{t} = \sum_{n=0}^{\infty} \frac{(-1)^n(\sqrt{t})^{2n+1}}{(2n+1)!} = \sum_{n=0}^{\infty} \frac{(-1)^n t^{n+1/2}}{(2n+1)!}$$

Therefore, the Laplace transform of this expression leads to

$$\begin{aligned}\mathscr{L}\{\sin\sqrt{t};p\} &= \sum_{n=0}^{\infty} \frac{(-1)^n}{(2n+1)!}\mathscr{L}\{t^{n+1/2};p\}\\ &= \sum_{n=0}^{\infty} \frac{(-1)^n}{(2n+1)!}\frac{\Gamma(n+3/2)}{p^{n+3/2}}\end{aligned}$$

where we are recalling Exam. 4.3. By using property (G8) in Sec. 1.2, we can write

$$\Gamma(n+3/2) = \frac{(2n+2)!}{2^{2n+2}(n+1)!}\sqrt{\pi} = \frac{(2n+1)!}{2^{2n+1}n!}\sqrt{\pi}$$

and thus we find that

$$\mathscr{L}\{\sin\sqrt{t};p\} = \frac{1}{2p}\sqrt{\frac{\pi}{p}}\sum_{n=0}^{\infty}\frac{(-1)^n}{n!}\left(\frac{1}{4p}\right)^n$$

We recognize this last series as that of an exponential function, and hence

$$\mathscr{L}\{\sin\sqrt{t};p\} = \frac{1}{2p}\sqrt{\frac{\pi}{p}}e^{-1/4p}, \qquad \mathrm{Re}(p) > 0$$

4.4.1 *Transforms Involving Bessel Functions*

Bessel functions arise in a variety of applications involving the Laplace transform and therefore it is important to know certain transform relations of these functions.

Example 4.15: Find the Laplace transform of $t^{\nu/2}J_\nu(2\sqrt{t})$.

Solution: Here we find it convenient to first express the given function in a power series; thus,

$$\begin{aligned}t^{\nu/2}J_\nu(2\sqrt{t}) &= t^{\nu/2}\sum_{n=0}^{\infty}\frac{(-1)^n\, t^{n+\nu/2}}{n!\Gamma(n+\nu+1)}\\ &= \sum_{n=0}^{\infty}\frac{(-1)^n\, t^{n+\nu}}{n!\Gamma(n+\nu+1)}\end{aligned}$$

By taking the Laplace transform of this series, we get

$$\mathscr{L}\{t^{\nu/2}J_\nu(2\sqrt{t});p\} = \sum_{n=0}^{\infty} \frac{(-1)^n}{n!\Gamma(n+\nu+1)} \mathscr{L}\{t^{n+\nu};p\}$$

$$= \sum_{n=0}^{\infty} \frac{(-1)^n}{n!p^{n+\nu+1}}$$

$$= \frac{1}{p^{\nu+1}} \sum_{n=0}^{\infty} \frac{(-1)^n}{n!}\left(\frac{1}{p}\right)^n$$

from which we deduce

$$\mathscr{L}\{t^{\nu/2}J_\nu(2\sqrt{t});p\} = \frac{1}{p^{\nu+1}} e^{-1/p}, \qquad \operatorname{Re}(p) > 0$$

The scaling property, Eq. (4.8), allows us to generalize the result of Exam. 4.15 to

$$\mathscr{L}\{t^{\nu/2}J_\nu(2\sqrt{at});p\} = (a^{\nu/2}/p^{\nu+1})\, e^{-a/p}, \qquad a > 0, \operatorname{Re}(p) > 0 \tag{4.21}$$

The special case $\nu = 0$ then yields

$$\mathscr{L}\{J_0(2\sqrt{at});p\} = (1/p)e^{-a/p}, \qquad a > 0, \operatorname{Re}(p) > 0 \tag{4.22}$$

Example 4.16: Find the Laplace transform of $t^\nu J_\nu(t)$, $\nu > -1/2$.

Solution: Again relying on the series definition of the Bessel function, we have

$$t^\nu J_\nu(t) = \sum_{n=0}^{\infty} \frac{(-1)^n t^{2n+2\nu}}{n!\Gamma(n+\nu+1)2^{2n+\nu}}$$

and thus

$$\mathscr{L}\{t^\nu J_\nu(t);p\} = \sum_{n=0}^{\infty} \frac{(-1)^n}{n!\Gamma(n+\nu+1)2^{2n+\nu}} \mathscr{L}\{t^{2n+2\nu};p\}$$

$$= \sum_{n=0}^{\infty} \frac{(-1)^n\, \Gamma(2n+2\nu+1)}{n!\Gamma(n+\nu+1)2^{2n+\nu}p^{2n+2\nu+1}}$$

By using the duplication formula of the gamma function [see (G7) in Sec. 1.2], it can be shown that

$$\frac{\Gamma(2n+2\nu+1)}{\Gamma(n+\nu+1)} = \frac{1}{\sqrt{\pi}} 2^{2n+2\nu}\, \Gamma(n+\nu+1/2)$$

(see Prob. 25 in Exer. 1.2). Hence, our transform relation reduces to

$$\mathscr{L}\{t^\nu J_\nu(t);p\} = \frac{2^\nu}{\sqrt{\pi}\, p^{2\nu+1}} \sum_{n=0}^{\infty} \frac{(-1)^n\Gamma(n+\nu+1/2)}{n!}\left(\frac{1}{p^2}\right)^n$$

This last series can be summed by using properties of the binomial series

$$(1 + x)^{-a} = \sum_{n=0}^{\infty} \binom{-a}{n} x^n = \sum_{n=0}^{\infty} \frac{(-1)^n \Gamma(n + a)}{n!\Gamma(a)} x^n, \qquad |x| < 1$$

where we are using the fact that

$$\binom{-a}{n} = \frac{(-1)^n a(a + 1) \cdots (a + n - 1)}{n!} = \frac{(-1)^n \Gamma(n + a)}{n!\Gamma(a)}$$

Therefore, we see that our Laplace transform series is a binomial series for which

$$\mathscr{L}\{t^\nu J_\nu(t);p\} = \frac{2^\nu \Gamma(\nu + 1/2)}{\sqrt{\pi}\, p^{2\nu+1}} (1 + 1/p^2)^{-\nu-1/2}, \qquad \text{Re}(p) > 1$$

or

$$\mathscr{L}\{t^\nu J_\nu(t);p\} = \frac{2^\nu\, \Gamma(\nu + 1/2)}{\sqrt{\pi}(p^2 + 1)^{\nu+1/2}}, \qquad \nu > -1/2, \text{Re}(p) > 1$$

When $\nu = 0$ in the result of Exam. 4.16, we obtain the special case

$$\mathscr{L}\{J_0(t);p\} = 1/\sqrt{p^2 + 1}, \qquad \text{Re}(p) > 1 \tag{4.23}$$

Also, by using Eq. (4.11) together with the relation $J_0'(t) = -J_1(t)$, we have

$$\mathscr{L}\{J_0'(t);p\} = p\, \mathscr{L}\{J_0(t);p\} - J_0(0)$$

or

$$-\mathscr{L}\{J_1(t);p\} = (p/\sqrt{p^2 + 1}) - 1$$

from which we deduce

$$\mathscr{L}\{J_1(t);p\} = (\sqrt{p^2 + 1} - p)/\sqrt{p^2 + 1}, \qquad \text{Re}(p) > 1 \tag{4.24}$$

EXERCISES 4.4

In Probs. 1–8, verify the Laplace transform relation involving error functions.

1. $\mathscr{L}\{\text{erf}(t/2a);p\} = (1/p)e^{a^2p^2} \text{erfc}(ap), \qquad a > 0$
2. $\mathscr{L}\{\text{erf}(1/\sqrt{t});p\} = (1/p)(1 - e^{-2\sqrt{p}})$
3. $\mathscr{L}\{\text{erf}(\sqrt{t});p\} = 1/p\sqrt{p + 1}$
4. $\mathscr{L}\{e^{at}\text{erf}(\sqrt{at});p\} = \sqrt{\frac{a}{p}}\left(\frac{1}{p - a}\right), \qquad a > 0$

5. $\mathscr{L}\{e^{at}\text{erfc}(\sqrt{at});p\} = \dfrac{1}{\sqrt{p}(\sqrt{p} + \sqrt{a})}, \qquad a > 0$

6. $\mathscr{L}\{t\ \text{erf}(2\sqrt{t});p\} = \dfrac{3p + 8}{p^2(p + 4)^{3/2}}$

7. $\mathscr{L}\{1/\sqrt{\pi t} - ae^{a^2 t}\text{erfc}(a\sqrt{t});p\} = \dfrac{1}{\sqrt{p} + a}, \qquad a > 0$

8. $\mathscr{L}\left\{e^{b(bt+a)}\text{erfc}\left(b\sqrt{t} + \dfrac{a}{2\sqrt{t}}\right);p\right\} = \dfrac{e^{-a\sqrt{p}}}{\sqrt{p}(\sqrt{p} + b)}$

In Probs. 9–15, use infinite series to derive the given Laplace transform relation.

9. $\mathscr{L}\{(\sin^2 t)/t;p\} = (1/4)\log(1 + 4/p^2)$

10. $\mathscr{L}\left\{(1/\sqrt{\pi t}) \cos 2\sqrt{at};p\right\} = \dfrac{1}{\sqrt{p}} e^{-a/p}, \qquad a > 0$

11. $\mathscr{L}\left\{\dfrac{\sin at}{t};p\right\} = \tan^{-1} a/p, \qquad a > 0$

12. $\mathscr{L}\{\text{erfc}(1/\sqrt{t});p\} = (1/p)e^{-2\sqrt{p}}$

13. $\mathscr{L}\{I_0(at);p\} = \dfrac{1}{\sqrt{p^2 - a^2}}, \qquad a > 0$

14. $\mathscr{L}\{t^{\nu/2}I_\nu(2\sqrt{at});p\} = \dfrac{a^{\nu/2}}{p^{\nu+1}}e^{a/p}, \qquad a > 0$

15. $\mathscr{L}\{t^\nu I_\nu(t);p\} = \dfrac{2^\nu\,\Gamma(\nu + \frac{1}{2})}{\sqrt{\pi}(p^2 - 1)^{\nu+1/2}}, \qquad \nu > -1/2$

In Probs. 16–22, verify the Laplace transform relation involving Bessel functions.

16. $\mathscr{L}\{e^{-at}J_0(bt);p\} = 1/\sqrt{p^2 - 2ap + a^2 + b^2}, \qquad b > 0$

17. $\mathscr{L}\{tJ_0(at);p\} = p/(p^2 + a^2)^{3/2}, \qquad a > 0$

18. $\mathscr{L}\{te^{-t}J_0(t);p\} = (p - 1)/(p^2 - 2p + 2)^{3/2}$

19. $\mathscr{L}\{tJ_1(t);p\} = 1/(p^2 + 1)^{3/2}$

20. $\mathscr{L}\{t^n J_n(at);p\} = \dfrac{(2n)!}{2^n\,n!}(p^2 + a^2)^{-n-1/2}, \qquad a > 0,\ n = 1,2,3,\ldots$

21. $\mathscr{L}\{t^{n+1}J_n(at);p\} = \dfrac{(2n + 1)!}{2^n\,n!}\,p(p^2 + a^2)^{-n-3/2}, \qquad a > 0,\ n = 1,2,3,\ldots$

22. $\mathscr{L}\{J_\nu(t);p\} = (\sqrt{p^2+1} - p)^\nu/\sqrt{p^2+1}$

23. Show that

(a) $\mathscr{L}\{J_0(t)\sin t;p\} = \dfrac{1}{\sqrt{p}\,\sqrt[4]{p^2+4}}\sin\left(\dfrac{1}{2}\tan^{-1}\dfrac{2}{p}\right)$

(b) $\mathscr{L}\{J_0(t)\cos t;p\} = \dfrac{1}{\sqrt{p}\,\sqrt[4]{p^2+4}}\cos\left(\dfrac{1}{2}\tan^{-1}\dfrac{2}{p}\right)$

24. Evaluate $\mathscr{L}\left\{\dfrac{1 - J_0(t)}{t};p\right\}$

25. If $F(p) = \mathscr{L}\{f(t);p\}$, show that

$$\mathscr{L}\left\{\int_0^\infty \frac{t^x f(x)}{\Gamma(x+1)}\,dx;p\right\} = \frac{F(\log p)}{p\log p}$$

In Probs. 26–28, use Laplace transforms to verify the Bessel function relation.

26. $\displaystyle\int_0^\infty J_0(2\sqrt{xt})\cos x\,dx = \sin t$

27. $\displaystyle\int_0^\infty J_0(2\sqrt{xt})\sin x\,dx = \cos t$

28. $\displaystyle\int_0^\infty J_0(2\sqrt{xt})J_0(x)\,dx = J_0(t)$

29. Use the result of Prob. 31 in Exer. 4.3 to deduce the value of $\displaystyle\int_0^\infty J_0(x)\,dx$.

30. If $L_n(t)$, $n = 0,1,2,\ldots$, are the Laguerre polynomials of Prob. 33 in Exer. 4.3, use Laplace transforms to show that

$$\sum_{n=0}^\infty \frac{L_n(t)}{n!} = e\,J_0(2\sqrt{t})$$

31. Show that

$$\mathscr{L}\{1 - C(\sqrt{2/\pi t}) - S(\sqrt{2/\pi t});p\} = (1/p)e^{-\sqrt{p}}\cos\sqrt{p}$$

where $C(x)$ and $S(x)$ are the Fresnel integrals (see Sec. 1.3.2).

32. Using the sifting property of the impulse function (see Sec. 1.5.2), show that

(a) $\mathscr{L}\{\delta(t);p\} = 1$

(b) $\mathscr{L}\{\delta(t-a);p\} = e^{-ap}, \qquad a > 0$

4.5 *The Inverse Laplace Transform*

Thus far we have concerned ourselves only with the problem of finding the transform function $F(p)$, given the function $f(t)$. However, the use of Laplace transforms is effective in applications only if we can also solve the inverse problem of finding $f(t)$, given the function $F(p)$. In symbols, we write the *inversion formula*

$$f(t) = \mathscr{L}^{-1}\{F(p);t\} \tag{4.25}$$

First of all, there is the question of whether a given function $F(p)$ of a particular class of functions actually represents the Laplace transform of some function $f(t)$. In this regard we have the following theorem.

Theorem 4.11. If f is piecewise continuous on $t \geq 0$ and is $0(e^{c_0 t})$, and if $F(p)$ is the Laplace transform of $f(t)$, then

$$\lim_{|p|\to\infty} F(p) = 0$$

The proof of Theor. 4.11 follows that of Theor. 4.1 and is left to the exericises (see Prob. 37 in Exer. 4.5). The real significance of Theor. 4.11 is that if $F(p)$ is any function for which $\lim_{|p|\to\infty} F(p) \neq 0$, then it does not represent the Laplace transform of any piecewise continuous function of exponential order. This condition rules out many functions as possible Laplace transforms, such as polynomials in p, e^p, $\cos p$, and so forth.

Given a particular function $f(t)$, we know that its transform $F(p)$ is uniquely determined as a consequence of the properties of integrals. Moreover, $F(p)$ is an analytic function even if $f(t)$ has certain discontinuities. The situation for the inverse transform, however, is not the same. For instance, if $f(t)$ and $g(t)$ are two functions that are identical except for a finite number of points, they will have the same transform, say $F(p)$, since their integrals are identical. Therefore, we can claim that either $f(t)$ or $g(t)$ is the inverse transform of $F(p)$. That is to say, the inverse transform of a given function $F(p)$ is uniquely determined only up to an additive *null function.** This result is known as *Lerche's theorem.* Null functions are normally of little consequence in applications and so the difficulty of finding unique inverse Laplace transforms is mostly of academic interest. If we can find a continuous function $f(t)$ that is the inverse transform of $F(p)$, that is the one we use.

When constructing inverse Laplace transforms, we find in many routine problems that the desired inverse transforms of $F(p)$ can be obtained

* A null function $n(t)$ is one for which $\int_0^t n(u)\,du = 0$ for all t.

directly from existing tables of transforms (see Appendix C). For instance, in Exams. 4.1 and 4.2 we derived the transform relations

$$\mathscr{L}\{e^{at};p\} = 1/(p - a)$$

and

$$\mathscr{L}\{\cos at;p\} = p/(p^2 + a^2)$$

and hence, we immediately have the inverse transform relations

$$\mathscr{L}^{-1}\left\{\frac{1}{p - a};t\right\} = e^{at}$$

and

$$\mathscr{L}^{-1}\left\{\frac{p}{p^2 + a^2};t\right\} = \cos at$$

Also, many of the operational properties used in finding the transform itself likewise can be used in constructing the inverse transform. For example, the *linearity property* (Theor. 4.3) and *shifting property* (Theor. 4.4) become, respectively,

$$\mathscr{L}^{-1}\{C_1F(p) + C_2G(p);t\} = C_1f(t) + C_2g(t) \tag{4.26}$$

and

$$\mathscr{L}^{-1}\{F(p - a);t\} = e^{at}\,\mathscr{L}^{-1}\{F(p);t\} \tag{4.27}$$

Example 4.17: Find $\mathscr{L}^{-1}\{(p - 5)/(p^2 + 6p + 13);\ t\}$

Solution: Completing the square in the denominator, we get

$$\frac{p - 5}{p^2 + 6p + 13} = \frac{p - 5}{(p + 3)^2 + 4} = \frac{(p + 3) - 8}{(p + 3)^2 + 4}$$

Then, using (4.26) and (4.27), we obtain

$$\begin{aligned}\mathscr{L}^{-1}\left\{\frac{p - 5}{p^2 + 6p + 13};t\right\} &= \mathscr{L}^{-1}\left\{\frac{(p + 3) - 8}{(p + 3)^2 + 4};t\right\}\\ &= e^{-3t}\mathscr{L}^{-1}\left\{\frac{p - 8}{p^2 + 4};t\right\}\\ &= e^{-3t}\left[\mathscr{L}^{-1}\left\{\frac{p}{p^2 + 4};t\right\} - 4\,\mathscr{L}^{-1}\left\{\frac{2}{p^2 + 4};t\right\}\right]\end{aligned}$$

which gives the result

$$\mathscr{L}^{-1}\left\{\frac{p - 5}{p^2 + 6p + 13};t\right\} = e^{-3t}(\cos 2t - 4 \sin 2t)$$

4.5.1 *Partial Fractions*

In many cases of practical importance we wish to find the inverse transform of a *rational function*, i.e., a function having the form

$$F(p) = R(p)/Q(p)$$

where $R(p)$ and $Q(p)$ are polynomials in p. The inverse transform in such cases can often be found quite easily be representing $F(p)$ in terms of its *partial fractions*. The partial fraction representation is the same as that found in the calculus, for example, as a means of integrating certain rational functions. Here it is assumed that $R(p)$ and $Q(p)$ have no common factors and that the degree of $R(p)$ is lower than that of $Q(p)$. Let us illustrate the technique with some examples.

Example 4.18: Find $\mathscr{L}^{-1}\{2/(p + 1)\,(p^2 + 1); t\}$

Solution: Using partial fraction expansions, we write

$$\frac{2}{(p + 1)(p^2 + 1)} = \frac{A}{p + 1} + \frac{Bp + C}{p^2 + 1}$$

and clearing fractions yields

$$2 = A(p^2 + 1) + (Bp + C)(p + 1)$$

Setting $p = -1$, we find $A = 1$, and equating like coefficients of p^2 and p^0 gives the equations

$$0 = A + B,$$
$$2 = A + C,$$

from which we deduce $B = -1$ and $C = 1$. Thus, we find

$$\begin{aligned}\mathscr{L}^{-1}\left\{\frac{2}{(p + 1)(p^2 + 1)};t\right\} &= \mathscr{L}^{-1}\left\{\frac{1}{p + 1};t\right\} \\ &\quad - \mathscr{L}^{-1}\left\{\frac{p}{p^2 + 1};t\right\} + \mathscr{L}^{-1}\left\{\frac{1}{p^2 + 1};t\right\} \\ &= e^{-t} - \cos t + \sin t\end{aligned}$$

Example 4.19: Find $\mathscr{L}^{-1}\{(p + 1)/p^2(p + 2)^3; t\}$

Solution: Let us write

$$\frac{p + 1}{p^2(p + 2)^3} = \frac{A}{p} + \frac{B}{p^2} + \frac{C}{p + 2} + \frac{D}{(p + 2)^2} + \frac{E}{(p + 2)^3}$$

and by clearing fractions, we have

$$p + 1 = Ap(p + 2)^3 + B(p + 2)^3 + Cp^2(p + 2)^2 + Dp^2(p + 2) + Ep^2$$

Solving for the constants yields $C = -A = 1/16$, $B = 1/8$, $D = 0$, and $E = -1/4$. Hence,

$$\mathscr{L}^{-1}\left\{\frac{p + 1}{p^2(p + 2)^3};t\right\} = -\frac{1}{16}\mathscr{L}^{-1}\left\{\frac{1}{p};t\right\} + \frac{1}{8}\mathscr{L}^{-1}\left\{\frac{1}{p^2};t\right\} + \frac{1}{16}\mathscr{L}^{-1}\left\{\frac{1}{p + 2};t\right\} - \frac{1}{4}\mathscr{L}^{-1}\left\{\frac{1}{(p + 2)^3};t\right\}$$

$$= -\frac{1}{16} + \frac{1}{8}t + \frac{1}{16}e^{-2t} - \frac{1}{8}t^2 e^{-2t}$$

In Sec. 4.6 we will develop more systematic means of finding the inverse Laplace transform of rational functions. Specifically, we will use methods of complex variables to derive what are known as Heaviside's expansion theorems. These theorems provide an approach to determining the unknown constants in partial fraction expansions that is more sophisticated than clearing fractions and matching like terms, and so on, as we have done here.

Finally, we wish to present one more example involving partial fractions that also makes use of a previously derived property of the Laplace transform.

Example 4.20: Find $\mathscr{L}^{-1}\{\log(1 + 1/p^2); t\}$

Solution: Letting $F(p) = \log(1 + 1/p^2)$, we first observe that

$$F'(p) = -\frac{2}{p(p^2 + 1)} = -2\left(\frac{1}{p} - \frac{p}{p^2 + 1}\right)$$

where we have used a partial fraction expansion. Now recalling the property [see Eq. (4.14)]

$$\mathscr{L}\{tf(t);p\} = -F'(p)$$

which can also be written as

$$\mathscr{L}^{-1}\{F'(p);t\} = -tf(t)$$

we see that in our particular case

$$\mathscr{L}^{-1}\{F'(p);t\} = -2(1 - \cos t) = -tf(t)$$

Hence, we deduce that

$$f(t) = 2(1 - \cos t)/t$$

which is the inverse Laplace transform of $F(p) = \log(1 + 1/p^2)$.

4.5.2 *Series Method*

In constructing Laplace transforms, we previously found it useful in certain situations to express the function $f(t)$ in terms of its power series and take the transform termwise. This same technique also proves fruitful in constructing inverse Laplace transforms. That is, if $F(p)$ has a convergent Laurent series expansion of the form

$$F(p) = \frac{a_0}{p} + \frac{a_1}{p^2} + \frac{a_2}{p^3} + \frac{a_3}{p^4} + \ldots \tag{4.28}$$

then under suitable conditions we can invert (4.28) to obtain*

$$f(t) = a_0 + a_1 t + a_2 \frac{t^2}{2!} + a_3 \frac{t^3}{3!} + \ldots \tag{4.29}$$

In some cases the resulting series (4.29) may be summable to a known function.

Example 4.21: Find the inverse Laplace transform of $(1/p)e^{-1/p}$.

Solution: Using infinite series, we have

$$\frac{1}{p} e^{-1/p} = \frac{1}{p} \sum_{n=0}^{\infty} \frac{(-1)^n}{n!\, p^n} = \sum_{n=0}^{\infty} \frac{(-1)^n}{n!\, p^{n+1}}$$

Inverting the series term by term yields

$$\mathcal{L}^{-1}\left\{\frac{1}{p} e^{-1/p};t\right\} = \sum_{n=0}^{\infty} \frac{(-1)^n}{n!} \mathcal{L}^{-1}\left\{\frac{1}{p^{n+1}};t\right\}$$
$$= \sum_{n=0}^{\infty} \frac{(-1)^n t^n}{(n!)^2}$$

and thus we see that

$$\mathcal{L}^{-1}\left\{\frac{1}{p} e^{-1/p};t\right\} = J_0(2\sqrt{t})$$

4.5.3 *Convolution Theorem*

It is often expedient to resolve a Laplace transform into the product of two transforms, i.e., $F(p)G(p)$, when the inverse transform of both $F(p)$

* See also Watson's lemma in Sec. 4.7 (Theor. 4.16).

and $G(p)$ are known. This same situation arises when one of them represents an arbitrary function — typically an input to some physical system. The inversion of the product $F(p)G(p)$ can then be obtained by application of what we call the *convolution theorem.*

In order to derive the convolution theorem, let us begin by writing the product of the transforms of $f(t)$ and $g(t)$ as the iterated integral

$$F(p)G(p) = \int_0^\infty e^{-px} f(x)\,dx \cdot \int_0^\infty e^{-pu} g(u)\,du$$
$$= \int_0^\infty \int_0^\infty e^{-p(x+u)} f(x)g(u)\,dx\,du$$

The change of variable $x = t - u$ then leads to

$$F(p)G(p) = \int_0^\infty \int_u^\infty e^{-pt} f(t-u)g(u)\,dt\,du \tag{4.30}$$

We can interpret the integrals in (4.30) as an iterated integral over the region $u \le t < \infty$, $0 \le u < \infty$, as shown in Fig. 4.5. If we change the order of integration, we find that the region of integration is characterized by $0 \le u \le t$, $0 \le t < \infty$, and thus

$$F(p)G(p) = \int_0^\infty \int_0^t e^{-pt} f(t-u)g(u)\,du\,dt$$
$$= \int_0^\infty e^{-pt} \int_0^t f(t-u)g(u)\,du\,dt$$

We now define the *Laplace convolution integral* of $f(t)$ and $g(t)$, i.e.,

$$(f * g)(t) = \int_0^t f(t-u)g(u)\,du \tag{4.31}$$

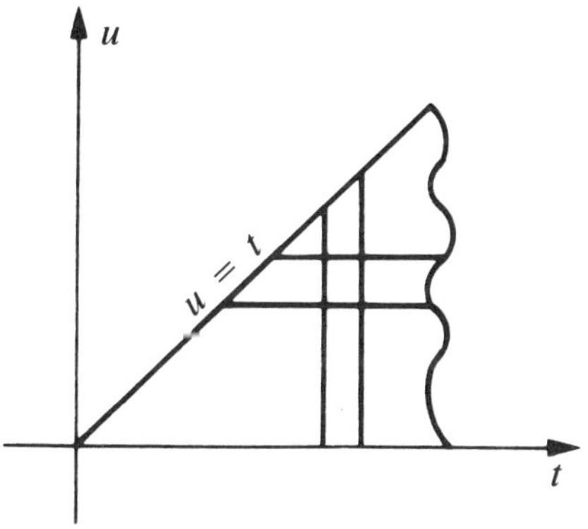

Figure 4.5 Region of integration

In terms of the convolution integral, we have shown that

$$F(p)G(p) = \mathcal{L}\{(f * g)(t);p\} \tag{4.32}$$

and by formally taking the inverse Laplace transform of (4.32), we deduce that

$$\mathcal{L}^{-1}\{F(p)G(p);t\} = (f * g)(t) = \int_0^t f(t - u)g(u)\,du \tag{4.33}$$

which is our intended result.

The integral in (4.33) arises in many applications where it represents a superposition of effects of magnitude $g(u)$, occurring at time $t = u$, for which $f(t - u)$ is the *influence function* or response of a system to a unit impulse delivered at time $t = u$.

In summary, we have the following theorem.

Theorem 4.12 (*Convolution theorem*). If $f(t)$ and $g(t)$ are piecewise continuous functions on $t \geq 0$ and are $0(e^{c_0 t})$, and if $F(p)$ and $G(p)$ are the Laplace transforms, respectively, of $f(t)$ and $g(t)$, then the inverse Laplace transform of the product $F(p)G(p)$ is given by the formula

$$\mathcal{L}^{-1}\{F(p)G(p);t\} = (f * g)(t)$$

Once again we point out that the conditions stated in Theor. 4.12 are more stringent than required for the validity of (4.33). That is, there exist functions satisfying (4.33) that are not piecewise continuous on $t \geq 0$. Some examples illustrating this point are included in the exercises (see also Exam. 4.23).

Example 4.22: Find $\mathcal{L}^{-1}\{1/p^2(p^2 + k^2);t\}$.

Solution: Let us select $F(p) = 1/p^2$ and $G(p) = 1/(p^2 + k^2)$, whose inverse transforms are

$$\mathcal{L}^{-1}\{1/p^2;t\} = f(t) = t$$

and

$$\mathcal{L}^{-1}\{1/(p^2 + k^2);t\} = g(t) = (1/k)\sin kt$$

Thus, using the convolution theorem (4.33), we write

$$\mathcal{L}^{-1}\{1/p^2(p^2 + k^2);t\} = (f * g)(t) = \int_0^t (t - u)(1/k)\sin ku\,du$$

which leads to the result

$$\mathcal{L}^{-1}\{1/p^2(p^2 + k^2);t\} = (1/k^3)(kt - \sin kt)$$

The convolution integral (4.31) satisfies certain properties which often prove useful in practice. For example, by making the change of variable $v = t - u$ in (4.31), we find

$$(f * g)(t) = -\int_t^0 f(v)g(t - v)dv = \int_0^t g(t - v)f(v)dv$$

from which we deduce the *commutative* relation

$$(f * g)(t) = (g * f)(t) \tag{4.34}$$

We also write this as simply $f * g = g * f$. From definition, it is clear that

$$f * (Cg) = (Cf) * g = C(f * g) \tag{4.35}$$

where C is a constant, and

$$f * (g + k) = f * g + f * k \tag{4.36}$$

which is the *distributive* relation. The convolution integral (4.31) also satisfies the *associative* relation

$$f * (g * k) = (f * g) * k \tag{4.37}$$

but this is a little more difficult to prove.

Our final example in this section illustrates how the convolution theorem can be used in a less direct manner to derive a certain result.

Example 4.23: Prove that

$$J_o(t) = \frac{2}{\pi}\int_0^1 \frac{\cos tx}{\sqrt{1 - x^2}}\,dx$$

Solution: Recalling Eq. (4.23), we have

$$\mathscr{L}\{J_o(t);p\} = \frac{1}{\sqrt{p^2 + 1}} = \frac{1}{\sqrt{p + i}} \cdot \frac{1}{\sqrt{p - i}}$$

Using the fact that

$$\mathscr{L}^{-1}\{1/\sqrt{p + a};t\} = (1/\sqrt{\pi t})e^{-at}$$

which follows from the shift property (Theor. 4.4) applied to the result in Exam. 4.3, we obtain by way of the convolution theorem,

$$\begin{aligned} J_o(t) &= \mathscr{L}^{-1}\left\{\frac{1}{\sqrt{p + i}} \cdot \frac{1}{\sqrt{p - i}};t\right\} \\ &= \frac{1}{\pi}\int_0^t u^{-1/2} e^{-iu} (t - u)^{-1/2} e^{i(t - u)}\, du \\ &= \frac{1}{\pi}\int_0^t \frac{e^{i(t - 2u)}}{\sqrt{u(t - u)}}\, du \end{aligned}$$

By letting $u = tv$, this becomes

$$J_0(t) = \frac{1}{\pi}\int_0^1 \frac{e^{it(1-2v)}}{\sqrt{v(1-v)}}\,dv$$

and then by introducing $x = 1 - 2v$, we have

$$\begin{aligned} J_0(t) &= \frac{1}{\pi}\int_{-1}^1 \frac{e^{itx}}{\sqrt{1-x^2}}\,dx \\ &= \frac{1}{\pi}\int_{-1}^1 \frac{\cos tx}{\sqrt{1-x^2}}\,dx + \frac{i}{\pi}\int_{-1}^1 \frac{\sin tx}{\sqrt{1-x^2}}\,dx \end{aligned}$$

Finally, by using properties of even and odd functions, we conclude that the integral involving $\sin tx$ vanishes and the remaining integral leads to

$$J_0(t) = \frac{2}{\pi}\int_0^1 \frac{\cos tx}{\sqrt{1-x^2}}\,dx$$

EXERCISES 4.5

In Probs. 1–10, determine the inverse Laplace transform using the table in Appendix C and various operational properties.

1. $F(p) = \dfrac{1}{2p + 3}$

2. $F(p) = \dfrac{3p + 7}{p^2 + 5}$

3. $F(p) = \dfrac{2p}{(p - 3)^5}$

4. $F(p) = \dfrac{1}{p^2 - 6p + 10}$

5. $F(p) = \dfrac{p}{p^2 - 6p + 13}$

6. $F(p) = \dfrac{5p - 2}{3p^2 + 4p + 8}$

7. $F(p) = \dfrac{2p + 3}{4p^2 + 4p + 5}$

8. $F(p) = \dfrac{p^2}{(p + 2)^4}$

9. $F(p) = \dfrac{e^{-5p}}{(p - 3)^4}$

10. $F(p) = \dfrac{(p + 1)e^{-\pi p}}{p^2 + p + 1}$

In Probs. 11–20, evaluate the inverse Laplace transform by the method of partial fractions.

11. $F(p) = \dfrac{1}{p(p + 1)}$

12. $F(p) = \dfrac{1}{(p - 1)(p + 2)(p + 4)}$

13. $F(p) = \dfrac{p^2}{(p + 2)^3}$

14. $F(p) = \dfrac{3p - 2}{p^3(p^2 + 4)}$

15. $F(p) = \dfrac{p+1}{(p^2-4p)(p+5)^2}$

16. $F(p) = \dfrac{1}{p^4-1}$

17. $F(p) = \dfrac{p}{(p^2+a^2)(p^2+b^2)}$

18. $F(p) = \dfrac{4p^2-16}{p^3(p+2)^2}$

19. $F(p) = \dfrac{p^2-3}{(p+2)(p-3)(p^2+2p+5)}$

20. $F(p) = \dfrac{3p^2-6p+7}{(p^2-2p+5)^2}$

21. Given that $\mathscr{L}^{-1}\{F(p);t\} = f(t)$, show for constants, a, b, and k that
(a) $\mathscr{L}^{-1}\{F(kp);t\} = (1/k)f(t/k), \quad k > 0$
(b) $\mathscr{L}^{-1}\{F(ap+b);t\} = (1/a)e^{-bt/a}f(t/a), \quad a > 0$

22. Given that $f(t) = t \sin t$ is the inverse Laplace transform of $F(p) = 2p/(p^2+1)^2$, use Theor. 4.7 to evaluate

$$\mathscr{L}^{-1}\{1/(p^2+1)^2;t\}$$

In Probs. 23–28, use infinite series to find the inverse Laplace transform of the given function.

23. $F(p) = \log(1 + 1/p)$

24. $F(p) = \log(1 + 1/p^2)$

25. $F(p) = \log\dfrac{p+1}{p-1}$

26. $F(p) = \log\dfrac{p-a}{p-b}$

27. $F(p) = 1/p\sqrt{p+4}$

28. $F(p) = (1/\sqrt{p})e^{-1/p}$

In Probs. 29–36, use the convolution theorem to find the inverse Laplace transform of the given function.

29. $1/p\sqrt{p+4}$

30. $2/(p+1)(p^2+1)$

31. $p/(p^2+a^2)^2$

32. $1/p^2(p+1)^2$

33. $1/\sqrt{p}(p-1)$

34. $1/(p-1)p\sqrt{p}$

35. $1/(p+1)^2(p^2+4)$

36. $1/(\sqrt{p}-1)$

37. Prove Theor. 4.11.

38. Show that
(a) $1 * 1 * 1 = t^2/2$
(b) $t * t * t = t^5/5!$
(c) $t^{m-1} * t^n = \dfrac{(m-1)!n!t^{m+n}}{(m+n)!}$; $m,n = 1,2,3,\ldots$

39. Starting with $f(t) = \displaystyle\int_0^t u^{x-1}(t-u)^{y-1}\,du$, $x,y > 0$,

(a) use the convolution theorem to show that

$$F(p) = \Gamma(x)\Gamma(y)/p^{x+y}$$

(b) From the result in (a), establish the formula

$$\int_0^1 u^{x-1}(1-u)^{y-1}\,du = \Gamma(x)\Gamma(y)/\Gamma(x+y), \qquad x>0,\ y>0$$

40. Show that

(a) $\int_0^t J_0(t-u)J_0(u)\,du = \sin t$

(b) $\int_0^t \cos(t-u)\sin u\,du = (1/2)t\sin t$

(c) $\int_0^t \sin(t-u)J_0(u)\,du = (1/2)tJ_1(t)$

(d) $\int_0^t J_1(t-u)J_0(u)\,du = J_0(t) - \cos t$

41. Show that

$$\frac{1}{\pi}\int_0^t \frac{e^{2au}}{\sqrt{u(t-u)}}\,du = e^{at}I_0(at)$$

42. Show that

$$\mathscr{L}^{-1}\left\{\frac{\sqrt{p}}{p-1};t\right\} = \frac{1}{\sqrt{\pi t}} + e^t\,\mathrm{erf}(t)$$

43. Use the result of Prob. 42 to evaluate $\mathscr{L}^{-1}\{1/(1+\sqrt{p});t\}$

44. Prove that

$$f*(g*k) = (f*g)*k$$

4.6 *Complex Inversion Formula*

The techniques introduced in Sec. 4.5 for evaluating inverse Laplace transforms are adequate for a wide variety of routine applications involving the Laplace transform. Mostly they make use of tabulated results coupled with operational properties of the transform. More complicated problems, however, may lead to transforms that are not in tables, and in such cases we may need more powerful direct methods for constructing the necessary inverse Laplace transforms. Some of the methods of complex variables can be particularly useful in deriving more direct inversion formulas.

In Sec. 4.1 we gave an heuristic argument that led to the integral formula

$$f(t) = \frac{1}{2\pi i}\int_{c-i\infty}^{c+i\infty} e^{pt}F(p)\,dp \tag{4.38}$$

where $f(t)$ and $F(p)$ are Laplace transform pairs. In order to provide a more rigorous derivation of this formula, let us assume that both $f(t)$ and $f'(t)$ are continuous functions on $t \geq 0$ and that $f(t)$ is $0(e^{c_o t})$. Then, based on Theors. 4.1 and 4.2, we know that the Laplace integral

$$F(p) = \int_0^\infty e^{-pu} f(u)\, du \tag{4.39}$$

converges absolutely and uniformly in the half-plane $\text{Re}(p) \geq c_2 > c_o$ and that $F(p)$ is analytic in the half-plane $\text{Re}(p) > c_o$. Substituting the integral representation (4.39) for $F(p)$ in the following integral, we obtain

$$\int_{c-i\lambda}^{c+i\lambda} e^{pt} F(p)\, dp = \int_{c-i\lambda}^{c+i\lambda} e^{pt} \int_0^\infty e^{-pu} f(u)\, du\, dp \tag{4.40}$$

Our goal at this point is to show that by allowing $\lambda \to \infty$, we can derive Eq. (4.38) from Eq. (4.40). Because of the requirements we have imposed on $f(t)$, we can interpret (4.40) as an iterated integral and interchange the order of integration. Doing so yields

$$\begin{aligned}\int_{c-i\lambda}^{c+i\lambda} e^{pt} F(p)\, dp &= \int_0^\infty f(u) \int_{c-i\lambda}^{c+i\lambda} e^{p(t-u)}\, dp\, du \\ &= 2i \int_0^\infty f(u) e^{c(t-u)} \frac{\sin \lambda(t-u)}{t-u}\, du \\ &= 2ie^{ct} \int_{-t}^\infty f(t+x) e^{-c(t+x)} \frac{\sin \lambda x}{x}\, dx\end{aligned}$$

where in the last step we have made the change of variable $u = t + x$. If we now invoke Lemma 2.2, it follows that

$$\begin{aligned}\lim_{\lambda\to\infty} \frac{1}{2\pi i} \int_{c-i\lambda}^{c+i\lambda} e^{pt} F(p)\, dp &= e^{ct} f(t) e^{-ct} \\ &= f(t), \qquad t > 0\end{aligned} \tag{4.41}$$

which is the same as (4.38).

Let us summarize this result in the form of an inversion theorem.

Theorem 4.13 (*Inversion theorem*). If $F(p)$ is an analytic function of the complex variable p in the half-plane $\text{Re}(p) > c$, and further, if $F(p)$ is $0(p^{-k})$,* where k is real and $k > 1$, then the inversion integral

$$f(t) = \frac{1}{2\pi i} \int_{c-i\infty}^{c+i\infty} e^{pt} F(p)\, dp$$

converges to the real function $f(t)$, which is independent of c and whose

---

* By saying $F(p)$ is $0(p^{-k})$, we mean there exists a positive real constant M such that $|p^k F(p| < M$ whenever $|p|$ is sufficiently large.

Laplace transform is $F(p)$ for $\text{Re}(p) > c$. Also, the function $f(t)$ is $0(e^{c_0 t})$ and is continuous everywhere, and $f(t) = 0$ when $t \leq 0$.

The conditions stated in Theor. 4.13 are quite severe in that they exclude, for example, the simple function $1/p$, which is $0(p^{-k})$ where $k = 1$. Also, these conditions are not satisfied by transforms of functions that are discontinuous or for which $f(0) \neq 0$. Nonetheless, by using a Fourier integral theorem and stating conditions on the function $f(t)$ rather than on $F(p)$, these conditions can be relaxed so that the inversion integral formula is valid in nearly all practical cases of interest to us.* We do note that the conditions stated in Theor. 4.13 ensure the existence of the inverse Laplace transform of $F(p)$, and moreover, ensure that the inverse Laplace transform $f(t)$ is that function for which $\mathscr{L}\{f(t);p\} = F(p)$.

In order to better understand the conditions stated in Theor. 4.13, let us consider the function $F(p) = e^{p^2}$, which is analytic everywhere in the complex plane. Therefore, integration of $e^{pt}e^{p^2}$ along the imaginary axis $p = iy$ yields

$$\frac{1}{2\pi i}\int_{-i\infty}^{i\infty} e^{pt}e^{p^2}\,dp = \frac{1}{2\pi}\int_{-\infty}^{\infty} e^{ity}\,e^{-y^2}\,dy$$

The integral on the right can be interpreted as a Fourier transform of the function e^{-y^2}, and based on the result of Exam. 2.6 in Chap. 2, we deduce that

$$\begin{aligned}\frac{1}{2\pi i}\int_{-i\infty}^{i\infty} e^{pt}e^{p^2}\,dp &= \frac{1}{\sqrt{2\pi}}\,\mathscr{F}\{e^{-y^2};t\} \\ &= \frac{1}{2\sqrt{\pi}}\,e^{-t^2/4} \end{aligned} \tag{4.42}$$

However, the Laplace transform of this last function is (see Prob. 23 in Exer. 4.2)

$$\mathscr{L}\left\{\frac{1}{2\sqrt{\pi}}\,e^{-t^2/4};p\right\} = \frac{1}{2}\,e^{p^2}\,\text{erfc}(p) \neq F(p) \tag{4.43}$$

What we are illustrating here is that the inversion integral of $F(p)$ may exist without representing the function $f(t)$ whose Laplace transform is $F(p)$. The reason it happens in this particular case is because the function $F(p) = e^{p^2}$ is *not* $0(p^{-k})$ for any positive k. Thus we have violated a condition of Theor. 4.13.

---

* See, for instance, R. V. Churchill, *Operational Mathematics*, New York: McGraw-Hill, 1972, Chap. 6.

The complex inversion integral for the inverse Laplace transform can often be evaluated quite readily through use of the theory of residues. Let us suppose that $F(p)$ is an analytic function in the complex p-plane, except for a finite number of *isolated singularities* $a_1, a_2, \ldots, a_N$. By integrating the function

$$e^{pt} F(p)$$

around the closed contour shown in Fig. 4.6, where c and R are selected such that no singularity a_k lies to the right of $\text{Re}(p) = c$ and the radius R of the circular arc C_R is large enough to enclose all singularities of $F(p)$, we have the result

$$\oint_C e^{pt}F(p)dp = \int_A^B e^{pt}F(p)dp + \int_{C_R} e^{pt}F(p)dp$$

$$= 2\pi i \sum_{k=1}^{N} \text{Res}\{e^{pt}F(p);a_k\} \quad (4.44)$$

The integral along C_R can be split into three parts, i.e.,

$$\int_{C_R} e^{pt}F(p)dp = \int_{BJ} e^{pt} F(p)dp + \int_{JKL} e^{pt} F(p)dp$$

$$+ \int_{LA} e^{pt} F(p)dp \quad (4.45)$$

Based upon Theorem A.6 in Appendix A, it follows that

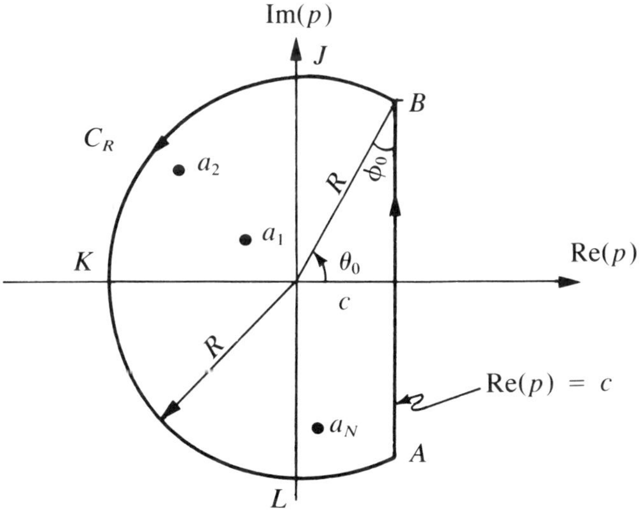

Figure 4.6 Contour of integration

$$\lim_{R\to\infty} \int_{JKL} e^{pt} F(p)dp = 0 \tag{4.46}$$

However, since the arcs BJ and LA do not lie in the second and/or third quadrants, the vanishing of the remaining two integrals in (4.45) in the limit must be treated separately. Under the assumption of Theor. 4.13, $F(p)$ is of order $0(p^{-k})$ and therefore it follows that there exists a constant M such that

$$|F(p)| < M/R^k, \qquad M > 0,\, k > 1 \text{ on } C_R$$

We now consider

$$\begin{aligned} |I_1| &= \left| \int_{BJ} e^{pt} F(p)dp \right| \\ &\le \int_{BJ} |e^{pt}|\, |F(p)|\, |dp| \\ &\le (M/R^k) \int_{BJ} |e^{pt}|\, |dp| \end{aligned}$$

Along the arc BJ we set $p = Re^{i\theta}$, $\theta_0 \le \theta \le \pi/2$, so that

$$\begin{aligned} |e^{pt}| &= |e^{Rt(\cos\theta + i\sin\theta)}| = e^{Rt\cos\theta} \\ |dp| &= Rd\theta \end{aligned}$$

and therefore

$$\begin{aligned} |I_1| &\le \frac{M}{R^{k-1}} \int_{\theta_0}^{\pi/2} e^{Rt\cos\theta}\, d\theta \\ &\le \frac{M}{R^{k-1}} \int_0^{\phi_0} e^{Rt\sin\phi}\, d\phi \end{aligned}$$

where $\phi = \pi/2 - \theta$ and $\phi_0 = \sin^{-1}(c/R)$. Using the obvious inequality

$$\sin\phi \le \sin\phi_0 = c/R$$

we deduce that

$$\begin{aligned} |I_1| &\le \frac{M}{R^{k-1}} \int_0^{\phi_0} e^{ct}\, d\phi \\ &\le \frac{M}{R^{k-1}} e^{ct} \sin^{-1}(c/R) \end{aligned}$$

As $R \to \infty$, we can use the approximation

$$\sin^{-1}(c/R) \simeq c/R, \qquad R >> c$$

to find

$$\lim_{R\to\infty} |I_1| \le \lim_{R\to\infty} (Mc/R^k)\, e^{ct} = 0 \tag{4.47}$$

Hence, the integral along the arc BJ vanishes in the limit $R \to \infty$. Similar arguments can be used to show that the integral in (4.45) along LA also vanishes as $R \to \infty$. Based on the above results, we deduce that

$$\lim_{R\to\infty} \int_{C_R} e^{pt} F(p)\, dp = 0 \tag{4.48}$$

and Eq. (4.44) reduces to

$$f(t) = \mathscr{L}^{-1}\{F(p);t\} = \sum_{k=1}^{N} \text{Res}\{e^{pt}F(p);a_k\} \tag{4.49}$$

If $F(p) = R(p)/Q(p)$, e.g., when both $R(p)$ and $Q(p)$ are polynomials, then Eq. (4.49) can be expressed in more explicit forms. For instance, if $Q(p)$ has simple zeros at $p = a_1, a_2, \ldots, a_N$, then (4.49) becomes

$$f(t) = \mathscr{L}^{-1}\left\{\frac{R(p)}{Q(p)};t\right\} = \sum_{k=1}^{N} \frac{R(a_k)}{Q'(a_k)}\, e^{a_k t} \tag{4.50}$$

This last result is due essentially to Heaviside and is widely known as the *Heaviside expansion theorem*. Other similar expansion theorems involving higher-order poles can also be readily obtained.

Example 4.24: Find the inverse Laplace transform of $p/(p^2 + a^2)$.

Solution: Although the inverse transform of this function is familiar to us, we wish to calculate it again using (4.50). Here we see that $Q(p) = p^2 + a^2$ has simple zeros at $p = \pm ia$. Also, $R(p)/Q'(p) = 1/2$ and hence,

$$\begin{aligned}\mathscr{L}^{-1}\left\{\frac{p}{p^2+a^2};t\right\} &= \text{Res}\left\{\frac{pe^{pt}}{p^2+a^2};ia\right\} + \text{Res}\left\{\frac{pe^{pt}}{p^2+a^2};-ia\right\}\\ &= \tfrac{1}{2}(e^{iat} + e^{-iat})\\ &= \cos at\end{aligned}$$

We can formally extend the result (4.49) to the case in which $F(p)$ has infinitely many isolated singularities by allowing N to become infinite. In this case, we obtain

$$f(t) = \mathscr{L}^{-1}\{F(p);t\} = \sum_{k=1}^{\infty} \text{Res}\{e^{pt}F(p);a_k\} \tag{4.51}$$

Example 4.25: Find the inverse Laplace transform of

$$F(p) = \frac{\cosh x\sqrt{p}}{p \cosh \sqrt{p}}, \quad 0 < x < 1$$

Solution: At first it might appear that $p = 0$ is a branch point of $F(p)$ because of the presence of $\sqrt{p}$. That this is not the case can be seen by writing

$$F(p) = \frac{\cosh x\sqrt{p}}{p\cosh\sqrt{p}} = \frac{1 + (x\sqrt{p})^2/2! + (x\sqrt{p})^4/4! + \dots}{p[1 + (\sqrt{p})^2/2! + (\sqrt{p})^4/4! + \dots]}$$

$$= \frac{1 + x^2p/2! + x^4p^2/4! + \dots}{p(1 + p/2! + p^2/4! + \dots)}$$

from which it is clear that $p = 0$ is not a branch point but a simple pole of $F(p)$. In addition, there are infinitely many other simple poles given by the roots of the transcendental equation

$$\cosh\sqrt{p} = 0$$

The solutions of this equation are given by

$$\sqrt{p} = (n - 1/2)\pi i,\ n = 1,2,3,\dots$$

or

$$p = a_n = -(n - 1/2)^2\pi^2,\ n = 1,2,3,\dots$$

Calculating the residues at $p = 0$ and $p = a_n$ yields

$$\text{Res}\{e^{pt}F(p);0\} = \left.\frac{\cosh x\sqrt{p}}{\cosh\sqrt{p}}\right|_{p=0} = 1$$

and

$$\text{Res}\{e^{pt}F(p);a_n\} = \left.\frac{e^{pt}\cosh x\sqrt{p}}{\cosh\sqrt{p} + \frac{1}{2}\sqrt{p}\sinh\sqrt{p}}\right|_{p=a_n}$$

$$= \frac{4(-1)^n}{\pi(2n-1)}\cos[(n - 1/2)\pi x]\, e^{-(n-1/2)^2\pi^2 t}$$

Hence, we deduce that

$$\mathcal{L}^{-1}\left\{\frac{\cosh x\sqrt{p}}{p\cosh\sqrt{p}};t\right\}$$

$$= 1 + \frac{4}{\pi}\sum_{n=1}^{\infty}\frac{(-1)^n}{(2n-1)}\cos[(n - 1/2)\pi x]e^{-(n-1/2)^2\pi^2 t}$$

4.6.1 *Multivalued Functions*

The method we have presented for evaluating inverse Laplace transforms when $F(p)$ has isolated singularities can be modified to include the case when $F(p)$ has branch points. We do this by deforming the contour of

Fig. 4.6 to exclude the branch point(s) and then take the limit as the radius of the small circle around the branch point(s) tends to zero. Let us illustrate the procedure with an example.

Example 4.26: Find the inverse Laplace transform of $F(p) = e^{-a\sqrt{p}}$, $a > 0$.

Solution: The function $F(p)$ has a branch point at the origin $p = 0$. Hence we integrate the function

$$e^{pt}\,F(p)$$

around the contour shown in Fig. 4.7. Along C_2 we write

$$p = xe^{i\pi}$$

where x is real and positive, and on C_3 we write

$$p = xe^{-i\pi}$$

Therefore, from the residue calculus, we find

$$\int_A^B e^{pt-a\sqrt{p}}\,dp + \int_{C_1} e^{pt-a\sqrt{p}}\,dp - \int_R^\rho e^{-xt-ia\sqrt{x}}\,dx$$

$$+ \int_{C_\rho} e^{pt-a\sqrt{p}}\,dp - \int_\rho^R e^{-xt+ia\sqrt{x}}\,dx + \int_{C_4} e^{pt-a\sqrt{p}}\,dp = 0$$

It can be shown that the integrals along C_1, C_ρ, and C_4 all tend to zero as $\rho \to 0$ and $R \to \infty$. Hence, it follows that

$$\mathscr{L}^{-1}\{e^{-a\sqrt{p}};t\} = \frac{1}{2\pi i}\int_0^\infty e^{-xt+ia\sqrt{x}}\,dx - \frac{1}{2\pi i}\int_0^\infty e^{-xt-ia\sqrt{x}}\,dx$$

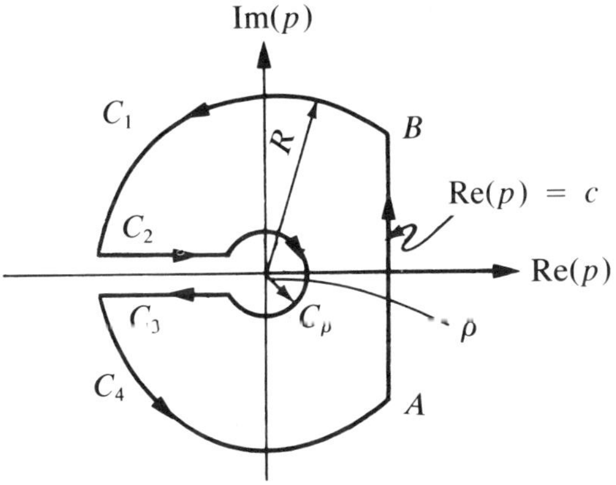

Figure 4.7 Contour of integration

$$= \frac{1}{\pi} \int_0^\infty e^{-xt} \sin(a\sqrt{x})\, dx$$

The change of variable $u = x^2$ further leads to

$$\mathscr{L}^{-1}\{e^{-a\sqrt{p}};t\} = \frac{2}{\pi} \int_0^\infty u e^{-u^2 t} \sin au\, du$$

$$= \sqrt{2/\pi}\, \mathscr{F}_S\{ue^{-u^2 t};a\}$$

From Exam. 2.6 in Chap. 2 we know that

$$\mathscr{F}\{e^{-u^2 t};a\} = (2t)^{-1/2} e^{-a^2/4t}$$

and thus

$$\mathscr{F}\{ue^{-u^2 t};a\} = -i \frac{d}{da}\left[(2t)^{-1/2} e^{-a^2/4t}\right]$$

$$= ia(2t)^{-3/2} e^{-a^2/4t}$$

However, for $a > 0$ the Fourier and Fourier sine transforms are related by

$$\mathscr{F}\{ue^{-u^2 t};a\} = i\, \mathscr{F}_S\{ue^{-u^2 t};a\}$$

and so we finally deduce the intended result*

$$\mathscr{L}^{-1}\{e^{-a\sqrt{p}};t\} = a(4\pi t^3)^{-1/2} e^{-a^2/4t}, \qquad a > 0$$

EXERCISES 4.6

In Probs. 1–12, use the complex inversion formula to evaluate the inverse Laplace transform of the given function having isolated singularities.

1. $F(p) = \dfrac{1}{(p+1)(p-3)^2}$

2. $F(p) = \dfrac{p}{(p+1)^3(p-1)^2}$

3. $F(p) = \dfrac{p^2}{(p^2+4)^2}$

4. $F(p) = \dfrac{p^2}{p^4+4}$

5. $F(p) = \dfrac{1}{p^2 \cosh p}$

6. $F(p) = \dfrac{\cosh xp}{p \sinh p}, \quad 0 < x < 1$

7. $F(p) = \dfrac{\sinh x\sqrt{p}}{\sinh \sqrt{p}},$
$0 < x < 1$

8. $F(p) = \dfrac{\sinh x\sqrt{p}}{\sqrt{p} \cosh \sqrt{p}},$
$0 < x < 1$

---

* See also Prob. 22(a) for another derivation of the Fourier sine transform $\mathscr{F}_S\{ue^{-u^2 t};a\}$.

9. $F(p) = \dfrac{1}{p(e^p + 1)}$

10. $F(p) = \dfrac{\sinh xp}{p^2 \cosh p}, \quad 0 < x < 1$

11. $F(p) = \dfrac{1}{p^2} - \dfrac{1}{p \sinh p}$

12. $F(p) = \dfrac{1}{p \cosh \sqrt{p}}$

13. Show that

$$\mathscr{L}^{-1}\left\{\frac{I_0(r\sqrt{p})}{I_0(b\sqrt{p})};t\right\} = \frac{2}{b}\sum_{n=1}^{\infty}\frac{k_n J_0(k_n r)}{J_1(k_n b)} e^{-k_n^2 t}$$

where the k_n are solutions of $J_0(k_n b) = 0$, $n = 1,2,3,\ldots$

14. Using the result of Exam. 4.26,

(a) determine the inverse Laplace transform of $(1/p)e^{-a\sqrt{p}}$ by way of the convolution theorem.

(b) Differentiate the result in (a) with respect to a to deduce that

$$\mathscr{L}^{-1}\left\{\frac{1}{\sqrt{p}} e^{-a\sqrt{p}};t\right\} = \frac{1}{\sqrt{\pi t}} e^{-a^2/4t}, \qquad a > 0$$

In Probs. 15–20, use the complex inversion formula to evaluate the inverse Laplace transform of the given function with branch points.

15. $F(p) = \dfrac{1}{\sqrt{p - a}}$

16. $F(p) = \dfrac{1}{p\sqrt{p + a^2}}$

17. $F(p) = \dfrac{1}{\sqrt{p}(p - a^2)}$

18. $F(p) = \dfrac{1}{\sqrt{p} + a}$

19. $F(p) = \log(1 + 1/p)$

20. $F(p) = \log(1 + 1/p^2)$

21. Considering the integral

$$I(t) = \int_0^\infty e^{-bu^2} \cos tu \, du, \qquad t \geq 0,\ b > 0$$

as a function of the variable t,

(a) express $\cos tu$ in a Maclaurin series and perform termwise integration to find a series representation for $I(t)$.

(b) Sum the series in (a) to deduce the result

$$I(t) - \frac{1}{2}\sqrt{\frac{\pi}{b}}\, e^{-t^2/4b}$$

22. Using the result of Prob. 21 and

(a) differentiating with respect to t, show that

$$\int_0^\infty u e^{-t^2u^2/2} \sin au \, du = \frac{a}{t}\sqrt{\frac{\pi}{2t}}\, e^{-a^2/2t}, \qquad t > 0$$

(b) integrating with respect to t, show that

$$\int_0^\infty (1/u)e^{-tu^2} \sin au \, du = (\pi/2) \operatorname{erf}(a/2\sqrt{t})$$

23. Using contour integration around the branch point $p = 0$, show that

(a) $\mathscr{L}^{-1}\left\{\frac{1}{p} e^{-a\sqrt{p}};t\right\} = 1 - \frac{1}{\pi}\int_0^\infty \frac{1}{x} e^{-xt} \sin a\sqrt{x} \, dx, \qquad a > 0$

(b) Use the result of Prob. 22(b) to arrive at

$$\mathscr{L}^{-1}\left\{\frac{1}{p} e^{-a\sqrt{p}};t\right\} = \operatorname{erfc}(a/2\sqrt{t}), \qquad a > 0$$

24. From the defining integral of the Laplace transform, show that

(a) $\mathscr{L}\left\{\frac{1}{\sqrt{\pi t}} + e^t \operatorname{erf}(\sqrt{t});p\right\} = \frac{\sqrt{p}}{p - 1}$

(b) Apply the complex inversion formula to $\sqrt{p}/(p - 1)$ and use the result in (a) to deduce that

$$\mathscr{L}\left\{\frac{\sqrt{t}}{t + 1};p\right\} = \sqrt{\frac{\pi}{p}} - \pi e^p \operatorname{erfc}(\sqrt{p})$$

(c) Finally, establish the relation

$$\mathscr{L}\left\{\frac{\sqrt{t}}{t + 1};p\right\} = \sqrt{\frac{\pi}{p}} - \mathscr{L}\left\{\frac{1}{\sqrt{t}(t + 1)};p\right\}$$

and thus conclude that

$$\mathscr{L}\left\{\frac{1}{\sqrt{t}(t + 1)};p\right\} = \pi e^p \operatorname{erfc}(\sqrt{p})$$

Hint: Let $t = u^2$ in the transform integral.

4.7 *Additional Topics*

In earlier sections we have found that the transform function $F(p)$ has certain useful properties, most of which can be established by relying on the Laplace transform integral for the definition of $F(p)$. For example, under appropriate conditions on the inverse transform $f(t)$, we have found that $F(p)$ is analytic (Theor. 4.2) and that $F(p) \to 0$ as $|p| \to \infty$ (Theor. 4.11). In addition, certain operational properties of this function were developed in Sec. 4.3. Here we will discuss briefly further asymptotic properties of $F(p)$ and extend the definition of the Laplace transform to functions defined on the entire real line rather than only on the positive real line.

4.7.1. *Asymptotic Properties of F(p)*

The behavior of a function $f(t)$ in the neighborhood of the origin is, in some sense, reflected in the behavior of its Laplace transform $F(p)$ as $|p| \to \infty$. For instance, if $f(t)$ is a polynomial of degree k, i.e., if

$$f(t) = a_0 + a_1 t + a_2 t^2 + \ldots + a_k t^k, \qquad k = 1,2,3,\ldots$$

then its Laplace transform is

$$F(p) = \frac{a_0}{p} + \frac{a_1}{p^2} + \frac{2a_2}{p^3} + \ldots + \frac{k!a_k}{p^{k+1}}$$

From these equations, we immediately deduce that

$$\lim_{|p|\to\infty} pF(p) = a_0 = f(0) \tag{4.52}$$

Not only is this particular property true of all polynomials, but applies to a relatively large class of functions.

Theorem 4.14 (*Initial Value Theorem*). If $f(t)$ is piecewise continuous on $t \geq 0$ and is $0(e^{c_0 t})$, then its Laplace transform $F(p)$ satisfies

$$\lim_{|p|\to\infty} pF(p) = \lim_{t\to 0} f(t) = f(0)$$

Proof: We will present the proof only for the stronger case where $f(t)$ is *continuous* on $t \geq 0$.

By Eq. (4.11) in Sec. 4.3,

$$\mathscr{L}\{f'(t);p\} = \int_0^\infty e^{-pt} f'(t)\,dt = pF(p) - f(0)$$

If $f'(t)$ is piecewise continuous and of exponential order, then its Laplace transform satisfies (Theor. 4.11)

$$\lim_{|p|\to\infty} \int_0^\infty e^{-pt} f'(t)\,dt = 0$$

Hence, we deduce that

$$\lim_{|p|\to\infty} [pF(p) - f(0)] = 0$$

or

$$\lim_{|p|\to\infty} pF(p) = \lim_{t\to 0} f(t) = f(0) \qquad \blacksquare$$

A related result is given in the next theorem.

Theorem 4.15 (***Final Value Theorem***). If $f(t)$ is piecewise continuous on $t \geq 0$ and is $0(e^{c_0 t})$, and if the integral

$$\int_0^\infty f'(t)\,dt$$

exists, then the transform $F(p)$ satisfies

$$\lim_{p\to 0} pF(p) = \lim_{t\to\infty} f(t) = f(\infty)$$

Proof: As in the proof of Theor. 4.14, we will present the proof here only for the case when $f(t)$ is continuous on $t \geq 0$.

Again we start with the relation

$$\mathscr{L}\{f'(t);p\} = \int_0^\infty e^{-pt} f'(t)\,dt = pF(p) - f(0)$$

Noting that

$$\begin{aligned}\lim_{p\to 0} \int_0^\infty e^{-pt} f'(t)\,dt &= \int_0^\infty f'(t)\,dt \\ &= \lim_{t\to\infty} \int_0^t f'(u)\,du \\ &= \lim_{t\to\infty} f(t) - f(0)\end{aligned}$$

we deduce that

$$\lim_{t\to\infty} f(t) - f(0) = \lim_{p\to 0} pF(p) - f(0)$$

or, upon simplification,

$$\lim_{p\to 0} pF(p) = f(\infty)$$

■

Example 4.27: Find the Laplace transform of the cosine integral $\mathrm{Ci}(t)$, defined by

$$\mathrm{Ci}(t) = \int_\infty^t \frac{\cos u}{u}\,du, \qquad t > 0$$

Solution: Although this transform can be found directly from Theors. 4.7 and 4.9, we wish to illustrate another technique using the final value theorem (Theor. 4.15).

If we set $f(t) = \mathrm{Ci}(t)$, then $f'(t) = (\cos t)/t$ and thus

$$tf'(t) = \cos t$$

Taking the Laplace transform, using Theorem 4.8 on the left-hand side, we obtain

$$\frac{d}{dp}[pF(p) - f(0)] = \frac{p}{p^2 + 1}$$

and upon integration, this leads to

$$pF(p) - f(0) = \int \frac{p}{p^2 + 1}\,dp$$

or

$$pF(p) = (1/2)\log(1 + p^2) + C$$

where $f(0)$ has been absorbed in the constant of integration to give us the constant C. By the final value theorem,

$$\lim_{p\to 0} pF(p) = \lim_{t\to\infty} f(t) = 0$$

Thus, we deduce that $C = 0$ and consequently,

$$\mathscr{L}\{\mathrm{Ci}(t);p\} = F(p) = (1/2p)\log(1 + p^2)$$

A result related to Theors. 4.14 and 4.15 is *Watson's lemma* below, which defines an asymptotic expansion of $F(p)$ ($|p| \to \infty$) for a certain class of functions $f(t)$.

Theorem 4.16 (*Watson's lemma*). If $f(t)$ is $0(e^{c_0 t})$ and if, in some neighborhood of $t = 0$, the function $f(t)$ has the Maclaurin series expansion

$$f(t) = \sum_{n=0}^{\infty} \frac{a_n}{n!} t^n, \qquad |t| < R$$

then the transform function $F(p)$ has the asymptotic series

$$F(p) \sim \sum_{n=0}^{\infty} \frac{a_n}{p^{n+1}}, \qquad |p| \to \infty$$

The result of Watson's lemma follows formally by applying the Laplace transform termwise to the Maclaurin series expansion of $f(t)$.* Let us illustrate the use of this theorem in the following example.

Example 4.28: Find an asymptotic expansion for the complementary error function $\mathrm{erfc}(x)$.

---

* For a rigorous proof of Watson's lemma, see I. N. Sneddon, *The Use of Integral Transforms*, New York: McGraw-Hill, 1972, pp. 188–190.

Solution: In Exam. 4.11 in Sec. 4.4, we have shown that

$$\mathscr{L}\{\mathrm{erf}(t);p\} = (1/p)e^{p^2/4}\,\mathrm{erfc}(p/2)$$

Now the error function has the Maclaurin series expansion [see Eq. (1.20) in Sec. 1.3]

$$\mathrm{erf}(t) = \frac{2}{\sqrt{\pi}}\sum_{n=0}^{\infty}\frac{(-1)^n t^{2n+1}}{n!(2n+1)}$$

so based upon Watson's lemma, we deduce that

$$\frac{1}{p}e^{p^2/4}\,\mathrm{erfc}(p/2) \sim \frac{2}{\sqrt{\pi}}\sum_{n=0}^{\infty}\frac{(-1)^n(2n)!}{n!p^{2n+2}}$$

$$\sim \frac{1}{\pi p}\sum_{n=0}^{\infty}\frac{(-1)^n\,\Gamma(n+1/2)}{(p/2)^{2n+1}}$$

where we have used the duplication formula of the gamma function. Finally, setting $x = p/2$, we get the desired asymptotic expansion

$$\mathrm{erfc}(x) \sim \frac{e^{-x^2}}{\pi}\sum_{n=0}^{\infty}\frac{(-1)^n\,\Gamma(n+1/2)}{x^{2n+1}}, \qquad |x| \to \infty$$

4.7.2 *Two-Sided Laplace Transform*

If the function $f(t)$ is defined over the entire real axis, we may consider the integral

$$\mathscr{L}_+\{f(t);p\} = \int_{-\infty}^{\infty} e^{-pt}f(t)\,dt = F_+(p) \tag{4.53}$$

known as the *two-sided Laplace transform* or *bilateral Laplace transform*. For pure imaginary p, say $p = is$, we see that (4.53) becomes

$$F_+(is) = \int_{-\infty}^{\infty} e^{-ist}f(t)\,dt \tag{4.54}$$

which is a multiple of our definition for the Fourier transform of $f(t)$. Thus, in a formal sense, there is a relation between the two-sided Laplace transform and the Fourier transform.

If the integral in (4.53) converges, it will usually do so only for restricted values of p in a vertical strip of the complex p-plane. To better understand this, let us rewrite the integral in two parts as

$$\int_{-\infty}^{\infty} e^{-pt}f(t)\,dt = \int_{-\infty}^{0} e^{-pt}f(t)\,dt + \int_{0}^{\infty} e^{-pt}f(t)\,dt$$

$$= \int_{0}^{\infty} e^{+pt}f(-t)\,dt + \int_{0}^{\infty} e^{-pt}f(t)\,dt$$

We see, therefore, that the two-sided Laplace transform is simply a sum of two ordinary Laplace transforms, i.e.,

$$\mathscr{L}_+\{f(t);p\} = \mathscr{L}\{f(-t);-p\} + \mathscr{L}\{f(t);p\} \tag{4.55}$$

The second transform in (4.55) exists in the plane $\text{Re}(p) > c_1$, where c_1 is the abscissa of convergence. Similarly, the first transform will exist in the plane $\text{Re}(-p) > -c_2$, or $\text{Re}(p) < c_2$, where $-c_2$ is the abscissa of convergence in this case. The two-sided Laplace transform of $f(t)$ exists only if these two half-planes overlap. That is, it will exist only in the strip $c_1 < \text{Re}(p) < c_2$. If $c_2 < c_1$, the two-sided Laplace transform does not exist, and if $c_2 = c_1$, the strip contracts to the vertical line $\text{Re}(p) = c_1$. Finally, if $c_2 = c_1 = 0$, the two-sided Laplace transform is then actually a Fourier transform [see Eq. (4.54)].

Example 4.29: Find the two-sided Laplace transform of e^{-t^2}.

Solution: Here $f(t) = f(-t) = e^{-t^2}$, for which $c_1 = -\infty$ and $c_2 = \infty$. Thus the two-sided Laplace transform exists for all values of $\text{Re}(p)$.

By definition,

$$\begin{aligned}\mathscr{L}_+\{e^{-t^2};p\} &= \int_{-\infty}^{\infty} e^{-pt}\, e^{-t^2}\, dt \\ &= e^{p^2/4} \int_{-\infty}^{\infty} e^{-(t+p/2)^2}\, dt\end{aligned}$$

from which we deduce

$$\mathscr{L}_+\{e^{-t^2};p\} = \sqrt{\pi}\, e^{p^2/4}$$

The inversion theorem for the two-sided Laplace transform has the same form as for the ordinary Laplace transform, except that for the two-sided transform the interval of convergence must also be established in order to uniquely establish the inverse transform. For example, consider the two functions

$$f(t) = \begin{cases} 0, & t<0 \\ e^{-2t} - e^{-t}, & t>0 \end{cases}$$

and

$$g(t) = \begin{cases} e^{-t}, & t<0 \\ e^{-2t}, & t>0 \end{cases}$$

whose two-sided Laplace transforms are, respectively,

$$F_+(p) = \frac{1}{p+2} - \frac{1}{p+1}, \qquad \text{Re}(p) > -1$$

and

$$G_+(p) = \frac{1}{p+2} - \frac{1}{p+1}, \qquad -2 < \text{Re}(p) < -1$$

Except for the strip of convergence specified in each case, these are identical transforms. This means that to use the inversion formula to find the inverse transforms of $F_+(p)$ and $G_+(p)$, we must set up contours over different vertical strips in the complex p-plane for each case. Specifically, these particular inversion formulas are

$$f(t) = \frac{1}{2\pi i}\int_{c-i\infty}^{c+i\infty} e^{pt}\, F_+(p)\, dp, \qquad c > -1$$

and

$$g(t) = \frac{1}{2\pi i}\int_{c-i\infty}^{c+i\infty} e^{pt}\, G_+(p)\, dp, \qquad -2 < c < -1$$

For a more detailed discussion of the two-sided Laplace transform, the reader is advised to consult W. R. LePage, *Complex Variables and the Laplace Transform for Engineers*, New York: Dover, 1980, or D. V. Widder, *The Laplace Transform*, Princeton: Princeton University Press, 1941.

EXERCISES 4.7

1. Verify the initial value theorem for the following functions:
(a) $f(t) = 5 + 4\cos 2t$
(b) $f(t) = (3t - 2)^3$
(c) $f(t) = \text{erf}(t)$

2. Verify the final value theorem for the following functions:
(a) $f(t) = 3 + e^{-2t}(\cos t + \sin t)$
(b) $f(t) = 1 + e^{-t^2}$
(c) $f(t) = \text{erfc}(1/\sqrt{t})$

In Probs. 3 and 4, use the technique of Exam. 4.27 to find the Laplace transform of the given function.

3. $\text{Si}(t) = \int_0^t \frac{\sin u}{u}\, du, \qquad t > 0$ **4.** $\text{Ei}(t) = \int_t^\infty \frac{e^{-u}}{u}\, du, \qquad t > 0$

5. Use the final value theorem to evaluate the following limit:

$$\lim_{t\to\infty} t^{\nu/2} J_\nu(2\sqrt{t})$$

6. If $f(t)$ and $f'(t)$ are continuous functions on $t \geq 0$, $f''(t)$ is piecewise continuous on $t \geq 0$, and all three functions are of exponential order c_0, show that $f(0) = 0$ and

$$\lim_{|p|\to\infty} p^2F(p) = f'(0)$$

7. Given the function

$$F(p) = \int_0^\infty \frac{e^{-px}}{1 + x^2}\, dx$$

(a) show that its inverse Laplace transform leads to the Maclaurin expansion

$$f(t) = \sum_{n=0}^{\infty} (-1)^n t^{2n}, \quad |t| < 1$$

(b) Use Watson's lemma to deduce the asymptotic expansion

$$F(p) \sim \sum_{n=0}^{\infty} \frac{(-1)^n(2n)!}{p^{2n+1}}, \qquad |p| \to \infty$$

In Probs. 8–10, use the technique of Prob. 7 to derive the given asymptotic formula.

8. $$\int_0^\infty \frac{e^{-px}}{1 + x}\, dx \sim \sum_{n=0}^{\infty} \frac{(-1)^n\, n!}{p^{n+1}}, \qquad |p| \to \infty$$

9. $$\int_0^\infty e^{-px} \cos x\, dx \sim \sum_{n=0}^{\infty} \frac{(-1)^n}{p^{2n+1}}, \qquad |p| \to \infty$$

10. $$\frac{1}{\Gamma(a)} \int_0^\infty e^{-px} x^{a-1} (1 + x)^{c-a-1}\, dx \sim \frac{1}{p^a} \sum_{n=0}^{\infty} \frac{(-1)^n(a)_n(1 + a - c)_n}{n!\, p^n}, \qquad |p| \to \infty$$

where $(a)_n = \Gamma(a + n)/\Gamma(a)$, $\quad n = 0,1,2,\ldots$.

In Probs. 11–14, find the two-sided Laplace transform of the given function and state the strip of convergence.

11. $f(t) = \begin{cases} e^{bt}, & t < 0 \\ e^{at}, & t > 0 \end{cases}$

$a < b$

12. $f(t) = e^{-|t|}$

13. $f(t) = \dfrac{\sin t}{t}$

14. $f(t) = e^{-a^2t^2}$

15. By integrating the function $F_+(p) = e^{p^2/4a^2}$ along the imaginary axis of the complex p-plane, find the inverse transform of this two-sided Laplace transform.

5

Applications Involving Laplace Transforms

5.1 *Introduction*

Like the Fourier transform, the Laplace transform is used in a variety of applications. Perhaps the most common usage of the Laplace transform is in the solution of initial value problems. However, there are other situations for which the properties of the Laplace transform are also very useful, such as in the evaluation of certain integrals and in the solution of certain integral equations. In this chapter we will briefly discuss applications of the Laplace transform in all of the above named areas.

5.2 *Evaluating Integrals*

An interesting application of Laplace transforms involves the evaluation of certain integrals, particularly those containing a free parameter. In some cases we simply recognize the integral as a special case of a Laplace transform for a particular value of the transform variable p. Other integrals may be solved by first taking the Laplace transform of the integrand with respect to a free parameter (not the variable of integration). The resulting integral is hopefully easier to evaluate than the original, and by applying the inverse Laplace transform we obtain our desired result. This latter procedure is direct and often simple, but it requires the interchange of two limit operations, so some caution should be exercised in its usage.

Example 5.1: Evaluate the integrals

$$I = \int_0^\infty \frac{\sin t}{t}\,dt$$

$$J = \int_0^\infty e^{-t}\frac{\sin t}{t}\,dt$$

Solution: Both of these integrals are special cases of known Laplace transform integrals. For example, consider the Laplace transform relation (recall Exam. 4.9 in Chap. 4)

$$\mathscr{L}\left\{\frac{\sin t}{t};p\right\} = \int_0^\infty e^{-pt}\frac{\sin t}{t}dt = \tan^{-1}\frac{1}{p}$$

By setting $p = 0$ and $p = 1$ in this result, we obtain, respectively,

$$I = \mathscr{L}\left\{\frac{\sin t}{t};p=0\right\} = \tan^{-1}\infty = \frac{\pi}{2}$$

$$J = \mathscr{L}\left\{\frac{\sin t}{t};p=1\right\} = \tan^{-1}1 = \frac{\pi}{4}$$

Example 5.2: Evaluate the integral

$$\int_0^\infty \frac{\cos tx}{x^2+1}\,dx \qquad t>0$$

Solution: Let us define the integral by $f(t)$ and take the Laplace transform with respect to t. This action leads to

$$\begin{aligned} F(p) &= \int_0^\infty \frac{p}{(x^2+1)(x^2+p^2)}\,dx \\ &= \frac{p}{p^2-1}\int_0^\infty \left(\frac{1}{x^2+1} - \frac{1}{x^2+p^2}\right)dx \\ &= \frac{p}{p^2-1}\left(\frac{\pi}{2} - \frac{\pi}{2p}\right) \\ &= \frac{\pi/2}{p+1}, \end{aligned}$$

and thus by taking the inverse Laplace transform, we obtain*

$$\mathscr{L}^{-1}\{F(p);t\} = f(t) = (\pi/2)e^{-t}, \quad t>0$$

* We might also recognize the integral $f(t)$ as a multiple of the Fourier cosine transform of the function $1/(x^2+1)$.

EXERCISES 5.2

In Probs. 1–10, use known Laplace transforms or transform properties to evaluate the given integral.

1. $\int_0^\infty te^{-2t}\cos t\,dt$

2. $\int_0^\infty t^2e^{-3t}\sin t\,dt$

3. $\int_{-\infty}^\infty \frac{\sinh t}{t}\,dt$

4. $\int_0^\infty \frac{e^{-3t}-e^{-6t}}{t}\,dt$

5. $\int_0^\infty J_0(t)\,dt$

6. $\int_0^\infty tJ_0(t)\,dt$

7. $\int_0^\infty \frac{\cos 6t - \cos 4t}{t}\,dt$

8. $\int_0^\infty e^{-t}\operatorname{erf}(\sqrt{t})\,dt$

9. $\int_0^\infty xe^{-x^2}J_0(ax)\,dx$

10. $\int_0^\infty xe^{-x^2}\operatorname{erfc}(x)\,dx$

In Probs. 11–16, use the technique illustrated in Exam. 5.2 to evaluate the given integral.

11. $\int_0^\infty e^{-tx^2}\,dx, \qquad t > 0$

12. $\int_0^\infty \frac{x\sin tx}{x^2+1}\,dx, \qquad t > 0$

13. $\int_0^\infty \exp(-x^2 - t^2/x^2)\,dx,$
$t > 0$

14. $\int_0^\infty \frac{\sin tx}{x}\,dx, \qquad t > 0$

15. $\int_0^\infty \frac{\sin tx}{\sqrt{x}}\,dx, \qquad t > 0$

16. $\int_{-\infty}^\infty \cos tx^2\,dx, \qquad t > 0$

In Probs. 17–20, introduce a parameter t somewhere in the integrand and then use the method of Exam. 5.2 to verify the given integral relation.

17. $\int_{-\infty}^\infty \sin x^2\,dx = \pi$

18. $\int_0^\infty x\cos x^3\,dx = \frac{\pi}{3\sqrt{3}\Gamma(1/3)}$

19. $\int_0^\infty \frac{\sin x}{x^\alpha}\,dx = \frac{\pi}{2\Gamma(\alpha)\sin \alpha\pi/2}, \qquad 0 < \alpha < 1$

20. $\int_0^\infty \frac{\cos x}{x^\alpha}\,dx = \frac{\pi}{2\Gamma(\alpha)\cos \alpha\pi/2}, \qquad 0 < \alpha < 1$

5.3 *Solution of ODEs*

In solving linear ordinary differential equations (ODEs) by the Laplace transform method, we first convert the equation in the unknown function $y(t)$ into an equation in $Y(p)$, and then if possible, solve for $Y(p)$. The inversion of $Y(p)$ will then give the solution $y(t)$ of the original ODE. This technique clearly has advantages only if the equation for $Y(p)$ is easier to solve than the equation for $y(t)$, and if $Y(p)$ is also invertable. In the case of an equation with constant coefficients, the transformed equation for $Y(p)$ turns out to be an algebraic one, and the Laplace transform method is therefore a powerful tool for solving this class of ODEs.

Because the appearance of the forms $y(0)$, $y'(0)$, $y''(0)$, and so on, in the transforms of the derivatives of $y(t)$, the Laplace transform method is best suited to *initial value problems,* i.e., those where the auxiliary conditions are all imposed at $t = 0$. Furthermore, the solution arises in the Laplace transform method with the initial conditions automatically built into it, unlike the conventional approach where one constructs the solution by adding a particular integral to the complementary function and then imposes the auxiliary conditions on it.

Let us illustrate the Laplace transform technique on some typical initial value problems.

Example 5.3: Solve the initial value problem

$$y'' - 6y' + 9y = t^2e^{3t}, \qquad y(0) = 2, y'(0) = 6$$

Solution: By introducing the Laplace transforms

$$\begin{aligned}
\mathscr{L}\{y(t);p\} &= Y(p) \\
\mathscr{L}\{y'(t);p\} &= pY(p) - y(0) \\
&= pY(p) - 2 \\
\mathscr{L}\{y''(t);p\} &= p^2Y(p) - py(0) - y'(0) \\
&= p^2Y(p) - 2p - 6
\end{aligned}$$

and

$$\mathscr{L}\{t^2e^{3t};p\} = 2!/(p - 3)^3$$

we find that the given initial value problem is transformed into the algebraic equation

$$[p^2Y(p) - 2p - 6] - 6[pY(p) - 2] + 9Y(p) = 2/(p - 3)^3$$

or

$$(p^2 - 6p + 9)Y(p) = 2(p - 3) + 2/(p - 3)^3$$

Solving for $Y(p)$, we obtain

$$Y(p) = \frac{2}{p-3} + \frac{2}{(p-3)^5}$$

and by taking the inverse Laplace transform, we have

$$y(t) = \mathscr{L}^{-1}\{Y(p);t\} = \mathscr{L}^{-1}\left\{\frac{2}{p-3};t\right\} + \mathscr{L}^{-1}\left\{\frac{2}{(p-3)^5};t\right\}$$

from which we deduce

$$y(t) = 2e^{3t} + (1/12)t^4e^{3t}$$

Example 5.4: Solve the initial value problem

$$y'' + 4y = f(t), \qquad y(0) = 1,\ y'(0) = 0$$

where

$$f(t) = \begin{cases} 4t, & 0 \le t \le 1 \\ 4, & t > 1 \end{cases}$$

Solution: By first writing the forcing function $f(t)$ in terms of the Heaviside unit function, we have

$$\begin{aligned} f(t) &= 4t[1 - h(t-1)] + 4h(t-1) \\ &= 4t - 4(t-1)h(t-1) \end{aligned}$$

the Laplace transform of which is (recall Theor. 4.5)

$$\mathscr{L}\{f(t);p\} = \frac{4}{p^2} - \frac{4}{p^2}e^{-p}$$

In this case the transformed initial value problem leads to

$$[p^2Y(p) - p] + 4Y(p) = \frac{4}{p^2} - \frac{4}{p^2}e^{-p}$$

with solution

$$Y(p) = \frac{p}{p^2+4} + \frac{4}{p^2(p^2+4)} - \frac{4}{p^2(p^2+4)}e^{-p}$$

Using a partial fraction expansion on the last two terms on the right-hand side of the above expression, we obtain

$$Y(p) = \frac{p}{p^2+4} + \frac{1}{p^2} - \frac{1}{p^2+4} - \left(\frac{1}{p^2} - \frac{1}{p^2+4}\right)e^{-p}$$

the inversion of which yields

$$y(t) = \cos 2t + t - \tfrac{1}{2}\sin 2t - [(t-1) - \tfrac{1}{2}\sin 2(t-1)]h(t-1)$$

The above examples illustrate the basic procedure used in the method of Laplace transforms. And although these problems can be solved by other techniques, the Laplace transform method offers the advantage of solving the problem directly without first producing the general solution of the DE. Moreover, in the case of Exam. 5.4 we found the solution without splitting the problem into two problems, one over each interval where $f(t)$ is defined, as required by more conventional methods. Forcing functions of this nature, as well as discontinuous or impulsive ones, are commonplace in circuit analysis problems and in certain problems involving mechanical vibrations.

While in general the Laplace transform method works best on constant-coefficient equations, there are some variable-coefficient equations which also lend themselves to the transform method. Consider the following examples.

Example 5.5: Solve the initial value problem

$$ty'' + y' + ty = 0, \qquad y(0) = 1, y'(0) = 0$$

Solution: Upon taking the Laplace transform of the given equation, we find

$$-\frac{d}{dp}[p^2Y(p) - p] + [pY(p) - 1] - \frac{d}{dp}Y(p) = 0$$

which reduces to

$$(p^2 + 1)\frac{dY}{dp} + pY = 0$$

Here we see that the transformed problem is another DE rather than an algebraic equation. However, since the largest power of t occurring in the given equation is unity, the transformed equation is a first-order DE whereas the original DE was second order. The general solution of this first-order linear DE is readily found to be

$$Y(p) = A/\sqrt{p^2 + 1}$$

where A is an arbitrary constant. The inverse Laplace transform of this result gives us

$$y(t) = AJ_0(t)$$

where $J_0(t)$ is the Bessel function of order zero. The initial conditions require the choice $A = 1$.

Example 5.6: Solve the initial value problem

$$y'' + ty' + y = 0, \qquad y(0) = 1, y'(0) = 0$$

Solution: The transformed problem becomes

$$[p^2y(p) - p] - \frac{d}{dp}[pY(p) - 1] + Y(p) = 0$$

or

$$\frac{dY}{dp} - pY = -1$$

The solution of this first-order linear DE is

$$Y(p) = e^{p^2/2}\left(A - \int e^{-p^2/2}\,dp\right)$$
$$= e^{p^2/2}[A - \sqrt{2\pi}\,\text{erf}(p/\sqrt{2})]$$

where A is an arbitrary constant and erf(x) is the error function. To determine the value of A, we use the initial value theorem (Theor. 4.14) which requires that

$$y(0) = 1 = \lim_{p\to\infty} pY(p)$$

Hence, we find that $A = \sqrt{2\pi}$ and therefore

$$Y(p) = \sqrt{2\pi}\,e^{p^2/2}\,\text{erfc}(p/\sqrt{2})$$

the inversion of which yields (recall Prob. 23 in Exer. 4.2)

$$y(t) = e^{-t^2/2}$$

Notice that in solving variable-coefficient DEs, the solution obtained by the Laplace transform method involved arbitrary constants that had to be resolved by use of the initial conditions. This is quite distinct from solutions of constant-coefficient DEs obtained through the transform method.

5.3.1 *Impulse Response Function*

To better understand the transform method and its relation to standard solution techniques, let us apply the Laplace transform to the general initial value problem

$$y'' + ay' + by = f(t), \qquad y(0) = k_0,\ y'(0) = k_1 \tag{5.1}$$

where a and b are known constants. The coefficient of y'' has been set to unity for mathematical convenience. If we introduce the Laplace transforms $\mathscr{L}\{y(t);p\} = Y(p)$ and $\mathscr{L}\{f(t);p\} = F(p)$, then (5.1) reduces to the algebraic equation

$$[p^2Y(p) - pk_0 - k_1] + a[pY(p) - k_0] + bY(p) = F(p)$$

or

$$(p^2 + ap + b)Y(p) = (p + a)k_0 + k_1 + F(p) \tag{5.2}$$

Solving for $Y(p)$, we have

$$Y(p) = \frac{(p + a)k_0 + k_1}{p^2 + ap + b} + \frac{F(p)}{p^2 + ap + b} \tag{5.3}$$

and by taking the inverse Laplace transform, we obtain

$$y(t) = \underbrace{\mathscr{L}^{-1}\left\{\frac{(p + a)k_0 + k_1}{p^2 + ap + b};t\right\}}_{y_H(t)} + \underbrace{\mathscr{L}^{-1}\left\{\frac{F(p)}{p^2 + ap + b};t\right\}}_{y_P(t)} \tag{5.4}$$

Here it is interesting to observe that the solution (5.4) has naturally split into two parts — the function $y_H(t)$, which is a solution of the initial value problem

$$y'' + ay' + by = 0, \qquad y(0) = k_0,\ y'(0) = k_1 \tag{5.5}$$

and $y_P(t)$, which satisfies

$$y'' + ay' + by = f(t), \qquad y(0) = 0,\ y'(0) = 0 \tag{5.6}$$

We can physically interpret the function $y_H(t)$ as the response of the system described by (5.1) entirely due to the initial conditions in the absence of an external disturbance $f(t)$. On the other hand, the function $y_P(t)$ represents the response of the same system which is at rest until time $t = 0$, at which time it is subject to the external input $f(t)$. By separating the solution in this fashion we can use the Laplace transform as an effective tool for analyzing the basic characteristics of a system in response to each of the input parameters.

In network analysis as well as other areas of application the analyst is often interested only in the system response to an external stimulus when the system is "at rest." That is the part of the solution above that we have designated by $y_P(t)$. If we assume that $y_H(t) = 0$, then we can represent the response of a system to the input $f(t)$ by

$$\begin{aligned} y(t) &= \mathscr{L}^{-1}\left\{\frac{F(p)}{p^2 + ap + b};t\right\} \\ &= \int_0^t g(t - u)f(u)\,du \end{aligned} \tag{5.7}$$

where we have used the convolution theorem and defined

$$g(t) = \mathscr{L}^{-1}\left\{\frac{1}{p^2 + ap + b};t\right\} \tag{5.8}$$

Equation (5.7) represents the response of any linear system, characterized by $y'' + ay' + b$, to the general input $f(t)$. The function $g(t)$ is called the *response function* of the system. Expressed as $g(t - u)$, it is also called the *one-sided Green's function* in much of the literature.*

Example 5.7: Construct the one-sided Green's function for the system described by $y'' - 2y' + 5y = f(t)$.

Solution: By use of Eq. (5.8), we first construct the response function

$$g(t) = \mathscr{L}^{-1}\left\{\frac{1}{p^2 - 2p + 5}; t\right\} = \mathscr{L}^{-1}\left\{\frac{1}{(p - 1)^2 + 4}; t\right\}$$

which yields

$$g(t) = \tfrac{1}{2}e^t \sin 2t$$

Hence, the one-sided Green's function is simply

$$g(t - u) = \tfrac{1}{2}e^{t-u} \sin 2(t - u)$$

If the forcing function to a system is the impulse function $\delta(t)$, then Eq. (5.7) yields

$$y(t) = \int_0^t g(t - u)\delta(u)\, du = g(t) \tag{5.9}$$

Hence, we see that the response function $g(t)$ is actually the response of the system to a unit impulse. For this reason, it is often called the *impulse response function* of the system. According to (5.7), all other solutions for general forcing functions $f(t)$ are simply superpositions of $f(u)$ with the "fundamental solution" $g(t - u)$.

Example 5.8: Use the impulse response function to find a general solution of

$$y'' - 2y' + 5y = f(t), \qquad y(0) = y_0,\ y'(0) = v_0$$

Solution: From Exam. 5.7, we know that the impulse response function is

$$g(t) = \tfrac{1}{2}\, e^t \sin 2t$$

Hence, based on Eq. (5.4) the general solution is given by

---

* See Chap. 2 in L. C. Andrews, *Elementary Partial Differential Equations with Boundary Value Problems,* Orlando: Academic Press, 1986.

$$y(t) = \mathscr{L}^{-1}\left\{\frac{y_0p + (v_0 - 2y_0)}{p^2 - 2p + 5};t\right\} + \frac{1}{2}\int_0^t e^{t-u}\sin 2(t-u)f(u)\,du$$

$$= \mathscr{L}^{-1}\left\{\frac{y_0(p-1) + (v_0 - y_0)}{(p-1)^2 + 4};t\right\} + \frac{1}{2}\int_0^t e^{t-u}\sin 2(t-u)f(u)\,du$$

so that upon taking the inverse Laplace transform, we arrive at the solution

$$y(t) = e^t[y_0\cos 2t + \tfrac{1}{2}(v_0 - y_0)\sin 2t] + \tfrac{1}{2}\int_0^t e^{t-u}\sin 2(t-u)f(u)\,du$$

The above expression represents a solution to the problem for any set of initial conditions and input function $f(t)$.

EXERCISES 5.3

In Probs. 1–15, use the Laplace transform to solve the given initial value problem.

1. $y'' - y = e^t \cos t, \quad y(0) = 0,\ y'(0) = 0$

2. $y'' + 2y' + y = 3te^{-t}, \quad y(0) = 4,\ y'(0) = 2$

3. $y'' - 4y' + 4y = t, \quad y(0) = 1,\ y'(0) = 0$

4. $y'' - 3y' + 2y = 4e^{2t}, \quad y(0) = -3,\ y'(0) = 5$

5. $y'' + 2y' + 5y = e^{-t}\sin t, \quad y(0) = 0,\ y'(0) = 1$

6. $y''' - 3y'' + 3y' - y = t^2e^t, \quad y(0) = 1,\ y'(0) = 0,\ y''(0) = -2$

7. $2y''' + 3y'' - 3y' - 2y = e^{-t}, \quad y(0) = 0,\ y'(0) = 0,\ y''(0) = 1$

8. $y''' - y'' + 4y' - 4y = t, \quad y(0) = 0,\ y'(0) = 0,\ y''(0) = 1$

9. $y'' + 4y = f(t) = \begin{cases} \cos 4t, & 0 \le t \le \pi \\ 0, & t > \pi \end{cases} \quad y(0) = 0,\ y'(0) = 1$

10. $y'' + 4y = \sin t - h(t - 2\pi)\sin(t - 2\pi), \quad y(0) = 0,\ y'(0) = 0$

11. $y'' + ty' - y = 0, \quad y(0) = 0,\ y'(0) = 1$

12. $y'' + aty' - 2ay = 1, \quad y(0) = 0,\ y'(0) = 1$

13. $ty'' + (2t + 3)y' + (t + 3)y = 3e^{-t},\ y(0) = 0$

14. $ty'' + (t - 1)y' + y = 0, \qquad y(0) = 0$

15. $t^2y'' - 2y = 2t, \qquad y(0) = 2$

In Probs. 16–21, use the Laplace transform to construct the impulse response function and the one-sided Green's function for the given differential operator M, where $D = d/dt$.

16. $M = (D - a)^2$

17. $M = (D - a)(D - b)$, $a \neq b$

18. $M = D^2 + 5$

19. $M = D^2 + 4D + 7$

20. $M = 4D^2 - 8D + 5$

21. $M = D^2 - D - 2$

22. The small motions $y(t)$ of an *undamped* spring-mass system are governed by the initial value problem

$$my'' + ky = f(t), \qquad y(0) = y_0,\ y'(0) = v_0$$

where m is the mass, k is the spring constant, and $f(t)$ is an external (driving) force. Show that the impulse response function of this system is

$$g(t) = (1/\omega_0)\sin \omega_0 t, \qquad \omega_0 = \sqrt{k/m}$$

23. Using the impulse response function given in Prob. 22, find the response of the spring-mass system given that
(a) $f(t) = P$ (constant)
(b) $f(t) = P \cos \omega t, \qquad \omega \neq \omega_0$
(c) $f(t) = P \cos \omega_0 t$

24. When resistive forces are taken into account for the spring-mass system in Prob. 22, the motions are called *damped*. In such cases the governing DE is modified to

$$my'' + cy' + ky = f(t)$$

where c is a positive constant. Determine the impulse response function for each of the following cases of damping:
(a) *underdamped* $(c^2 < 4mk)$
(b) *critically damped* $(c^2 = 4mk)$
(c) *overdamped* $(c^2 > 4mk)$

25. Determine the impulse response function for each of the following differential operators ($D = d/dt$):
(a) $M = D^n$, $n = 2,3,4, \ldots$
(b) $M = D^2(D^2 - 1)$
(c) $M = D^4 - 1$
(d) $M = D^3 - 6D^2 + 11D - 6$

5.4 *Solutions of PDEs*

The Laplace transform is especially well-suited for solving initial-boundary value problems for which some auxiliary conditions are prescribed at $t = 0$. Such problems arise naturally in the solution of the heat equation and the wave equation, the independent variable t generally being interpreted as the time variable. The Laplace transform, however, is not generally appropriate in the solution of potential problems. For example, the Laplace transform of $u_{xx}(x,y)$, $x > 0$, leads to

$$\mathscr{L}\{u_{xx}(x,y); x \rightarrow p\} = p^2U(p,y) - pu(0,t) - u_x(0,t) \tag{5.10}$$

To use this expression would require knowledge of both u and u_x at the boundary $x = 0$. Yet, prescribing both u and u_x on the boundary would generally lead to an *ill-posed problem* (i.e., not usually solvable). In heat conduction problems this is equivalent to prescribing both the temperature and heat flux at the boundary, which may not be compatible. Even if either u or u_x is left undetermined until later in the problem, the problem would almost surely become unwieldy at some point.

Since we have previously discussed physical situations leading to the heat equation and wave equation (see Chap. 3), here we will simply illustrate the solution technique of the Laplace transform.

5.4.1 *Heat Conduction*

Let us start by considering a very long homogeneous rod, one end of which is exposed to a time-varying heat reservoir. If we assume the initial temperature distribution is 0°C along the rod, we have the mathematical problem described by*

$$\begin{aligned} & u_{xx} = a^{-2}u_t, \quad 0 < x < \infty,\ t > 0 \\ \text{B.C.:}\quad & u(0,t) = f(t), \quad u(x,t) \rightarrow 0 \text{ as } x \rightarrow \infty \\ \text{I.C.:}\quad & u(x,0) = 0, \quad 0 < x < \infty \end{aligned} \tag{5.11}$$

By applying the Laplace transform to the PDE and boundary conditions, we have

$$\begin{aligned} & U_{xx} - (p/a^2)\,U = 0, \quad 0 < x < \infty \\ \text{B.C.:}\quad & U(0,p) = F(p), \quad U(x,p) \rightarrow 0 \text{ as } x \rightarrow \infty \end{aligned} \tag{5.12}$$

where $U(x,p) = \mathscr{L}\{u(x,t);t\rightarrow p\}$ and $F(p) = \mathscr{L}\{f(t);p\}$. The general solution

* The problem described by (5.11) can also be solved by applying the Fourier sine transform to the variable x, but the Laplace transform is an easier tool to use in this case.

of this second-order linear DE is

$$U(x,p) = A(p)e^{x\sqrt{p}/a} + B(p)e^{-x\sqrt{p}/a} \qquad (5.13)$$

where $A(p)$ and $B(p)$ are arbitrary functions of p. However, to satisfy the condition $U(x,p) \to 0$ as $x \to \infty$, we must choose $A(p) = 0$. The remaining boundary condition demands that $B(p) = F(p)$, and thus

$$U(x,p) = F(p)e^{-x\sqrt{p}/a} \qquad (5.14)$$

From Exam. 4.26 in Chap. 4, we have that

$$\mathscr{L}^{-1}\{e^{-x\sqrt{p}/a};p\to t\} = \frac{x}{2a\sqrt{\pi}t^{3/2}}e^{-x^2/4a^2t} \qquad (5.15)$$

and therefore, through use of the convolution theorem, we arrive at the result

$$u(x,t) = \frac{x}{2a\sqrt{\pi}}\int_0^t \frac{f(\tau)}{(t-\tau)^{3/2}} \exp\left[-\frac{x^2}{4a^2(t-\tau)}\right] d\tau \qquad (5.16)$$

An alternate form of the solution (5.16) can be derived by making the change of variable $z = x/2a\sqrt{t-\tau}$, which leads to

$$u(x,t) = \frac{2}{\sqrt{\pi}}\int_{x/2a\sqrt{t}}^{\infty} f(t - x^2/4a^2z^2)e^{-z^2}\, dz \qquad (5.17)$$

In particular, when the temperature at the end is given by $f(t) = T_0$ (constant), we find that (5.17) reduces to

$$u(x,t) = T_0 \operatorname{erfc}(x/2a\sqrt{t}) \qquad (5.18)$$

where erfc(x) is the complementary error function (see Sec. 1.3.1).

Suppose we now consider a homogeneous rod of unit length where the initial temperature of the rod is zero. It is assumed that the end $x = 0$ is maintained at zero temperature while the end at $x = 1$ is kept at constant temperature T_0. The problem is characterized by

$$u_{xx} = a^{-2}u_t, \qquad 0 < x < 1,\ t > 0$$

$$\text{B.C.:} \quad u(0,t) = 0, \qquad u(1,t) = T_0 \qquad (5.19)$$

$$\text{I.C.:} \quad u(x,0) = 0, \qquad 0 < x < 1$$

Application of the Laplace transform to (5.19) leads to the transformed problem

$$U_{xx} - (p/a^2)\,U = 0, \qquad 0 < x < 1$$

$$\text{B.C.:} \quad U(0,p) = 0, \qquad U(1,p) = T_0/p \qquad (5.20)$$

the general solution of which is*

$$U(x,p) = A(p)\cosh(x\sqrt{p}/a) + B(p)\sinh(x\sqrt{p}/a) \tag{5.21}$$

Imposing the boundary conditions requires that $A(p) = 0$ and $B(p) = T_0/p\sinh(\sqrt{p}/a)$; hence,

$$U(x,p) = T_0 \frac{\sinh(x\sqrt{p}/a)}{p\sinh(\sqrt{p}/a)} \tag{5.22}$$

In order to invert (5.22), we use the method of residues from Sec. 4.6. The function $U(x,p)$ has a simple pole at $p = 0$ and simple poles at $p = -n^2\pi^2a^2$, $n = 1,2,3,\ldots$. Therefore, we find that

$$\mathscr{L}^{-1}\{U(x,p);p\to t\} = T_0\left[\operatorname{Res}\left\{\frac{e^{pt}\sinh(x\sqrt{p}/a)}{p\sinh(\sqrt{p}/a)};p = 0\right\}\right.$$
$$\left. + \sum_{n=1}^{\infty} \operatorname{Res}\left\{\frac{e^{pt}\sinh(x\sqrt{p}/a)}{p\sinh(\sqrt{p}/a)};p = -n^2\pi^2a^2\right\}\right]$$

from which we deduce

$$u(x,t) = T_0\left[x + \frac{2}{\pi}\sum_{n=1}^{\infty}\frac{(-1)^n}{n}\sin(n\pi x)e^{-n^2\pi^2a^2t}\right]^{\dagger} \tag{5.23}$$

5.4.2 *Mechanical Vibrations*

Let $u(x,t)$ denote the transverse displacement of a semiinfinite stretched string, one end of which is fixed far out on the x axis and the other end looped around the point $x = 0$. The string is presumed to be at rest initially with the looped end later moved in some prescribed manner perpendicular to the x axis, i.e., $u(0,t) = f(t)$. Because the string is initially at rest, it follows that $f(0) = 0$. If no external forces are acting on the string, the above conditions are characterized by the boundary value problem

$$u_{xx} = c^{-2}u_{tt}, \quad 0 < x < \infty,\ t > 0$$
$$\text{B.C.:} \quad u(0,t) = f(t), \quad u(x,t) \to 0 \text{ as } x \to \infty \tag{5.24}$$
$$\text{I.C.:} \quad u(x,0) = 0, \quad u_t(x,0) = 0,\ 0 < x < \infty$$

---

* As a general rule, we use hyperbolic functions in the general solution when the domain is finite and exponential functions [see (5.13)] when the domain is infinite.

† The standard method of solving this problem is by use of separation of variables, e.g., see L. C. Andrews, *Elementary Partial Differential Equations with Boundary Value Problems*, Orlando: Academic Press, 1986.

The Laplace transform applied to (5.24) yields

$$U_{xx} - (p/c)^2\,U = 0, \qquad 0 < x < \infty \tag{5.25}$$

$$\text{B.C.:} \quad U(0,p) = F(p), \qquad U(x,p) \to 0 \text{ as } x \to \infty$$

the solution of which is

$$U(x,p) = F(p)e^{-xp/c} \tag{5.26}$$

The translation property of the Laplace transform (Theor. 4.5) enables us to invert (5.26) directly, from which we get

$$u(x,t) = f(t - x/c)h(t - x/c) = \begin{cases} 0, & t \le x/c \\ f(t - x/c), & t > x/c \end{cases} \tag{5.27}$$

The interpretation of this solution is that a point on the string x units from the origin remains at rest until time $t = x/c$, and then it executes the same motion as the loop at the point $x = 0$. In other words, the displacement which is imposed on the end $x = 0$ propagates down the string with velocity c.

Suppose the semiinfinite string is now fixed at $x = 0$ and subject to the external force $f(x,t) = -f_0\delta(t - x/v)$, which is a concentrated load moving with speed v according to $x = vt$. If we assume the string is initially at rest, the resulting motion of the string is governed by

$$c^2u_{xx} = u_{tt} + f_0\delta(t - x/v), \qquad 0 < x < \infty,\ t > 0$$

$$\text{B.C.:} \quad u(0,t) = 0, \qquad u(x,t) \to 0 \text{ as } x \to \infty \tag{5.28}$$

$$\text{I.C.:} \quad u(x,0) = 0, \qquad u_t(x,0) = 0,\ 0 < x < \infty$$

The transformed problem of (5.28) reads

$$U_{xx} - (p/c)^2\,U = (f_0/c^2)\,e^{-xp/v}, \qquad 0 < x < \infty \tag{5.29}$$

$$\text{B.C.:} \quad U(0,t) = 0, \qquad U(x,p) \to 0 \text{ as } x \to \infty$$

This is a nonhomogeneous ODE whose solution is readily found to be

$$U(x,p) = \begin{cases} \dfrac{f_0v^2}{(c^2 - v^2)p^2}\,(e^{-xp/v} - e^{-xp/c}), & v \ne c \\[2ex] -\dfrac{f_0x}{2cp}\,e^{-xp/c}, & v = c \end{cases} \tag{5.30}$$

and by use of the translation property, the inversion of this expression yields

$$u(x,t) = \begin{cases} \dfrac{f_0v^2}{c^2 - v^2}\left[\left(t - \dfrac{x}{v}\right)h\left(t - \dfrac{x}{v}\right) - \left(t - \dfrac{x}{c}\right)h\left(t - \dfrac{x}{c}\right)\right], & v \ne c \\[2ex] -\dfrac{f_0x}{2c}\,h\left(t - \dfrac{x}{c}\right), & v = c \end{cases} \tag{5.31}$$

For our final example involving mechanical vibrations, let us consider a semiinfinite beam which is initially at rest along the x axis and then at time $t = 0$ given a transverse displacement b at the end $x = 0$. The subsequent displacements are solutions of

$$u_{xxxx} + a^{-2}u_{tt} = 0, \qquad 0 < x < \infty,\ t > 0$$

$$\text{B.C.:} \quad \begin{cases} u(0,t) = b, & u_{xx}(0,t) = 0 \\ u(x,t) \to 0 \text{ as } x \to \infty \end{cases} \tag{5.32}$$

$$\text{I.C.:} \quad u(x,0) = 0, \qquad u_t(x,0) = 0,\ 0 < x < \infty$$

By application of the Laplace transform, we obtain the transformed problem

$$U_{xxxx} + (p/a)^2\, U = 0, \qquad 0 < x < \infty$$

$$\text{B.C.:} \quad \begin{cases} U(0,p) = b/p,\ U_{xx}(0,p) = 0 \\ U(x,p) \to 0 \text{ as } x \to \infty \end{cases} \tag{5.33}$$

The general solution of this fourth-order ODE is

$$\begin{aligned} U(x,p) &= e^{-x\sqrt{p/2a}}\,[A(p)\cos(x\sqrt{p/2a}) + B(p)\sin(x\sqrt{p/2a})] \\ &\quad + e^{x\sqrt{p/2a}}\,[C(p)\cos(x\sqrt{p/2a}) + D(p)\sin(x\sqrt{p/2a})] \end{aligned} \tag{5.34}$$

To satisfy the boundedness condition in (5.33), we must set $C(p) = D(p) = 0$. The remaining boundary conditions in (5.33) lead to $A(p) = b/p$ and $B(p) = 0$; thus, our solution reduces to

$$U(x,p) = (b/p)e^{-x\sqrt{p/2a}}\cos(x\sqrt{p/2a}) \tag{5.35}$$

Now let us use Euler's formula $\cos z = \frac{1}{2}(e^{iz} + e^{-iz})$ so that we may rewrite (5.35) as*

$$U(x,p) = \frac{b}{2p}\,(e^{-x\sqrt{-ip/a}} + e^{-x\sqrt{ip/a}}) \tag{5.36}$$

Then, using the inverse transform relation (see Prob. 14 in Exer. 4.6)

$$\mathscr{L}^{-1}\{(1/p)e^{-a\sqrt{p}};t\} = \operatorname{erfc}(a/2\sqrt{t}) \tag{5.37}$$

we deduce that

$$\begin{aligned} u(x,t) &= \frac{b}{2}\left[\operatorname{erfc}\left(\frac{x}{2}\sqrt{\frac{-i}{at}}\right) + \operatorname{erfc}\left(\frac{x}{2}\sqrt{\frac{i}{at}}\right)\right] \\ &= b\left[1 - \frac{1}{2}\operatorname{erf}\left(\frac{x}{2}\sqrt{\frac{-i}{at}}\right) - \frac{1}{2}\operatorname{erf}\left(\frac{x}{2}\sqrt{\frac{i}{at}}\right)\right] \end{aligned} \tag{5.38}$$

---

* Note that $1 \pm i = \sqrt{\pm 2i}$.

Finally, recalling Prob. 13 in Exer. 1.3, we obtain the solution in terms of the Fresnel integrals, i.e.,

$$u(x,t) = b\,[1 - C(x/\sqrt{2a\pi t}) - S\,(x/\sqrt{2a\pi t})] \tag{5.39}$$

EXERCISES 5.4

1. If the boundary condition in (5.11) is

$$u(0,t) = f(t) = \begin{cases} T_1, & 0 < t + b \\ 0, & t \geq b \end{cases}$$

(a) show that the subsequent temperature distribution is

$$u(x,t) = \begin{cases} T_1 \operatorname{erfc}(x/2a\sqrt{t}), & 0 < t < b \\ T_1[\operatorname{erf}(x/2a\sqrt{t-b}) - \operatorname{erf}(x/2a\sqrt{t})], & t \geq b \end{cases}$$

(b) Show that $u(x,t)$ given in (a) is continuous at $t = b$.

2. Solve the problem described by (5.11) when $u(0,t) = f(t) = T_1/\sqrt{t}$.

3. Given the heat conduction problem

$$u_{xx} = a^{-2}u_t, \qquad 0 < x < \infty,\ t > 0$$

B.C.: $\quad u_x(0,t) = -f(t), \qquad u(x,t) \to 0$ as $x \to \infty$

I.C.: $\quad u(x,0) = 0, \qquad 0 < x < \infty$

show that its solution can be expressed as

$$u(x,t) = \frac{a}{\sqrt{\pi}} \int_0^t \frac{f(\tau)}{\sqrt{t-\tau}} \exp\left[\frac{x^2}{4a^2(t-\tau)}\right] d\tau$$

4. Show that the solution in Prob. 3 can be expressed in the alternate form

$$u(x,t) = \frac{x}{\sqrt{\pi}} \int_{x/2a\sqrt{t}}^{\infty} f(t - x^2/4a^2z^2)z^{-2}\, e^{-z^2}\, dz$$

5. For the special case $f(t) = K$ (constant), show that the solution of Prob. 3 is

$$u(x,t) = K\,[2a\sqrt{t/\pi}e^{-x^2/4a^2t} - x \operatorname{erfc}(x/2a\sqrt{t})]$$

Hint: Use Prob. 4 and integration by parts.

6. A semiinfinite conducting solid has initial temperature T_0. Radiation into a medium $x < 0$ at temperature zero is assumed to be such that the flux at the face $x = 0$ is proportional to the difference in tem-

peratures of the face $x = 0$ and the medium $x < 0$. Given that the mathematical formulation of the problem is

$$u_{xx} = u_t, \qquad 0 < x < \infty, t > 0$$

B.C.: $u_x(0,t) = -ku(0,t), \qquad u(x,t) \to 0 \text{ as } x \to \infty$

I.C.: $u(x,0) = T_0, \qquad 0 < x < \infty$

show that the temperatures inside the conducting solid are given by

$$u(x,t) = \frac{2kT_0}{\pi} \int_0^{\infty} \frac{e^{-tz^2}}{z} \left(\frac{z \cos xz - k \sin xz}{z^2 + k^2} \right) dz$$

In Probs. 7–10, use the Laplace transform to solve the given heat conduction boundary-value problem.

7. $u_{xx} = a^{-2}u_t, \qquad 0 < x < 1, t > 0$

B.C.: $u(0,t) = T_0, \qquad u_x(1,t) = 0$

I.C.: $u(x,0) = 0$

8. $u_{xx} = a^{-2}u_t, \qquad 0 < x < 1, t > 0$

B.C.: $u(0,t) = 0, \qquad u(1,t) = 0$

I.C.: $u(x,0) = T_0$

9. $0.25\, u_{xx} = u_t - 1, \qquad 0 < x < 10, t > 0$

B.C.: $u_x(0,t) = 0, \qquad u(10,t) = 20$

I.C.: $u(x,0) = 50$

10. $u_{xx} = u_t - 2x, \qquad 0 < x < 1, t > 0$

B.C.: $u(0,t) = 0, \qquad u(1,t) = 0$

I.C.: $u(x,0) = x(1 - x)$

11. Given the boundary-value problem

$$u_{xx} = u_t, \qquad 0 < x < 1, t > 0$$

B.C.: $u(0,t) = 0, \qquad u(1,t) = T_0$

I.C.: $u(x,0) = T_0, \qquad 0 < x < 1$

(a) show that the solution of the transformed problem can be expressed in the form

$$U(x,p) = \frac{T_0}{p} \left\{ 1 - e^{-\sqrt{p}} \left[\frac{1 - e^{-2(1-x)\sqrt{p}}}{1 - e^{-2\sqrt{p}}} \right] \right\}$$

(b) By expanding $(1 - e^{-2\sqrt{p}})^{-1}$ in a series of ascending powers of $e^{-2\sqrt{p}}$, show that

$$U(x,p) = \frac{T_0}{p}\left[1 - e^{-x\sqrt{p}} + e^{-(2-x)\sqrt{p}} - e^{-(2+x)\sqrt{p}} + \cdots\right]$$

(c) Inverting the series in (b) termwise, deduce that

$$u(x,t) = T_0\,\{\operatorname{erf}(x/2\sqrt{t}) + \operatorname{erfc}[(2-x)/2\sqrt{t}] - \operatorname{erfc}[(2+x)/2\sqrt{t}] + \cdots\}$$

12. A heat source of strength $q(t)h(t)$, where h is the Heaviside unit function, appears at the origin of a long rod at time $t = 0$ and moves along the positive x axis with constant speed v. The problem is characterized by

$$u_{xx} = u_t - \delta(x - vt)q(t)h(t), \qquad -\infty < x < \infty,\ t > 0$$

B.C.: $u(x,t) \to 0$ as $|x| \to \infty$

I.C.: $u(x,0) = 0, \qquad -\infty < x < \infty$

Using the Laplace transform, show that

$$u(x,t) = \frac{1}{2\sqrt{\pi}}\int_0^t q(\tau)(t - \tau)^{-1/2}\, e^{-(x - v\tau)^2/4(t - \tau)}\, d\tau$$

13. Solve the heat conduction problem

$$u_{xx} = a^{-2}u_t - \delta(x)\delta(t), \qquad -\infty < x < \infty,\ t > 0$$

B.C.: $u(x,t) \to 0$ as $|x| \to \infty$

I.C.: $u(x,0) = T_0$

14. Using the Laplace transform, find a bounded solution of the exterior temperature distribution problem for a sphere described by

$$u_{rr} + (2/r)u_r = u_t, \qquad 1 < r < \infty,\ t < 0$$

B.C.: $u(1,t) = T_1, \qquad u(r,t) \to 0$ as $r \to \infty$

I.C.: $u(r,0) = T_0$

15. The temperature distribution $u(r,t)$ in a thin circular plate, which is initially at 0°C, has its faces insulated and its boundary held at temperature T_1, is governed by the boundary value problem

$$u_{rr} + (1/r)u_r = u_t, \qquad 0 < r < 1,\ t < 0$$

B.C.: $u(1,t) = T_1$

I.C.: $u(r,0) = 0$

Use the Laplace transform to find a bounded solution.

Hint: Recall Prob. 13 in Exer. 4.6.

16. Given the boundary value problem

$$u_{xx} = u_{tt}, \qquad 0 < x < \infty, \, t > 0$$

B.C.: $u_x(0,t) = f(t), \qquad u(x,t) \to 0 \text{ as } x \to \infty$

I.C.: $u(x,0) = 0, \qquad u_t(x,0) = 0$

show that $u(x,t) = g(t - x)$, where $g(z) = 0$, $z > 0$, and

$$g(z) = -\int_0^z f(\tau)\, d\tau, \qquad z \geq 0$$

17. Consider the motions of a string fastened at the origin but whose far end is looped around a frictionless peg that exerts no vertical force on the loop. The string is initially supported at rest along the x axis and is released at time $t = 0$, moving downward under the action of gravity. Determine the subsequent displacements given that the problem is characterized by

$$c^2 u_{xx} = u_{tt} - g, \qquad 0 < x < \infty, \, t > 0 \; (g \text{ constant})$$

B.C.: $u(0,t) = 0, \qquad u(x,t) \to 0 \text{ as } x \to \infty$

I.C.: $u(x,0) = 0, \qquad u_t(x,0) = 0$

18. Show that the boundary-value problem

$$u_{xx} = u_{tt}, \qquad 0 < x < \infty, \, t > 0$$

B.C.: $u_x(0,t) = 0, \qquad u(x,t) \to 0 \text{ as } x \to \infty$

I.C.: $u(x,0) = e^{-x}, \qquad u_t(x,0) = 0$

has the solution

$$u(x,t) = \begin{cases} e^{-t} \cosh x, & x < t \\ e^{-x} \cosh t, & x > t. \end{cases}$$

In Probs. 19–25, use the Laplace transform to solve the given boundary-value problem.

19.

$$u_{xx} = c^{-2} u_{tt}, \qquad 0 < x < \infty, \, t > 0$$

B.C.: $u(0,t) = 0, \qquad u_x(x,t) \to 0 \text{ as } x \to \infty$

I.C.: $u(x,0) = 0, \qquad u_t(x,0) = v_0$

20.

$$u_{xx} = c^{-2} u_{tt}, \qquad 0 < x < \infty, \, t > 0$$

B.C.: $u(0,t) = 0, \qquad u(x,t) \to 0 \text{ as } x \to \infty$

I.C.: $u(x,0) = A, \qquad u_t(x,0) = 0$

21.

$$u_{xx} = c^{-2} u_{tt} - A \sin \pi x, \qquad 0 < x < 1, \, t > 0$$

B.C.: $u(0,t) = 0, \qquad u(1,t) = 0$

I.C.: $u(x,0) = 0, \qquad u_t(x,0) = 0$

22. $u_{xx} = c^{-2}u_{tt}, \quad 0 < x < 1,\ t > 0$

B.C.: $u(0,t) = 0, \quad u(1,t) = 1$

I.C.: $u(x,0) = 0, \quad u_t(x,0) = 0$

23. $u_{xx} = c^{-2}u_{tt}, \quad 0 < x < 1,\ t > 0$

B.C.: $u_x(0,t) = 0, \quad u(1,t) = 1$

I.C.: $u(x,0) = 0, \quad u_t(x,0) = 0$

24. $u_{xx} = c^{-2}u_{tt}, \quad 0 < x < 1,\ t > 0$

B.C.: $u(0,t) = 0, \quad u_x(1,t) = At^2$

I.C.: $u(x,0) = 0, \quad u_t(x,0) = 0$

25. $u_{xx} = c^{-2}u_{tt}, \quad 0 < x < 1,\ t > 0$

B.C.: $u(0,t) = 0, \quad u_x(1,t) = 0$

I.C.: $u(x,0) = 0, \quad u_t(x,0) = x$

5.5 *Linear Integral Equations*

Integral equations of the form

$$\int_0^t u(\tau)k(t - \tau)\, d\tau = f(t), \qquad t > 0 \tag{5.40}$$

are known as *Volterra equations of convolution type* (see also the discussion in Sec. 3.2.1). The Laplace transform provides a useful technique for the solution of such equations in which $f(t)$ and $k(t - \tau)$ are known functions and $u(t)$ is to be determined.

Using the result of Eq. (4.32) in Sec. 4.5.3 to take the Laplace transform of (5.40), we obtain

$$U(p)K(p) = F(p) \tag{5.41}$$

from which it follows that

$$U(p) = F(p)/K(p) \tag{5.42}$$

Inverting (5.42) leads to the solution

$$u(t) = \mathscr{L}^{-1}\left\{\frac{F(p)}{K(p)};t\right\} \tag{5.43}$$

If the function $1/K(p)$ has an inverse Laplace transform, say

$$\mathscr{L}^{-1}\left\{\frac{1}{K(p)};t\right\} = g(t) \tag{5.44}$$

then we can use the convolution theorem (Theorem 4.12) to express the solution (5.43) as

$$u(t) = \int_0^t f(\tau)g(t - \tau)\, d\tau, \qquad t > 0 \tag{5.45}$$

In some cases it may happen that the inverse transform (5.44) does not exist, but $1/pK(p)$ has an inverse transform. Thus, (5.42) becomes

$$U(p) = pF(p) \cdot L(p) \tag{5.46}$$

where $L(p) = 1/pK(p)$. Now if $f(t)$ is a differentiable function such that $f(0) = 0$, then

$$\mathscr{L}^{-1}\{pF(p);t\} = f'(t) \tag{5.47}$$

and the inversion of (5.46) yields

$$u(t) = \int_0^t f'(\tau)\ell(t - \tau)\, d\tau, \qquad t > 0 \tag{5.48}$$

where $\ell(t) = \mathscr{L}^{-1}\{L(p);t\}$.

Example 5.9: Solve the integral equation

$$\int_0^t u(\tau)J_0(t - \tau)\, d\tau = \sin t, \qquad t > 0$$

where $J_0(t)$ is the Bessel function.

Solution: Recalling that

$$\mathscr{L}\{J_0(t);p\} = 1/\sqrt{p^2 + 1}$$

we find that the Laplace transform applied to the integral equation leads to

$$U(p)/\sqrt{p^2 + 1} = 1/(p^2 + 1)$$

The solution of this transformed problem is

$$U(p) = 1/\sqrt{p^2 + 1}$$

and therefore we deduce that

$$u(t) = J_0(t)$$

By substituting $u(t) = J_0(t)$ into the original integral equation, we get the interesting integral formula

$$\int_0^t J_0(\tau)J_0(t - \tau)\, d\tau = \sin t$$

Example 5.10: Solve the integral equation

$$u(t) - \int_0^t e^{t-\tau} u(\tau)\, d\tau = f(t), \qquad t > 0$$

Solution: This is an integral equation of the second kind (see Sec. 3.2.1). By taking the Laplace transform of each term in the equation, we find

$$U(p) - \frac{U(p)}{p - 1} = F(p)$$

which leads to

$$U(p) = \frac{(p - 1)F(p)}{p - 2} = F(p) + \frac{F(p)}{p - 2}$$

Now taking the inverse transform of this last expression, we get the formal solution

$$u(t) = f(t) + \int_0^t e^{2(t-\tau)} f(\tau)\, d\tau$$

5.5.1 *The Tautochrone Problem*

The great Norwegian mathematician Niels Abel (1802–1829) studied a particular integral equation of the Volterra type which has several important applications. The most famous application is the *tautochrone problem,* which is to determine a curve passing through the origin for which the time required for a particle to slide down the curve is independent of the starting point. The particle is allowed to slide freely from rest under the action of gravity and the reaction of the curve on which it is constrained to move (see Fig. 5.1). It is this curve that is called the tautochrone.

If we assume the particle is initially at rest at $P(x,y)$, then its kinetic energy is zero and its potential energy is mgy, where m is the mass of

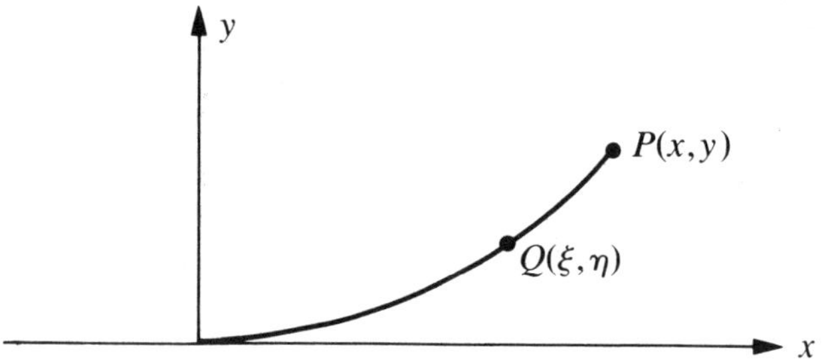

Figure 5.1 Tautochrone

the particle and g is the gravitational constant. At some intermediate point $Q(\xi,\eta)$, in accordance with the conservation of total energy, we can equate the gain in kinetic energy to the loss of potential energy, which leads to

$$\tfrac{1}{2}mv^2 = mg(y - \eta) \tag{5.49}$$

where v is the instantaneous speed of the particle. Solving for v, we obtain

$$v = ds/dt = \sqrt{2g(y - \eta)}$$

where s denotes the arclength along the curve. The time of travel from P to Q is thus given by the expression

$$t = \frac{1}{\sqrt{2g}} \int_P^Q \frac{ds}{\sqrt{y - \eta}} \tag{5.50}$$

From the calculus we have the arclength relation

$$ds = -\sqrt{1 + (d\xi/d\eta)^2}\, d\eta = -u(\eta)d\eta \tag{5.51}$$

where the negative sign reflects the fact that η is a decreasing variable from P to Q. Substituting this last expression into (5.50) yields

$$t = -\frac{1}{\sqrt{2g}} \int_y^\eta \frac{u(\eta)}{\sqrt{y - \eta}}\, d\eta$$

and therefore the total time of descent T from P to the origin is given by the Volterra integral equation

$$T = \frac{1}{\sqrt{2g}} \int_0^y \frac{u(\eta)}{\sqrt{y - \eta}}\, d\eta \tag{5.52}$$

where T is a fixed constant.

To solve (5.52) for the unknown function $u(y)$, we apply the Laplace transform to the variable y, which gives us

$$T\sqrt{2g}/p = \sqrt{\pi/p}\, U(p)$$

or

$$U(p) = T\sqrt{2g/\pi p} \tag{5.53}$$

The inverse Laplace transform of this expression is

$$u(y) = \frac{T}{\pi}\sqrt{\frac{2g}{y}} \tag{5.54}$$

Rewriting Eq. (5.51) in terms of x and y, and substituting (5.54) into the square of the resulting expression leads to

$$1 + \left(\frac{dx}{dy}\right)^2 = \frac{2gT^2}{\pi^2 y} = \frac{a}{y}$$

where $a = 2gT^2/\pi^2$. Separating variables yields

$$dx = \sqrt{\frac{a - y}{y}}\, dy$$

and by substituting $y = a \sin^2 \frac{1}{2}\theta$, this last expression simplifies to

$$dx = a \cos^2 \tfrac{1}{2}\theta \, d\theta = (a/2)(1 + \cos \theta) d\theta \tag{5.55}$$

Observing that $x = 0$ when $y = 0$, the solution of (5.55) leads to the set of parametric equations for the tautochrone

$$x = (a/2)(\theta + \sin \theta), \qquad y = (a/2)(1 - \cos \theta) \tag{5.56}$$

which represent a curve called a *cycloid*. This same curve is generated by a point P on a circle of radius $a/2$ as the circle rolls along the lower side of the line $y = a$.

Abel's integral equation (5.52) is a special case of the more general integral equation

$$\int_0^t \frac{u(\tau)}{(t - \tau)^\alpha} d\tau = f(t), \qquad t > 0 \tag{5.57}$$

where $f(t)$ is given and α is a constant such that $0 < \alpha < 1$. If we formally apply the Laplace transform to (5.57), we get

$$\Gamma(1 - \alpha)p^{\alpha - 1}U(p) = F(p)$$

or

$$U(p) = \frac{F(p)}{\Gamma(1 - \alpha)p^{\alpha - 1}} = \frac{pF(p)}{\Gamma(1 - \alpha)p^\alpha} \tag{5.58}$$

Assuming that $f(t)$ is differentiable and that $f(0) = 0$, then

$$\mathscr{L}^{-1}\{pF(p);t\} = f'(t) \tag{5.59}$$

Also, we have that

$$\mathscr{L}^{-1}\left\{\frac{1}{\Gamma(1 - \alpha)p^\alpha};t\right\} = \frac{t^{\alpha - 1}}{\Gamma(1 - \alpha)\Gamma(\alpha)}$$

$$= \frac{\sin \pi\alpha}{\pi} t^{\alpha - 1} \tag{5.60}$$

and thus we deduce that

$$u(t) = \frac{\sin \pi\alpha}{\pi} \int_0^t \frac{f'(\tau)}{(t - \tau)^{1 - \alpha}} d\tau, \qquad t > 0 \tag{5.61}$$

EXERCISES 5.5

In Probs. 1–10, we find a continuous solution of the given integral equation.

1. $\int_0^t \frac{u(\tau)}{\sqrt{t-\tau}}\,d\tau = \sqrt{t}$

2. $\int_0^t \frac{u(\tau)}{\sqrt{t-\tau}}\,d\tau = 1 + t + 3t^2$

3. $\int_0^t u(\tau)u(t-\tau)\,d\tau = 4\sin 2t$

4. $u(t) = t + \int_0^t (t-\tau)u(\tau)\,d\tau$

5. $u(t) = 4t - 3\int_0^t u(\tau)\sin(t-\tau)\,d\tau$

6. $u(t) + \int_0^t e^{-\tau}\,u(t-\tau)\,d\tau = 1$

7. $u(t) = a\sin t - 2\int_0^t u(\tau)\cos(t-\tau)\,d\tau$

8. $u(t) = t + \frac{1}{6}\int_0^t (t-\tau)^3 u(\tau)\,d\tau$

9. $\int_0^t \frac{u(\tau)}{(t-\tau)^{1/3}}\,d\tau = t(1+t)$

10. $u(t) = \frac{1}{2}\sin 2t + \int_0^t u(\tau)u(t-\tau)\,d\tau$

In Probs. 11–15, solve the given integrodifferential equation.

11. $\int_0^t u(\tau)\cos(t-\tau)\,d\tau = u'(t), \qquad u(0) = 1$

12. $\int_0^t u(\tau)\cos 2(t-\tau)\,d\tau = 1 - u'(t), \qquad u(0) = 3$

13. $\int_0^t u'(\tau)u(t-\tau)\,d\tau = 24t^3, \qquad u(0) = 0$

14. $\int_0^t u''(\tau)u'(t-\tau)\,d\tau = u'(t) - u(t), \qquad u(0) = 0,\ u'(0) = 0$

15. $f(t) = \int_0^t (t-\tau)^{-\alpha}\,u'(\tau)\,d\tau, \qquad 0 < \alpha < 1$

16. Show that

$$u(t) + \int_0^t u(\tau)\sin\omega(t-\tau) = f(t)$$

has the formal solution

$$u(t) = f(t) - \frac{\omega}{\Omega}\int_0^t f(\tau)\sin\Omega(t-\tau)\,d\tau, \qquad \Omega^2 = \omega(1+\omega)$$

17. Show that

$$f(t) = \int_0^t J_0(2\sqrt{t-\tau})u(\tau)\,d\tau, \qquad f(0) = 0$$

has the formal solution

$$u(t) = f'(0)I_0(2\sqrt{t}) + \int_0^t I_0(2\sqrt{t-\tau})f(\tau)\,d\tau$$

18. Solve the integral equation

$$\int_0^t J_n(t-\tau)u(\tau)\,d\tau = J_{n+1}(t)$$

(a) for the case $n = 0$.
(b) for the cases $n = 1,2,3,\ldots$

19. Show that

$$u(t) + \int_0^t \frac{u(\tau)}{\sqrt{t-\tau}}\,d\tau = f(t), \qquad u(0) = f(0) = 0$$

has a formal solution of the form

$$u(t) = \int_0^t f'(t-\tau)g(\tau)\,d\tau$$

and identify the function $g(t)$.

In Probs. 20–22, find formal solutions of the given integral equations.

20. $f(t) = \displaystyle\int_0^t \frac{u(s)}{(t^2 - s^2)^\alpha}ds, \qquad 0 < \alpha < 1$

21. $f(t) = 2\displaystyle\int_t^1 \frac{su(s)}{\sqrt{s^2 - t^2}}\,ds$

22. $f(t) = t\displaystyle\int_t^\infty \frac{u'(s)}{\sqrt{s-t}}\,ds$

6

The Mellin Transform

6.1 *Introduction*

Generally speaking, unlike the Fourier and Laplace transforms, we find that the Mellin transform is not very useful in a direct manner. It is quite effective, however, in the derivation of certain properties of integrals, in summing series, and in statistics. In this sense, we generally think of the Mellin transform as a sort of indirect tool in applications.

As in the case of the Laplace transform, the Mellin transform and its inversion formula can be formally derived from the Fourier integral theorem

$$g(u) = \frac{1}{2\pi}\int_{-\infty}^{\infty} e^{-i\xi u} \int_{-\infty}^{\infty} g(t)e^{i\xi t}\, dt\, d\xi \tag{6.1}$$

Let us begin by introducing the change of variables $x = e^t$, $y = e^u$, and $s = c + i\xi$, where c is a fixed constant. Then, after some algebraic manipulation, we find

$$g(\log y)y^{-c} = \frac{1}{2\pi i}\int_{c-i\infty}^{c+i\infty} y^{-s} \int_0^{\infty} g(\log x)x^{-c}x^{s-1}\, dx\, ds \tag{6.2}$$

If we now define

$$f(x) = g(\log x)x^{-c} \tag{6.3}$$

then (6.2) leads to the pair of transform formulas

$$F(s) = \int_0^\infty x^{s-1} f(x)\, dx \tag{6.4}$$

and

$$f(x) = \frac{1}{2\pi i} \int_{c-i\infty}^{c+i\infty} x^{-s} F(s)\, ds \tag{6.5}$$

We define (6.4) as the *Mellin transform* of $f(x)$, and (6.5) is the related *inversion formula*. We also use the notations

$$F(s) = \mathcal{M}\{f(x);s\} \tag{6.6}$$

and

$$f(x) = \mathcal{M}^{-1}\{F(s);x\} \tag{6.7}$$

respectively, for the Mellin transform and its inverse.

If the integrals

$$\int_0^\infty x^{a-1} f(x)\, dx, \qquad \int_0^\infty x^{b-1} f(x)\, dx$$

both converge for real a and b such that $a < b$, then the Mellin transform of $f(x)$ converges uniformly to $F(s)$ in any finite region interior to the infinite vertical strip $a < \sigma < b$, where $\sigma = \text{Re}\,(s)$. In such cases the transform $F(s)$ is analytic in this vertical strip. Also, if $x^{k-1}f(x)$ is absolutely integrable on the positive real axis for some $k > 0$, and if $F(s)$ is defined by (6.4), then the inversion formula (6.5) is valid for $c > k$.

6.2 *Evaluation of Mellin Transforms*

In this section we will calculate the Mellin transform of several functions and develop certain operational properties. In many cases we find that we can relate the desired Mellin transform to known results involving either the Fourier or Laplace transform.

Example 6.1: Find the Mellin transform of e^{-x}.

Solution: From definition,

$$\mathcal{M}\{e^{-x};s\} = \int_0^\infty x^{s-1} e^{-x}\, dx$$

but this is precisely the definition of the gamma function. Hence, we immediately deduce that

$$\mathcal{M}\{e^{-x};s\} = \Gamma(s), \qquad \text{Re}(s) > 0$$

Observe that, in Exam. 6.1, we could also relate the Mellin transform of e^{-x} to the Laplace transform of x^{x-1}, i.e.,

$$\mathcal{M}\{e^{-x};s\} = \mathcal{L}\{x^{s-1};\, p=1\} \tag{6.8}$$

Example 6.2: Find the Mellin transforms of $\cos x$ and $\sin x$.

Solution: Starting with $\cos x$, we have

$$\mathcal{M}\{\cos x;s\} = \int_0^\infty x^{s-1}\cos x\, dx$$

but the integral on the right is recognized as a multiple of a special case of the cosine transform

$$\begin{aligned}\mathcal{F}_C\{x^{s-1};\xi\} &= \sqrt{\frac{2}{\pi}}\int_0^\infty x^{s-1}\cos \xi x\, dx \\ &= \sqrt{\frac{2}{\pi}}\frac{\Gamma(s)}{\xi^s}\cos\left(\tfrac{1}{2}\pi s\right), \qquad 0 < \text{Re}(s) < 1\end{aligned}$$

[recall Eq. (2.68a) in Sec. 2.6.1]. Therefore we deduce that ($\xi = 1$)

$$\mathcal{M}\{\cos x;s\} = \Gamma(s)\cos\left(\tfrac{1}{2}\pi s\right), \qquad 0 < \text{Re}(s) < 1$$

Similarly, the result

$$\mathcal{M}\{\sin x;s\} = \Gamma(s)\sin\left(\tfrac{1}{2}\pi s\right), \qquad 0 < \text{Re}(s) < 1$$

follows directly from the sine transform given by Eq. (2.68b) in Sec. 2.6.1.

Example 6.3 Find the Mellin transform of $1/(1 + x)$.

Solution: From the defining integral,

$$\mathcal{M}\left\{\frac{1}{1+x};s\right\} = \int_0^\infty \frac{x^{s-1}}{1+x}\, dx$$

By imposing the restriction $0 < \text{Re}(s) < 1$, the integral is exactly that given by Eq. (1.13) in Sec. 1.2.2. Hence, it follows that

$$\mathcal{M}\left\{\frac{1}{1+x};s\right\} = \Gamma(s)\Gamma(1-s), \qquad 0 < \text{Re}(s) < 1$$

Also, by using properties of the gamma function, we can write this transform in the more convenient form

$$\mathcal{M}\left\{\frac{1}{1+x};s\right\} = \frac{\pi}{\sin \pi s}, \qquad 0 < \text{Re}(s) < 1$$

6.2.1 *Operational Properties*

The Mellin transform enjoys certain operational properties analogous to those of the Fourier and Laplace transforms. For example, if C_1 and C_2 are any constants, then we have the *linearity property*

$$\mathcal{M}\{C_1 f(x) + C_2 g(x);s\} = C_1 F(s) + C_2 G(s) \tag{6.9}$$

where $F(s)$ and $G(s)$ are the Mellin transforms, respectively, of $f(x)$ and $g(x)$. This property is a simple consequence of the linearity property of integrals.

By making the simple change of variable $x = t/a$ in the following integral

$$\begin{aligned}\mathcal{M}\{f(ax);s\} &= \int_0^\infty x^{s-1} f(ax)\,dx \\ &= \frac{1}{a^s}\int_0^\infty t^{s-1} f(t)\,dt\end{aligned}$$

we deduce the *scaling property*

$$\mathcal{M}\{f(ax);s\} = (1/a^s)F(s), \qquad a > 0 \tag{6.10}$$

where $F(s)$ is the Mellin transform of $f(x)$. Also, we find that

$$\mathcal{M}\{x^a f(x);s\} = \int_0^\infty x^{s+a-1} f(x)\,dx$$

which leads to the *translation property*

$$\mathcal{M}\{x^a f(x);s\} = F(s + a) \tag{6.11}$$

Similarly, it can be shown that

$$\mathcal{M}\{f(x^a);s\} = (1/a)F(s/a), \qquad a > 0 \tag{6.12}$$

and

$$\mathcal{M}\{(1/x)f(1/x);s\} = F(1 - s) \tag{6.13}$$

the proofs of which are left to the exercises (see Probs. 16 and 17 in Exer. 6.2).

If $f(x)$ is continuous on $x \geq 0$ and has a Mellin transform $F(s)$, then

$$\mathcal{M}\{f'(x);s\} = \int_0^\infty x^{s-1} f'(x)\, dx$$

$$= x^{s-1} f(x)\Big|_0^\infty - (s-1)\int_0^\infty x^{s-2} f(x)\, dx$$

Now, if it happens that σ_1 and σ_2 exist such that

$$\lim_{x\to 0} x^{s-1} f(x) = 0, \qquad \sigma_1 < \mathrm{Re}(s) < \sigma_2 \tag{6.14a}$$

$$\lim_{x\to \infty} x^{s-1} f(x) = 0, \qquad \sigma_1 < \mathrm{Re}(s) < \sigma_2 \tag{6.14b}$$

then the Mellin transform of $f'(x)$ leads to

$$\mathcal{M}\{f'(x);s\} = -(s-1)F(s-1), \qquad \sigma_1 < \mathrm{Re}(s) < \sigma_2 \tag{6.15}$$

provided $F(s - 1)$ exists in the stated vertical strip of the s plane. A second application of (6.15) yields

$$\mathcal{M}\{f''(x);s\} = -(s-1)\,\mathcal{M}\{f'(x);s-1\}$$

or

$$\mathcal{M}\{f''(x);s\} = (s-1)(s-2)F(s-2) \tag{6.16}$$

Continued application of (6.15) eventually leads to the general result

$$\mathcal{M}\{f^{(n)}(x);s\} = (-1)^n \frac{\Gamma(s)}{\Gamma(s-n)} F(s-n), \qquad n = 1,2,3,\ldots \tag{6.17}$$

provided

$$\lim_{x\to 0} x^{s-k-1} f^{(k)}(x) = 0, \qquad k = 1,2,\cdots,n-1 \tag{6.18}$$

The *convolution theorem* for the Mellin transform is given by

$$\mathcal{M}^{-1}\{F(s)G(s);x\} = \int_0^\infty f(x/u)g(u)\frac{du}{u} \tag{6.19}$$

the derivation of which we leave to the exercises (see Prob. 25 in Exer. 6.2). Additional properties of the Mellin transform are also provided in the exercises.

Example 6.4: Find the Mellin transform of $1/(1 + ax)^m$, $m > 0$.

Solution: Let us first find the Mellin transform of $f(x) = 1/(1 + x)^m$. From the defining integral

$$\mathcal{M}\{f(x);s\} = \int_0^\infty \frac{x^{s-1}}{(1+x)^m}\, dx$$

but recalling the definition of the beta function (see Prob. 21 in Exer. 1.2)

$$B(x,y) = \int_0^\infty \frac{t^{x-1}}{(1+t)^{x+y}}\,dt, \qquad x > 0, y > 0$$

we see immediately that

$$\mathcal{M}\{f(x);s\} = B(s, m - s) = \frac{\Gamma(s)\Gamma(m-s)}{\Gamma(m)}$$

Now using the scaling property (6.10), we have

$$\mathcal{M}\left\{\frac{1}{(1+ax)^m};s\right\} = \frac{\Gamma(s)\Gamma(m-s)}{a^s\Gamma(m)}, \qquad 0 < \mathrm{Re}(s) < m$$

Example 6.5: Find the Mellin transform of $x^{-\nu}J_\nu(ax)$, $a > 0$, $\nu > -1/2$.

Solution: In this case the Mellin transform

$$\mathcal{M}\{x^{-\nu}J_\nu(ax);s\} = \int_0^\infty x^{s-\nu-1}J_\nu(ax)\,dx$$

can be evaluated most easily in an indirect manner using known results from the Fourier cosine transform. First we recall the cosine transforms

$$\mathcal{F}_C\{x^{-\nu}J_\nu(ax);\xi\} = \frac{(a^2-\xi^2)^{\nu-1/2}\,h(a-\xi)}{2^{\nu-1/2}a^\nu\Gamma(\nu+1/2)}$$

and

$$\mathcal{F}_C\{x^{s-1};\xi\} = \sqrt{2/\pi}\,\Gamma(s)\cos(\tfrac{1}{2}\pi s)\xi^{-s}$$

[see Eqs. (2.66) and (2.68a) in Sec. 2.6]. Then, using the cosine transform relation

$$\int_0^\infty f(x)g(x)\,dx = \int_0^\infty F_C(\xi)G_C(\xi)\,d\xi$$

[see Eq. (2.90) in Sec. 2.7], we obtain

$$\int_0^\infty x^{s-\nu-1}J_\nu(ax)\,dx = \sqrt{\frac{2}{\pi}}\frac{\Gamma(s)\cos(\frac{1}{2}\pi s)}{2^{\nu-1/2}a^\nu\Gamma(\nu+1/2)}\int_0^a \xi^{-s}(a^2-\xi^2)^{\nu-1/2}\,d\xi$$

However, if $0 < \mathrm{Re}(s) < 1$, then

$$\begin{aligned}\int_0^a \xi^{-s}(a^2-\xi^2)^{\nu-1/2}\,d\xi &= \frac{1}{2}a^{2\nu-s}\int_0^1 t^{-(s+1)/2}(1-t)^{\nu-1/2}\,dt \\ &= \frac{1}{2}a^{2\nu-s}\frac{\Gamma(\frac{1}{2}-\frac{1}{2}s)\Gamma(\nu+\frac{1}{2})}{\Gamma(\nu-\frac{1}{2}s+1)},\end{aligned}$$

from which it follows that

$$\int_0^\infty x^{s-\nu-1}J_\nu(ax)\,dx = a^{\nu-s}\frac{\Gamma(s)\Gamma(\frac{1}{2}-\frac{1}{2}s)\cos(\frac{1}{2}\pi s)}{\sqrt{\pi}\,2^\nu\Gamma(\nu-\frac{1}{2}s+1)}$$

The integral on the left is the Mellin transform we seek. By use of properties of the gamma function, we can write

$$\Gamma(s)\Gamma(\tfrac{1}{2}-\tfrac{1}{2}s)\cos(\tfrac{1}{2}\pi s) = \frac{\pi\Gamma(s)}{\Gamma(\frac{1}{2}+\frac{1}{2}s)} = \sqrt{\pi}\,2^{s-1}\,\Gamma(\tfrac{1}{2}s)$$

and hence we deduce that

$$\mathcal{M}\{x^{-\nu}J_\nu(ax);s\} = \frac{a^{\nu-s}2^{s-\nu-1}\,\Gamma(\frac{1}{2}s)}{\Gamma(\nu-\frac{1}{2}s+1)}, \qquad a>0,\ \nu>-1/2$$

EXERCISES 6.2

In Probs. 1–15, evaluate the Mellin transform of the given function. When possible, use known integral results from previous chapters.

1. $f(x) = h(a-x), \quad a>0$

2. $f(x) = e^{-bx}, \quad b>0$

3. $f(x) = x^a e^{-bx}, \quad a,b>0$

4. $f(x) = e^{-b^2x^2}, \quad b>0$

5. $f(x) = \log(b/x), \quad b>0$

6. $f(x) = x^a\log(b/x), \quad a,b>0$

7. $f(x) = x^a/(1+x)^b, \quad a,b>0$

8. $f(x) = x^a(1-x)^{b-1}h(1-x), \quad a>0$

9. $f(x) = \dfrac{h(1-x)}{(x-1)^a}, \quad a>0$

10. $f(x) = \dfrac{[\sqrt{x^2+1}-x]^a}{\sqrt{x^2+1}}, \quad a>0$

Hint: Let $x = \frac{1}{2}y/\sqrt{y^2+1}$.

11. $f(x) = \dfrac{1}{1+x^a}, \quad a>0$

12. $f(x) = 1/(1+x^a)^m, \quad a,m>0$

13. $f(x) = J_\nu(x), \quad \nu>-1/2$

14. $f(x) = J_\nu(\sqrt{x}), \quad \nu>-1/2$

15. $f(x) = J_\nu(x^2), \quad \nu>-1/2$

Hint: In Probs. 14 and 15, use the result of Prob. 13 and Eq. (6.12).

In Probs. 16–28, verify the given operational property of the Mellin transform.

16. $\mathcal{M}\{f(x^a);s\} = (1/a)\,F(s/a), \quad a>0$

17. $\mathcal{M}\{(1/x)f(1/x);s\} = F(1-s)$

18. $\mathcal{M}\{(\log x)f(x);s\} = F'(s)$ **19.** $\mathcal{M}\{(x\, d/dx)f(x);s\} = -s\, F(s)$

20. $\mathcal{M}\{(x\, d/dx)^n f(x);s\} = (-s)^n F(s), \quad n = 1,2,3,\ldots$

21. $\mathcal{M}\{x^n f^{(n)}(x);s\} = (-1)^n \dfrac{\Gamma(s+n)}{\Gamma(s)} F(s), \quad n = 1,2,3,\ldots$

22. $\mathcal{M}\{x^2 f''(x) + xf'(x);s\} = s^2 F(s)$

23. $\mathcal{M}\left\{\int_0^x f(u)\, du;s\right\} = -(1/s)\, F(s+1)$

24. $\mathcal{M}\left\{\int_x^\infty f(u)\, du;s\right\} = (1/s)\, F(s+1)$

25. $\mathcal{M}^{-1}\{F(s)G(s);x\} = \int_0^\infty f(x/u)g(u)\, du/u$

26. $\mathcal{M}\{f(x)g(x);s\} = \dfrac{1}{2\pi i}\int_{c-i\infty}^{c+i\infty} F(p)G(s-p)\, dp$

27. $\mathcal{M}\left\{\int_0^\infty u^m f(u)g(xu)\, du;s\right\} = F(1+m-s)G(s)$

28. $\mathcal{M}\left\{\int_0^\infty u^m f(u)g(x/u)\, du;s\right\} = F(1+m+s)G(s)$

In Probs. 29–33, verify the given Mellin transform relation.

29. $\mathcal{M}^{-1}\{\Gamma(s)F(1-s);x\} = \mathcal{L}\{f(t);x\}$

30. $\mathcal{M}^{-1}\{\cos(\tfrac{1}{2}\pi s)\Gamma(s)F(1-s);x\} = \sqrt{\pi/2}\ \mathcal{F}_C\{f(t);x\}$

31. $\mathcal{M}^{-1}\{\sin(\tfrac{1}{2}\pi s)\Gamma(s)F(1-s);x\} = \sqrt{\pi/2}\ \mathcal{F}_S\{f(t);x\}$

32. $\mathcal{M}\{e^{-x\cos\phi}\cos(x\sin\phi);s\} = \Gamma(s)\cos(s\phi), \quad -\pi/2 < \phi < \pi/2, \quad \mathrm{Re}(s) > 0$

33. $\mathcal{M}\{e^{-x\cos\phi}\sin(x\sin\phi);s\} = \Gamma(s)\sin(s\phi), \quad -\pi/2 < \phi < \pi/2, \quad \mathrm{Re}(s) > -1$

34. Starting with the Dirichlet series

$$g(s) = \sum_{n=1}^{\infty} \frac{a_n}{n^s}$$

and the integral representation

$$\frac{1}{n^s} = \frac{1}{\Gamma(s)}\int_0^\infty e^{-nx}x^{s-1}\, dx$$

(a) show, by summing over all positive integers n, that

$$g(s)\Gamma(s) = \int_0^\infty x^{s-1} f(x)\,dx$$

where

$$f(x) = \sum_{n=1}^{\infty} a_n e^{-nx}$$

(b) From (a), deduce the inverse transform relation

$$f(x) = \frac{1}{2\pi i}\int_{c-i\infty}^{c+i\infty} x^{-s}\, g(s)\Gamma(s)\,ds$$

35. Show that

$$\mathcal{M}\left\{\frac{1}{e^x - 1};s\right\} = \zeta(s)\Gamma(s)$$

where $\zeta(s)$ is the *Riemann zeta function* defined by

$$\zeta(s) = \sum_{n=1}^{\infty} \frac{1}{n^s}, \qquad \mathrm{Re}(s) > 1$$

Hint: See Prob. 34.

36. Using Prob. 35, show that

(a) $\mathcal{M}\left\{\frac{1}{e^x - 1};s\right\} - \mathcal{M}\left\{\frac{1}{e^x + 1};s\right\} = 2^{1-s}\,\zeta(s)\Gamma(s)$

(b) $\mathcal{M}\left\{\frac{1}{e^x + 1};s\right\} = (1 - 2^{1-s})\zeta(s)$

37. Starting with the Mellin transform relation

$$\mathcal{M}\left\{\frac{1}{1 + x};s\right\} = \frac{\pi}{\sin \pi s}$$

use Prob. 23 to deduce that

$$\mathcal{M}\{\log(1 + x);s\} = \pi/(s \sin \pi s)$$

38. Starting with the Mellin transform relation

$$\mathcal{M}\left\{\frac{1}{1 + x^2};s\right\} = \frac{\pi}{2\sin(\frac{1}{2}\pi s)}$$

use Prob. 24 to deduce that

$$\mathcal{M}\left\{\left(\frac{\pi}{2} - \tan^{-1}x\right);s\right\} = \frac{\pi}{2s\cos(\frac{1}{2}\pi s)}$$

6.3 *Complex Variable Methods*

Up to this point we have evaluated Mellin transforms primarily by relating the transform integral to known results involving other integral transforms. A more direct evaluation of the Mellin transform may sometimes be accomplished by complex variable methods, particularly when the function $f(x)$ is a rational function. Also, since the inversion formula of the Mellin transform is formulated in the complex plane, it likewise lends itself to contour integration techniques similar to those used in evaluating inverse Laplace transforms. In this section we will briefly illustrate the method of residues as it applies to both the Mellin transform and its inversion formula.

6.3.1 *Mellin Transforms*

Suppose that $f(z)$ is a rational function having no poles on the positive real axis. Suppose further that there exists real constants σ_1 and σ_2 such that

$$\begin{aligned} \lim_{z\to 0} z^s f(z) &= 0, \qquad \sigma_1 < \text{Re}(s) < \sigma_2 \\ \lim_{|z|\to\infty} z^s f(z) &= 0, \qquad \sigma_1 < \text{Re}(s) < \sigma_2 \end{aligned} \tag{6.20}$$

If $f(z)$ has N poles at $z = a_1, a_2, \cdots, a_N$, then we begin by integrating the function $z^{s-1}f(z)$ around the contour shown in Fig. 6.1, which encloses all poles of $f(z)$. Based on the Residue Theorem (see Appendix A)

$$\oint_C z^{s-1} f(z)\, dz = 2\pi \text{i} \sum_{k=1}^{N} \text{Res}\{z^{s-1}f(z);a_k\}$$

or

$$\int_\rho^R x^{s-1}f(x)dx + \int_{C_R} z^{s-1}f(z)dz + \int_R^\rho (xe^{2\pi i})^{s-1}f(x)dx + \int_{C_\rho} z^{s-1}f(z)dz$$
$$= 2\pi i \sum_{k=1}^{N} \text{Res}\{z^{s-1}f(z);a_k\} \tag{6.21}$$

Because of the limit relations (6.20), it can be shown that the integral around the large circle of radius R vanishes in the limit as $R \to \infty$, as does the integral around the small circle of radius ρ as $\rho \to 0$. Hence, in the limit as $R \to \infty$ and $\rho \to 0$, we obtain

$$(1 - e^{2\pi i s}) \int_0^\infty x^{s-1}f(x)\, dx = 2\pi i \sum_{k=1}^{N} \text{Res}\{z^{s-1}f(z);a_k\}. \tag{6.22}$$

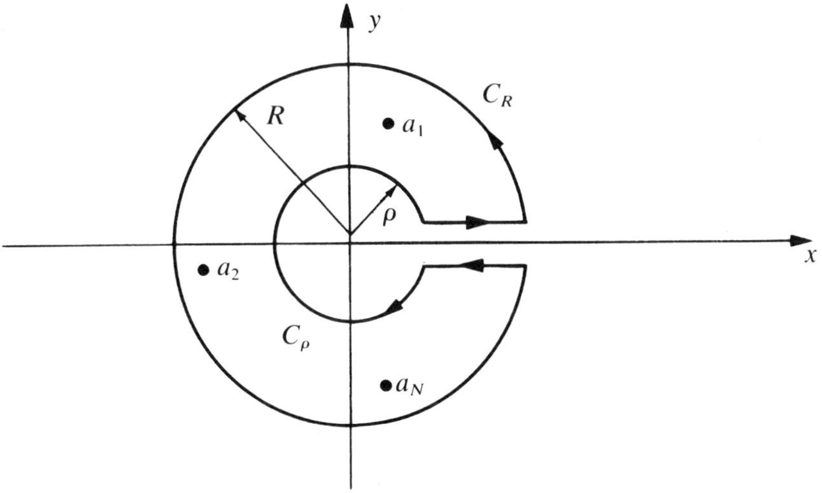

Figure 6.1 Contour of integration

However,

$$1 - e^{2\pi is} = e^{i\pi s}(e^{-i\pi s} - e^{i\pi s}) = -2ie^{i\pi s}\sin \pi s$$

and thus (6.22) becomes

$$\int_0^\infty x^{s-1} f(x)\, dx = -\frac{\pi e^{-i\pi s}}{\sin \pi s} \sum_{k=1}^{N} \text{Res}\{z^{s-1}f(z);a_k\} \tag{6.23}$$

Eq. (6.23) can be easily modified to include the case where $f(z)$ has poles on the positive real axis (see Prob. 8 in Exer. 6.3).

Example 6.6: Find the Mellin transform of $1/(1 + x^2)$.

Solution: We first note that conditions (6.20) hold if $0 < \text{Re}(s) < 2$. Also, the poles of $f(z) = 1/(1 + z^2)$ are simple poles at $z = \pm i$, and the residues at these poles are

$$\text{Res}\left\{\frac{z^{s-1}}{1 + z^2};i\right\} = \frac{i^{s-1}}{2i} = -\frac{1}{2}\, e^{i\pi s/2}$$

and

$$\text{Res}\left\{\frac{z^{s-1}}{1 + z^2};-i\right\} = \frac{(-i)^{s-1}}{-2i} = -\frac{1}{2}\, e^{3i\pi s/2}$$

where we have observed that $i = e^{i\pi/2}$ and $-i = e^{3i\pi/2}$. From (6.23), it now follows that

$$\int_0^\infty \frac{x^{s-1}}{1+x^2}\,dx = \frac{\pi e^{-i\pi s}}{\sin \pi s}\cdot\frac{1}{2}(e^{i\pi s/2} + e^{3i\pi s/2})$$

$$= \frac{\pi}{\sin \pi s}\cdot\frac{1}{2}(e^{-i\pi s/2} + e^{i\pi s/2})$$

$$= \frac{\pi \cos(\pi s/2)}{\sin \pi s}$$

One further simplification leads to our final result

$$\mathcal{M}\left\{\frac{1}{1+x^2};s\right\} = \frac{\pi}{2\sin(\pi s/2)}, \qquad 0 < \mathrm{Re}(s) < 2$$

Observe also that we could obtain this result from Exam. 6.3 and the operational property given by Eq. (6.12).

6.3.2 *Inverse Mellin Transforms*

In Exam. 6.1 we obtained the Mellin transform

$$\mathcal{M}\{e^{-x};s\} = \Gamma(s), \qquad \mathrm{Re}(s) > 0 \tag{6.24}$$

To formally recover the function e^{-x} from the transform function $\Gamma(s)$, we consider the inversion formula

$$\mathcal{M}^{-1}\{\Gamma(s);x\} = \frac{1}{2\pi i}\int_{c-i\infty}^{c+i\infty} x^{-s}\,\Gamma(s)\,ds \tag{6.25}$$

All of the poles of $\Gamma(s)$ occur in the left-half plane, viz., at $s = -n$ $(n = 0,1,2,\ldots)$. Thus, since $\Gamma(s)$ is analytic in the right-half plane, we can evaluate (6.25) by considering the same kind of contour used in evaluating inverse Laplace transforms [see Fig. 6.2(a)]. Due to the asymptotic behavior of $\Gamma(s)$ for large $|s|$, it can be shown that the integral along the large circular arc tends to zero for all x as the radius of the circular arc tends to infinity. Based on the calculus of residues, we then conclude that

$$\mathcal{M}^{-1}\{\Gamma(s);x\} = \sum_{n=0}^{\infty}\mathrm{Res}\{x^{-s}\Gamma(s); -n\} \tag{6.26}$$

Now, by writing [recall property (G9) in Sec. 1.2.2]

$$\Gamma(s) = \frac{\pi}{\Gamma(1-s)\sin \pi s} \tag{6.27}$$

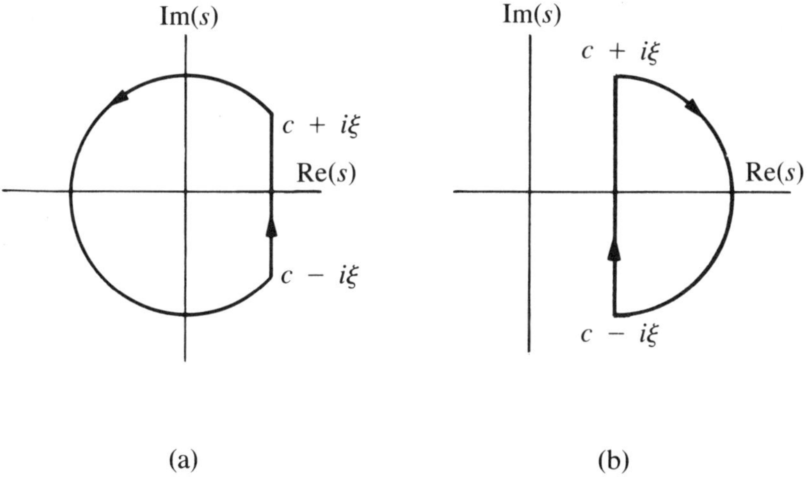

Figure 6.2 Contours of integration

we find

$$\operatorname{Res}\{x^{-s}\Gamma(s);-n\} = \left.\frac{x^{-s}}{\Gamma(1-s)\cos \pi s}\right|_{s=-n} = \frac{(-1)^n x^n}{n!}$$

and therefore it follows from (6.26) that

$$\mathcal{M}^{-1}\{\Gamma(s);x\} = \sum_{n=0}^{\infty}\frac{(-1)^n x^n}{n!} = e^{-x} \tag{6.28}$$

In general, the result we obtain by enclosing the poles of $F(s)$ in the left-half plane by a contour as shown in Fig. 6.2(a) will be restricted to the interval $0 < x < x_0$ where x_0 is determined such that the integral along the circular arc vanishes in the limit as the radius tends to infinity. If the number of poles along the negative real axis is infinite, this will lead to an *ascending* series in x like that in (6.28). In some cases the function $F(s)$ may also have poles along the positive real axis which we then enclose by a contour as shown in Fig. 6.2(b). This time the result obtained will be valid for $x > x_0$, and will be a *descending* series in x.

Example 6.7: Use the method of residues to find the inverse Mellin transform of

$$F(s) = \frac{\Gamma(s)\Gamma(m-s)}{a^s\Gamma(m)}, \qquad a > 0,\ 0 < \operatorname{Re}(s) < m$$

Solution: The function $F(s)$ has two sets of poles, viz., at $s = -n$ $(n = 0,1,2,\ldots)$ and at $s = m + n$ $(n = 0,1,2,\ldots)$. Thus, to find the inverse Mellin transform we must apply the residue theory to both contours shown in Fig. 6.2.

By first choosing a contour as shown in Fig. 6.2(a), we have that

$$\mathcal{M}^{-1}\{F(s);x\} = \sum_{n=0}^{\infty} \text{Res}\{x^{-s}F(s);-n\}$$

for some restricted values of x to be determined. Using Eq. (6.27), the residue at $s = -n$ is found to be

$$\begin{aligned}\text{Res}\{x^{-s}F(s);-n\} &= \left.\frac{x^{-s}\Gamma(m-s)}{a^s\Gamma(m)\cos\pi s}\right|_{s=-n}\\ &= \frac{(-1)^n\Gamma(m+n)}{n!\Gamma(m)}(ax)^n\end{aligned}$$

However, using the identity (see Prob. 10 in Exer. 1.2)

$$\frac{(-1)^n\Gamma(m+n)}{n!\Gamma(m)} = (-1)^n\binom{m+n-1}{n} = \binom{-m}{n}$$

we get the ascending binomial series

$$\mathcal{M}^{-1}\{F(s);x\} = \sum_{n=0}^{\infty}\binom{-m}{n}(ax)^n$$

It is known that this series converges for $|ax| < 1$,* and hence we have

$$\mathcal{M}^{-1}\{F(s);x\} = 1/(1+ax)^m, \qquad 0 < x < 1/a$$

Choosing a second contour as shown in Fig. 6.2(b), the residue calculus applied to the remaining poles of $F(s)$ on the positive real axis leads to

$$\mathcal{M}^{-1}\{F(s);x\} = -\sum_{n=0}^{\infty} \text{Res}\{x^{-s}F(s);m+n\}$$

where the negative sign in front of the summation is due to integrating in the negative direction along the closed contour in Fig. 6.2(b). Recalling Eq. (1.6) in Sec. 1.2, we find that

$$\begin{aligned}\text{Res}\{x^{-s}F(s);m+n\} &= \lim_{s\to m+n}\frac{(s-m-n)\Gamma(s)\Gamma(m-s)}{(ax)^s\Gamma(m)}\\ &= -\lim_{s\to m+n}\frac{\Gamma(s)\Gamma(m-s+n+1)}{(ax)^s(m-s)(m-s+1)\cdots(m-s+n-1)}\end{aligned}$$

---

*These are the same values of x for which the integral along the circular arc in Fig. 6.2(a) vanishes in the limit as the radius tends to infinity.

$$= -\frac{(-1)^n\Gamma(m+n)}{n!\Gamma(m)}$$

$$= -\binom{-m}{n}$$

Hence, we obtain the descending binomial series

$$\mathcal{M}^{-1}\{F(s);x\} = \frac{1}{(ax)^m}\sum_{n=0}^{\infty}\binom{-m}{n}(ax)^{-n}$$

which converges for $|ax| > 1$, and therefore

$$\mathcal{M}^{-1}\{F(s);x\} = 1/(1+ax)^m, \qquad x > 1/a$$

Combining results, we finally deduce that

$$\mathcal{M}^{-1}\left\{\frac{\Gamma(s)\Gamma(m-s)}{a^s\Gamma(m)};x\right\} = \frac{1}{(1+ax)^m}, \qquad 0 < x < \infty$$

in agreement with Exam. 6.4.

6.3.3 *Transforms in Polar Coordinates*

In the application of Mellin transforms in polar coordinates (r,θ) it is frequently necessary to consider inverse Mellin transforms of functions such as $F(s)\cos s\theta$ and $F(s)\sin s\theta$, where $F(s)$ is the Mellin transform of a real function $f(\xi)$ (e.g., see Sec. 6.4.3). The technique illustrated below for evaluating such inverse transform is based on a paper by W. J. Harrington,* who provides a more rigorous treatment and justification of the formal result.

Suppose that $F(s)$ is the Mellin transform of a real function $f(\xi)$. Then, we have formally

$$\mathcal{M}\{f(re^{i\theta});r\to s\} = \int_0^\infty r^{s-1}f(re^{i\theta})\,dr$$

$$= e^{-is\theta}\int_0^\infty \xi^{s-1}f(\xi)\,d\xi$$

where we have set $\xi = re^{i\theta}$. Hence, we deduce that

$$\mathcal{M}\{f(re^{i\theta});r\to s\} - e^{-is\theta}\,F(s)$$

*W. J. Harrington, "A property of Mellin transforms," *SIAM Review*, **9**, No. 3, 542–547, July (1967).

which leads to

$$\mathcal{M}^{-1}\{F(s)\cos s\theta; s \to r\} = \operatorname{Re}[f(re^{i\theta})] \tag{6.29a}$$

$$\mathcal{M}^{-1}\{F(s)\sin s\theta; s \to r\} = -\operatorname{Im}[f(re^{i\theta})] \tag{6.29b}$$

Example 6.8: Find the inverse Mellin transforms of

$$\frac{\pi \cos s\theta}{\sin s\pi} \text{ and } \frac{\pi \sin s\theta}{\sin s\pi}$$

Solution: We first recall Exam. 6.3, which gives us

$$\mathcal{M}^{-1}\left\{\frac{\pi}{\sin s\pi}; x\right\} = \frac{1}{1 + x}$$

Therefore, from (6.29a) it follows that

$$\mathcal{M}^{-1}\left\{\frac{\pi \cos s\theta}{\sin s\pi}; s \to r\right\} = \operatorname{Re}\left[\frac{1}{1 + re^{i\theta}}\right] = \frac{1 + r\cos\theta}{1 + 2r\cos\theta + r^2}$$

and similarly, from (6.29b),

$$\mathcal{M}^{-1}\left\{\frac{\pi \sin s\theta}{\sin s\pi}; s \to r\right\} = \frac{r\sin\theta}{1 + 2r\cos\theta + r^2}$$

EXERCISES 6.3

In Probs. 1–6, use residue calculus to evaluate the Mellin transform of the given function.

1. $f(x) = \dfrac{1}{(x + a)(x + b)}, \quad a,b > 0$

2. $f(x) = \dfrac{1}{(1 + x)^3}$

3. $f(x) = \dfrac{1}{1 + x^3}$

4. $f(x) = \dfrac{1}{1 + x^4}$

5. $f(x) = \dfrac{x}{(1 + x^2)^2}$

6. $f(x) = \dfrac{1}{(x^2 + a^2)(x^2 + b^2)}, \quad a \neq b,\ a,b > 0$

7. Show that Eq. (6.23) can be equivalently expressed in the form

$$\mathcal{M}\{f(x); s\} = \frac{\pi}{\sin \pi s} \sum_{k=1}^{N} \operatorname{Res}\{(e^{-i\pi}z)^{s-1} f(z); a_k\}$$

8. If $f(z)$ is a rational function satisfying conditions (6.20), and if $f(z)$

also has poles $b_1, b_2, \ldots, b_M$ on the positive real axis, show that

$$\mathcal{M}\{f(x);s\} = \frac{\pi}{\sin \pi s} \sum_{k=1}^{N} \text{Res}\{(e^{-i\pi}z)^{s-1}f(z);a_k\}$$

$$- \pi \cot \pi s \sum_{k=1}^{N} \text{Res}\{f(z);b_k\}$$

In Probs. 9 and 10, use the result of Prob. 8 to find the Mellin transform of the given function.

9. $f(x) = \dfrac{1}{1 - x}$ **10.** $f(x) = \dfrac{1}{4 - x^2}$

11. Show that

(a) $\mathcal{M}\{\text{erfc}(x);s\} = \Gamma\left(\dfrac{s+1}{2}\right)/\sqrt{\pi}\, s$

(b) Using residue methods, show that

$$\mathcal{M}^{-1}\left\{\frac{\Gamma\left(\dfrac{s+1}{2}\right)}{\sqrt{\pi}\, s};x\right\} = 1 - \frac{2}{\sqrt{\pi}} \sum_{n=0}^{\infty} \frac{(-1)^n x^{2n+1}}{n!(2n+1)} = \text{erfc}(x)$$

12. Show that

(a) $\mathcal{M}\{E_1(x);s\} = \Gamma(s)/s$
where $E_1(x)$ is the *exponential integral* defined by

$$E_1(x) = \int_x^{\infty} \frac{e^{-t}}{t}\, dt$$

(b) Using residue methods, find an ascending series for the inverse Mellin transform of $F(s) = \Gamma(s)/s$.

13. Show that

(a) $\mathcal{M}\{\text{Ci}(x);s\} = -\dfrac{\Gamma(s)}{s} \cos(\pi s/2)$

where $\text{Ci}(x)$ is the *cosine integral* defined by

$$\text{Ci}(x) = -\int_x^{\infty} \frac{\cos t}{t}\, dt$$

(b) Using residue methods, find an ascending series for the inverse Mellin transform of $F(s) = -\Gamma(s)\cos(\pi s/2)/s$.

14. Use residues to show that

(a) $$\mathcal{M}^{-1}\left\{\frac{\pi}{2s\cos(\pi s/2)};x\right\} = \frac{\pi}{2} - \sum_{n=0}^{\infty} \frac{(-1)^n x^{2n+1}}{2n+1}, \qquad 0 < x < 1$$

(b) $\mathcal{M}^{-1}\left\{\dfrac{\pi}{2s\cos(\pi s/2)};x\right\} = \sum_{n=0}^{\infty} \dfrac{(-1)^n}{(2n+1)x^{2n+1}}, \qquad x > 1$

(c) From (a) and (b), deduce that

$$\mathcal{M}^{-1}\left\{\frac{\pi}{2s\cos(\pi s/2)};x\right\} = \frac{\pi}{2} - \tan^{-1}x, \qquad x > 0$$

In Probs. 15–26, use residue methods to determine the inverse Mellin transform of the given function.

15. $F(s) = \dfrac{\Gamma(s/2)}{2a^s}$

16. $F(s) = \dfrac{\pi}{\sin \pi s}$

17. $F(s) = \dfrac{\pi}{2\sin(\pi s/2)}$

18. $F(s) = \Gamma(s)\cos(\pi s/2)$

19. $F(s) = \Gamma(s)\sin(\pi s/2)$

20. $F(s) = \dfrac{2^{s-1}\Gamma(s/2)}{\Gamma(1 - s/2)}$

21. $F(s) = \dfrac{\cos s\phi}{\sin s\alpha}$, $-\dfrac{\pi}{2} < \phi < \dfrac{\pi}{2}$

22. $F(s) = \dfrac{\sin s\phi}{\sin s\alpha}$, $-\dfrac{\pi}{2} < \phi < \dfrac{\pi}{2}$

23. $F(s) = \dfrac{\pi \cos s\theta}{s \sin s\pi}$

Hint: See Prob. 37 in Exer. 6.2.

24. $F(s) = \dfrac{\pi \sin s\theta}{s \sin s\pi}$

25. $F(s) = \dfrac{\cos s\theta}{\cos s\pi}$

26. $F(s) = \dfrac{\sin s\theta}{\cos s\pi}$

6.4 *Applications*

Although the Mellin transform is not nearly as versatile in applications as are the Fourier and Laplace transforms, there are some areas of application where it can be a useful tool. In particular, it is useful in the summation of certain series, in finding the distribution function for products of random variables, and in solving for the potential function in a wedge-shaped region. Our discussion of these applications, however, will be intentionally brief since any deeper treatment would require mathematical knowledge beyond the stated prerequisites.

6.4.1 *Summation of Series*

The discussion in this section is based on a paper by G. G. MacFarlane, who considered more general series than those presented here.*

Given that $f(x)$ and $F(s)$ are Mellin transform pairs, they are related by the inversion formula

$$f(x) = \frac{1}{2\pi i}\int_{c-i\infty}^{c+i\infty} x^{-s} F(s)\, ds$$

Replacing x by nx, where $n = 1,2,3,\ldots$, this inversion formula becomes

$$f(nx) = \frac{1}{2\pi i}\int_{c-i\infty}^{c+i\infty} n^{-s}x^{-s}F(s)\, ds, \qquad n = 1,2,3,\ldots \tag{6.30}$$

Now let us sum both sides of (6.30) over all positive integer values of n, interchanging the order of summation and integration on the right-hand side. This action leads to

$$\sum_{n=1}^{\infty} f(nx) = \frac{1}{2\pi i}\int_{c-i\infty}^{c+i\infty} x^{-s}\zeta(s)F(s)\, ds \tag{6.31}$$

where $\zeta(s)$ is the *Riemann zeta function* defined by

$$\zeta(s) = \sum_{n=1}^{\infty}\frac{1}{n^s}, \qquad \mathrm{Re}(s) > 1 \tag{6.32}$$

Thus we have the result

$$\sum_{n=1}^{\infty} f(nx) = \mathcal{M}^{-1}\{\zeta(s)F(s)\,;x\} \tag{6.33}$$

The importance of Eq. (6.33) is that we can replace the series on the left, which in some cases may converge very slowly, by the inverse Mellin transform on the right. This inverse transform leads to another infinite series, but in certain cases it may be summed exactly. In other cases this new series may converge faster than the original series, which is important for computational purposes. Proficiency in this technique of summing series requires knowledge of some properties of the zeta function, most of which were developed in 1859 by G. Riemann (1826–

*G. G. MacFarlane, "The application of Mellin transforms to the summation of slowly convergent series," *Phil. Mag.*, *(vii)*, **40**, 188 (1949).

1866). Some of the relations involving the zeta function that prove useful in our work here include the following:*

$$\zeta(0) = -1/2 \tag{6.34}$$

$$\zeta(1) = \infty \tag{6.35}$$

$$\zeta(2) = \pi^2/6 \tag{6.36}$$

$$\zeta(4) = \pi^4/90, \tag{6.37}$$

$$\zeta(-2n) = 0, \qquad n = 1,2,3,\ldots \tag{6.38}$$

$$\zeta'(0) = -\tfrac{1}{2}\log 2\pi \tag{6.39}$$

$$\pi^s\zeta(1-s) = 2^{1-s}\Gamma(s)\zeta(s)\cos(\pi s/2) \tag{6.40}$$

Example 6.9: Sum the series $\sum_{n=1}^{\infty}(\cos an)/n^2$.

Solution: The given series defines the function

$$f(x) = \frac{\cos ax}{x^2}$$

whose Mellin transform is

$$\begin{aligned}\mathcal{M}\{f(x);s\} &= \int_0^\infty x^{s-3}\cos ax\,dx\\ &= \mathcal{M}\{\cos ax;\, s-2\}\\ &= -(1/a^{s-2})\Gamma(s-2)\cos(\pi s/2)\end{aligned}$$

In deriving this transform we have used the result of Exam. 6.2 and the scaling property (6.10). Substituting this result into Eq. (6.33) yields

$$\sum_{n=1}^{\infty}\frac{\cos an}{n^2} = -\mathcal{M}^{-1}\left\{\frac{1}{a^{s-2}}\zeta(s)\Gamma(s-2)\cos(\pi s/2);x=1\right\}$$

Using property (6.40) of the zeta function, we can express this inverse Mellin transform in the more convenient form

$$\begin{aligned}\sum_{n=1}^{\infty}\frac{\cos an}{n^2} &= -\frac{a^2}{2}\mathcal{M}^{-1}\left\{\left(\frac{2\pi}{a}\right)^s\frac{\zeta(1-s)\Gamma(s-2)}{\Gamma(s)};x=1\right\}\\ &= -\frac{a^2}{2}\mathcal{M}^{-1}\left\{\left(\frac{2\pi}{a}\right)^s\frac{\zeta(1-s)}{(s-1)(s-2)};x=1\right\}\end{aligned}$$

*For a detailed discussion of the zeta function, see E. T. Whittaker and G. N. Watson, *A Course of Modern Analysis*, Cambridge: Cambridge University Press, 1965.

Thus, we have three simple poles at $s = 0, 1, 2$, the residues of which are

$$\text{Res}\{0\} = -1/2, \text{Res}\{1\} = \pi/a, \text{Res}\{2\} = -\pi^2/3a^2$$

The details of evaluating these residues are left to the exercises (see Probs. 1 and 2 in Exer. 6.4). Combining results, we finally obtain

$$\sum_{n=1}^{\infty} \frac{\cos an}{n^2} = -\frac{a^2}{2}\left(-\frac{1}{2} + \frac{\pi}{a} - \frac{\pi^2}{3a^2}\right) = \frac{a^2}{4} - \frac{\pi a}{2} + \frac{\pi^2}{6}$$

Observe that in Example 6.9 the limit $a \to 0^+$ leads to the special case

$$\sum_{n=1}^{\infty} \frac{1}{n^2} = \frac{\pi^2}{6} \tag{6.41}$$

6.4.2 *Products of Random Variables*

Unlike the distribution of sums of independent random variables, the distribution of products of independent random variables has received relatively little attention in the literature. This is particularly true of products involving more than two variables. It has been shown that the probability density function (PDF) for such products can be obtained through use of the Mellin transform.* The general theory concerning these problems is too complex for our purposes, but we can illustrate the utility of the Mellin transform by examining certain products involving only two random variables.

Let us define $Z = XY$, where X and Y are independent, positive, random variables. From probability theory it is known that the PDF of Z is related to the joint PDF of X and Y by

$$p_Z(z) = \frac{d}{dz} \iint_{D_z} p_{X,Y}(x,y)\, dy\, dx \tag{6.42}$$

where D_z is the region of the xy plane such that $xy \le z$. (In this section the variable z is considered to be real.) Since X and Y are independent random variables, we can express their joint density function as a product of their respective PDFs, and thus (6.42) becomes

$$p_Z(z) = \frac{d}{dz} \int_0^{\infty} \int_0^{z/x} p_X(x) p_Y(y)\, dy\, dx$$

*See M. D. Springer and W. E. Thompson, "The distribution of products of beta, gamma, and Gaussian random variables," *SIAM J. Appl. Math.* **18**, No. 4, 721–737, June 1970.

which simplifies to

$$p_Z(z) = \int_0^\infty p_X(x)p_Y(z/x)\frac{dx}{x} \tag{6.43}$$

We recognize (6.43) as the convolution integral of the Mellin transform [see Eq. (6.19)], and, therefore, we can express (6.43) in the alternate form

$$p_Z(z) = \mathcal{M}^{-1}\{F_1(s)F_2(s);z\} \tag{6.44}$$

where $F_1(s)$ and $F_2(s)$ are the Mellin transforms, respectively, of $p_X(x)$ and $p_Y(y)$.

Although we will not provide the details, we can generalize (6.44) to the case involving products of N random variables. That is, if

$$Z = \prod_{j=1}^{N} X_j \tag{6.45}$$

where $X_1, X_2, \ldots, X_N$ are independent, positive, random variables, then it can be shown that

$$p_Z(z) = \mathcal{M}^{-1}\left\{\prod_{j=1}^{N} F_j(s);z\right\} \tag{6.46}$$

where $F_j(s)$ is the Mellin transform of the PDF for X_j, $j = 1,2,\ldots,N$. Further generalizations to products involving both positive and negative random variables have been established, as well as generalizations to certain quotients of random variables.

Example 6.10: Find the PDF of the product $Z = XY$, where X and Y are independent Cauchy random variables, each having density function

$$p(x) = 2/\pi(1 + x^2), \qquad x > 0$$

Solution: Using (6.43), we find immediately (upon simplification)

$$p_Z(z) = \frac{4}{\pi^2}\int_0^\infty \frac{x\,dx}{(x^2 + 1)(x^2 + z^2)}$$

which we can interpret as a Mellin transform of the rational function

$$f(x) = 1/(x^2 + 1)(x^2 + z^2)$$

for the special value $s = 2$. Using the method of residues, it can be shown that (see Prob. 11 in Exer. 6.4)

$$\mathcal{M}\{f(x);s\} = \frac{\pi(z^{s-2} - 1)}{2(z^2 - 1)\sin(\pi s/2)}$$

Therefore, the result we need is

$$p_Z(z) = \lim_{s\to 2} \frac{2(z^{s-2} - 1)}{\pi(z^2 - 1)\sin(\pi s/2)}$$

which, through an application of Leibnitz's rule, leads to

$$p_Z(z) = \frac{4 \log z}{\pi^2(z^2 - 1)}, \qquad z > 0$$

We could equally well use Eq. (6.44) to find the PDF by first noting that

$$\mathcal{M}\{p_X(x);s\} = \mathcal{M}\{p_Y(y);s\} = 1/\sin(\pi s/2)$$

However, the details of this approach are left to the exercises (see Prob. 12 in Exer. 6.4).

Finally, observe that, because $p_X(x)$ and $p_Y(y)$ are even functions, we could extend our result to products involving Cauchy random variables that extend over $-\infty < x < \infty$, $-\infty < y < \infty$. The PDF for $Z = XY$ in this case assumes the form

$$p_Z(z) = \frac{\log z^2}{\pi^2(z^2 - 1)}, \qquad -\infty < z < \infty$$

the verification of which we leave to the reader.

6.4.3 *Distribution of Potential in a Wedge*

Let us consider the boundary value problem of determining a potential function in an infinite wedge-shaped region in the xy plane (see Fig. 6.3). In polar coordinates the problem is mathematically formulated by

$$r^2u_{rr} + ru_r + u_{\theta\theta} = 0, \qquad 0 < r < \infty,\ -\alpha < \theta < \alpha$$

$$\text{B.C.:} \quad \begin{cases} u(r,-\alpha) = f(r), & u(r,\alpha) = g(r) \\ u(r,\theta) \to 0 \text{ as } r \to \infty, & |\theta| < \alpha \end{cases} \tag{6.47}$$

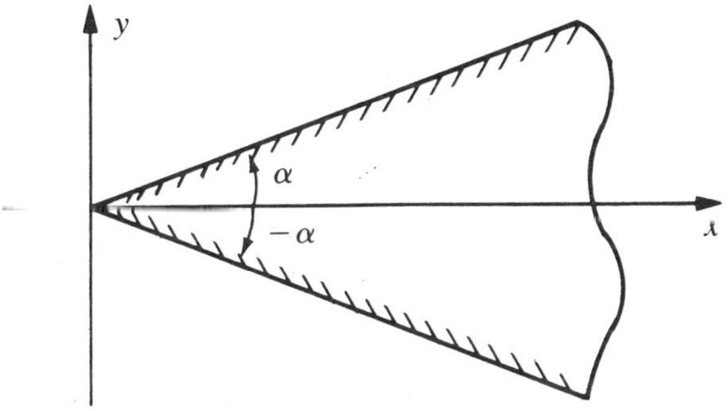

Figure 6.3 Infinite wedge

Among other areas of application, this problem describes the steady-state temperature distribution in an infinite wedge, given the temperature distribution along the boundaries of the wedge.

Using the differentiation property of the Mellin transform (see Prob. 22 in Exer. 6.2)

$$\mathcal{M}\{r^2 u_{rr} + r u_r ; r \to s\} = s^2 U(s, \theta) \tag{6.48}$$

we find that under the Mellin transform the potential problem (6.47) is transformed to

$$\begin{aligned} & U_{\theta\theta} + s^2 U = 0, \qquad -\alpha < \theta < \alpha \\ \text{B.C.:}\quad & U(s, -\alpha) = F(s), \qquad U(s, \alpha) = G(s) \end{aligned} \tag{6.49}$$

The general solution of the transformed DE is

$$U(s, \theta) = A(s)\cos s\theta + B(s)\sin s\theta$$

and by imposing the boundary conditions in (6.49), we obtain

$$U(s, \theta) = F(s)\frac{\sin s(\alpha - \theta)}{\sin 2\alpha s} + G(s)\frac{\sin s(\alpha + \theta)}{\sin 2\alpha s} \tag{6.50}$$

The inversion of this solution by the convolution theorem, given by Eq. (6.19), leads to the formal result

$$\begin{aligned} u(r, \theta) = \frac{r^m}{2\alpha}\cos(m\theta)\Bigg[& \int_0^\infty \frac{\xi^{m-1} f(\xi)\, d\xi}{\xi^{2m} + 2r^m \xi^m \sin(m\theta) + r^{2m}} \\ & + \int_0^\infty \frac{\xi^{m-1} g(\xi)\, d\xi}{\xi^{2m} - 2r^m \xi^m \sin(m\theta) + r^{2m}}\Bigg] \end{aligned} \tag{6.51}$$

where we are using the inverse Mellin transform relation (see Prob. 22 in Exer. 6.3)

$$\mathcal{M}^{-1}\left\{\frac{\sin s\phi}{\sin 2\alpha s}; s \to r\right\} = \frac{1}{2\alpha} \cdot \frac{r^m \sin(m\phi)}{1 + 2r^m \cos(m\phi) + r^{2m}} \tag{6.52}$$

where $m = \pi/2\alpha$.

EXERCISES 6.4

1. Given the function

$$F(s) = (1/a^{s-2})\,\zeta(s)\Gamma(s - 2)\cos(\pi s/2)$$

show that $\text{Res}\{F(s); 0\} = -1/2$.

2. Given the function

$$F(s) = \left(\frac{2\pi}{a}\right)^s \frac{\zeta(1-s)}{(s-1)(s-2)}$$

show that

(a) $\operatorname{Res}\{F(s);1\} = -\frac{2\pi}{a}\zeta(0) = \frac{\pi}{a}$

(b) $\operatorname{Res}\{F(s);2\} = \frac{4\pi^2}{a^2}\zeta(-1) = -\frac{\pi^2}{3a^2}$

3. Show that

$$\sum_{n=1}^{\infty} \frac{\sin an}{n} = \frac{\pi - a}{2}, \qquad 0 < a < 2\pi$$

4. Show that

$$\sum_{n=1}^{\infty} e^{-n^2} = -\frac{1}{2} + \sqrt{\pi}\sum_{k=0}^{\infty} e^{-\pi^2 k^2}$$

5. Show that

(a) $\displaystyle\sum_{n=1}^{\infty} \frac{(-1)^{n-1}}{n^s} = (1 - 2^{1-s})\zeta(s)$

(b) $\displaystyle\sum_{n=0}^{\infty} \frac{1}{(2n+1)^s} = (1 - 2^{-s})\zeta(s)$

6. From the results of Prob. 5, deduce that

(a) $\displaystyle\sum_{n=1}^{\infty} (-1)^{n-1} f(nx) = \mathcal{M}^{-1}\{(1 - 2^{1-s})\zeta(s)F(s);x\}$

(b) $\displaystyle\sum_{n=0}^{\infty} f[(2n+1)x] = \mathcal{M}^{-1}\{(1 - 2^{-s})\zeta(s)F(s);x\}$

In Probs. 7–10, use Eq. (6.33) or the results of Prob. 5 to sum the given series.

7. $\displaystyle\sum_{n=1}^{\infty} \frac{(-1)^{n-1}}{n^2} \cos an$

8. $\displaystyle\sum_{n=1}^{\infty} \frac{\sin an}{2n-1}$

9. $\displaystyle\sum_{n=1}^{\infty} \frac{(-1)^{n-1}}{n} \sin an$

10. $\displaystyle\sum_{n=0}^{\infty} \frac{J_1[(2n+1)a]}{2n+1}$

11. Show that

$$\mathcal{M}\left\{\frac{1}{(x^2+1)(x^2+z^2)};s\right\} = \frac{\pi(z^{s-2} - 1)}{2(z^2-1)\sin(\pi s/2)}$$

12. By using Eq. (6.44),
(a) show that the PDF of Exam. 6.10 can also be expressed as

$$p_Z(z) = \mathcal{M}^{-1}\left\{\frac{1}{\sin^2(\pi s/2)};z\right\}$$

(b) Using the method of residues, show that the inverse Mellin transform in (a) leads to the result of Exam. 6.10.

13. Given that X and Y both have the PDF $p(x) = (a + 1)x^a h(1 - x)$, $x > 0$,
(a) show that $Z = XY$ has the PDF

$$p_Z(z) = \frac{(a+1)^2}{2} z^a h(1 - z)\log\frac{1}{z}, \qquad z > 0$$

(b) If Z is a product of N independent random variables, all having the above density function, show that

$$p_Z(z) = \frac{(a+1)^{N-1}}{(N-1)!} z^a h(1 - z)\left(\log\frac{1}{z}\right)^{N-1}, \qquad z > 0$$

14. The PDF of $W = 1/Y$ is known to satisfy the Mellin transform relation

$$\mathcal{M}\{p_W(w);s\} = \mathcal{M}\{p_Y(y);-s+2\}$$

Use this relation to show that the PDF of $Z = X/Y$, where X and Y both have the PDF $p(x) = (a + 1)x^a h(1 - x)$, $x > 0$, is given by

$$p_Z(z) = \begin{cases} [(a+1)/2]\, z^a, & 0 \le z \le 1 \\ [(a+1)/2]\, z^{-a-2}, & z > 1. \end{cases}$$

15. Recall from Exam. 3.10 in Sec. 3.8 that the product of two independent, zero-mean, Gaussian random variables with equal variances has the PDF

$$p_Z(z) = (1/\pi)K_0(|z|)$$

where $K_0(x)$ is a modified Bessel function of the second kind. Use this result to deduce that the Bessel function $K_0(z)$ has the integral representation

$$K_0(z) = \frac{\pi}{2}\int_0^\infty e^{-t-z^2/4t}\frac{dt}{t}, \qquad z > 0$$

16. Use the result of Prob. 15 to deduce that the PDF of $Z = XY$, where X and Y have the same PDF $p(x) = e^{-x}h(x)$, is given by

$$p_Z(z) = (2/\pi)K_0(\sqrt{z}/2)h(z)$$

17. Use the result of Prob. 14 to find the PDF of $Z = X/Y$, where X and Y are positive independent random variables, each with PDF

$$p(x) = \sqrt{2/\pi}\, e^{-x^2/2}\, h(x)$$

18. For the special case of (6.47) where the boundary conditions are

$$u(r, -\alpha) = u(r, \alpha) = f(r)$$

(a) show that the solution of the transformed problem is

$$U(s, \theta) = F(s)\frac{\cos s\theta}{\cos s\alpha}$$

(b) Following the technique in Sec. 6.3.3, show that

$$\mathcal{M}^{-1}\left\{\frac{\cos s\theta}{\cos s\alpha}; s \to r\right\} = \frac{r^{m/2}}{\alpha}\cdot\frac{(1 + r^m)\cos(m\theta/2)}{1 + 2r^m\cos(m\theta) + r^{2m}},$$

where $m = \pi/\alpha$.

(c) Using the result in (b), find a formal solution of the potential problem by using the convolution theorem.

19. Find a formal solution of the problem described by (6.47) when the boundary conditions are prescribed by

$$u(r, \alpha) = u(r, -\alpha) = \begin{cases} 1, & 0 < r < a \\ 0, & r > a \end{cases}$$

20. Find the steady-state temperature distribution inside the infinite wedge-shaped region $0 < r < \infty$, $0 < \theta < \alpha$, if the boundary $\theta = 0$ is held at temperature zero while the other boundary is maintained at

$$u(r, \alpha) = \begin{cases} T_0, & 0 < r < a \\ 0, & r > a \end{cases}$$

21. Show that the integral equation

$$u(x) = f(x) + \int_0^\infty u(\xi) g(x/\xi)\frac{d\xi}{\xi}$$

has the formal solution

$$u(x) = \frac{1}{2\pi i}\int_{c-i\infty}^{c+i\infty}\frac{x^{-s}F(s)}{1 - G(s)}\,ds$$

22. If $g(x) = \lambda/(1 + x)$ in Prob. 21,

(a) show that

$$u(x) = f(x) + \frac{1}{2\pi i}\int_{c-i\infty}^{c+i\infty}\frac{(\lambda\pi)x^{-s}F(s)}{\sin s\pi - \lambda\pi}\,ds$$

(b) If $\lambda\pi = \sin\alpha\pi$, where $0 < \alpha < 1/2$, show that

$$\mathcal{M}^{-1}\left\{\frac{\sin\alpha\pi}{\sin s\pi - \sin\alpha\pi}; x\right\} = \frac{\tan\alpha\pi}{\pi}\left(\frac{x^{-\alpha} - x^{1+\alpha}}{1 - x^2}\right)$$

(c) From (a) and (b), deduce the formal solution

$$u(x) = f(x) + x\left(\frac{\tan\alpha\pi}{\pi}\right)\int_0^\infty f(\xi)\left[\frac{(x/\xi)^{-1-\alpha} - (x/\xi)^{\alpha}}{\xi^2 - x^2}\right] d\xi$$

23. Show that the integral equation

$$f(x) = \int_0^\infty g(x\xi)u(\xi)\, d\xi$$

has the formal solution

$$u(x) = \frac{1}{2\pi i}\int_{c-i\infty}^{c+i\infty} \frac{F(1 - s)}{G(1 - s)} x^{-s}\, ds$$

Hint: See Prob. 26 in Exer. 6.2.

24. Use the result of Prob. 23 to solve the Laplace integral equation

$$\frac{1}{1 + x} = \int_0^\infty e^{-x\xi} u(\xi)\, d\xi, \qquad x > 0$$

25. Show that the integral equation

$$u(x) = f(x) + \int_0^\infty g(x\xi)u(\xi)\, d\xi$$

has the formal solution

$$u(x) = \frac{1}{2\pi i}\int_{c-i\infty}^{c+i\infty} \frac{[F(s) + G(s)F(1 - s)]}{1 - G(s)G(1 - s)} x^{-s}\, ds$$

26. Use the result of Prob. 25 to solve the integral equation

$$u(x) = f(x) + \frac{1}{\sqrt{\pi}}\int_0^\infty e^{-x\xi} u(\xi)\, d\xi, \qquad x > 0$$

where

$$f(x) = \begin{cases} \dfrac{\log(1 + x)}{x}, & 0 < x < 1 \\[2ex] \dfrac{\log(1 + x)}{x} - \dfrac{\pi}{x}, & x > 1 \end{cases}$$

6.5 *Table of Mellin Transforms*

The following is a short table of Mellin transforms and their inverses.

Table 6.1 Mellin Transforms

No.	$f(x) = \dfrac{1}{2\pi i}\displaystyle\int_{c-i\infty}^{c+i\infty} x^{-s} F(s)\, ds$	$F(s) = \displaystyle\int_0^\infty x^{s-1} f(x)\, dx$
1	$\dfrac{1}{1+x}$	$\dfrac{\pi}{\sin \pi s}$
2	$\dfrac{1}{1-x}$	$\pi \cot \pi s$
3	$\dfrac{1}{(1+ax)^m}, \quad a > 0, m > 0$	$\dfrac{\Gamma(s)\Gamma(m-s)}{a^s\Gamma(m)}$
4	$\dfrac{1}{(1+x^a)^m}, \quad a > 0, m > 0$	$\dfrac{\Gamma(s/a)\Gamma(m-s/a)}{a\,\Gamma(m)}$
5	$e^{-ax}, \quad a > 0$	$\Gamma(\mathrm{s})/a^s$
6	$e^{-a^2x^2}, \quad a > 0$	$\Gamma(s/2)/2a^s$
7	$\cos ax, \quad a > 0$	$\dfrac{\Gamma(s)}{a^s}\cos(\pi s/2)$
8	$\sin ax, \quad a > 0$	$\dfrac{\Gamma(s)}{a^s}\sin(\pi s/2)$
9	$x^{-\nu}J_\nu(ax), \quad a > 0, \nu > -1/2$	$\dfrac{a^{\nu-s2s-\nu-1}\Gamma(s/2)}{\Gamma(\nu+1-s/2)}$
10	$J_\nu(ax), \quad a > 0, \nu > -1/2$	$\dfrac{2^{s-1}\,\Gamma\left(\dfrac{s+\nu}{2}\right)}{a^s\,\Gamma\left(\dfrac{\nu-s}{2}+1\right)}$

7

The Hankel Transform

7.1 *Introduction*

Hankel transforms arise naturally in solving boundary-value problems formulated in cylindrical coordinates. They also occur in other applications such as determining the oscillations of a heavy chain suspended from one end, first treated by D. Bernoulli. This latter problem is of some special historical significance since it was in this analysis of Bernoulli in 1703 that the Bessel function of order zero appeared for the first time.

To formally derive the Hankel transform and its inversion formula, we start with the two-dimensional Fourier transform pair (see Sec. 2.9.2)

$$F(\xi,\eta) = \frac{1}{2\pi}\int_{-\infty}^{\infty}\int_{-\infty}^{\infty} e^{i(x\xi+y\eta)} f(x,y)\,dx\,dy \tag{7.1}$$

and

$$f(x,y) = \frac{1}{2\pi}\int_{-\infty}^{\infty}\int_{-\infty}^{\infty} e^{-i(x\xi+y\eta)} F(\xi,\eta)\,d\xi\,d\eta \tag{7.2}$$

In applications involving these transform pairs, such as in optics, for instance, the problem often exhibits circular symmetry for natural reasons. When this happens, it may be expected that a simplification will result since one radial variable will suffice in place of two independent rectangular coordinates. Thus, let us assume that $f(x,y) \equiv f(r)$, where $r = (x^2 + y^2),^{1/2}$ and transform (7.1) and (7.2) into polar coordinate representations. In terms of the polar coordinates

$$x = r\cos\theta, \qquad y = r\sin\theta$$
$$\xi = \rho\cos\phi, \qquad \eta = \rho\sin\phi \tag{7.3}$$

we find that (7.1) becomes

$$F(\rho,\phi) = \frac{1}{2\pi}\int_{-\infty}^{\infty}\int_0^{2\pi} e^{i\rho r\cos(\theta-\phi)}\, rf(r)\, d\theta\, dr \tag{7.4}$$

where we have recognized that $x\xi + y\eta = \rho r \cos(\theta - \phi)$. Recalling property (J12) in Sec. 1.4.1, we have

$$\frac{1}{2\pi}\int_0^{2\pi} e^{i\rho r\cos(\theta-\phi)}\, d\theta = J_0(\rho r) \tag{7.5}$$

where $J_0(x)$ is the *Bessel function of order zero*. This result suggests that $F(\rho,\phi)$ is a function of ρ alone, i.e., $F(\rho,\phi) \equiv F(\rho)$, and in this case, (7.4) leads to

$$F(\rho) = \int_0^{\infty} rf(r)J_0(\rho r)\, dr \tag{7.6}$$

called the *Hankel transform of order zero*. Clearly, the substitutions (7.3) into Eq. (7.2) will produce the similar result

$$f(r) = \int_0^{\infty} \rho F(\rho)J_0(\rho r)\, d\rho \tag{7.7}$$

which represents the *inversion formula* for the Hankel transform of order zero.

When circular symmetry does not necessarily prevail, we define the general *Hankel transform of order* ν by

$$\mathscr{H}_\nu\{f(r);\rho\} = \int_0^{\infty} rf(r)J_\nu(\rho r)\, dr = F(\rho), \qquad \nu > -1/2^* \tag{7.8}$$

the corresponding *inversion formula* of which takes the form

$$\mathscr{H}_\nu^{-1}\{F(\rho);r\} = \int_0^{\infty} \rho F(\rho)J_\nu(\rho r)\, d\rho = f(r), \qquad \nu > -1/2 \tag{7.9}$$

A heuristic argument for the special case $\nu = n/2 - 1$ can be presented to derive (7.8) and (7.9), which parallels the above development for $n = 2$, by considering the multiple Fourier transform of order n applied to a radially symmetric function.†

---

*The restriction $\nu > -1/2$ may be extended to a larger interval in certain transform results.

†See I. N. Sneddon, *The Use of Integral Transforms*, New York: McGraw-Hill, 1972, pp. 79–83.

The basic requirement for the existence of the Hankel transform (7.8) is that the function $\sqrt{r}f(r)$ be piecewise continuous and absolutely integrable on the positive real line. The proof of the Hankel inversion formula (7.9) is similar to, but more complicated than, the corresponding proof of the Fourier inversion theorem. This is due primarily to the fact that the Hankel inversion formula relies on a good understanding of the properties of Bessel functions, which are more complicated than those of the corresponding kernels of the Fourier transform. The interested reader can consult Sneddon for the proof.*

7.2 *Evaluation of Hankel Transforms*

As in the case of the Mellin transform, we find that certain Hankel transforms are directly related to known transforms previously calculated. Also, many Hankel transforms and basic properties can be developed for general values of ν, although most applications involve only the cases $\nu = 0$ and $\nu = 1$. In this section we will determine the Hankel transform of certain functions and develop some of the fundamental operational properties of the transform. When possible, we will keep the value of ν arbitrary.

Example 7.1: Find the Hankel transform of $r^\nu h(a - r)$, $a > 0$.

Solution: From definition, we have

$$\mathcal{H}_\nu\{r^\nu h(a - r);\rho\} = \int_0^a r^{\nu+1} J_\nu(\rho r)\, dr$$
$$= \frac{1}{\rho^{\nu+2}} \int_0^{a\rho} x^{\nu+1} J_\nu(x)\, dx$$

the last step resulting from a change of variable. Now using Eq. (1.42) in Sec. 1.4.1, we deduce that

$$\mathcal{H}_\nu\{r^\nu h(a - r);\rho\} = (a^{\nu+1}/\rho)\, J_{\nu+1}(a\rho), \qquad a > 0,\ \nu > -1/2$$

Example 7.2: Find the Hankel transform of $r^{\nu-1}e^{-ar}$, $a > 0$.

Solution: Here we see that the Hankel transform can be expressed as a Laplace transform, i.e.,

---

*I. N. Sneddon, *The Use of Integral Transforms,* New York: McGraw-Hill, 1972, pp. 301–309.

$$\mathscr{H}_\nu\{r^{\nu-1}e^{-ar};\rho\} = \int_0^\infty r^\nu e^{-ar} J_\nu(\rho r)\, dr$$
$$= \mathscr{L}\{r^\nu J_\nu(\rho r); r \to a\}$$

Thus, recalling Exam. 4.16 in Sec. 4.4.1 and the scaling property of the Laplace transform, we immediately have the result

$$\mathscr{H}_\nu\{r^{\nu-1}e^{-ar};\rho\} = \frac{(2\rho)^\nu \Gamma(\nu + 1/2)}{\sqrt{\pi}(\rho^2 + a^2)^{\nu+1/2}}, \qquad a > 0, \nu > -1/2$$

The special case $\nu = 0$ in Exam. 7.2 leads to the results

$$\mathscr{H}_0\left\{\frac{1}{r} e^{-ar};\rho\right\} = \frac{1}{\sqrt{\rho^2 + a^2}}, \qquad a > 0 \tag{7.10}$$

and $(a \to 0^+)$

$$\mathscr{H}_0\left\{\frac{1}{r};\rho\right\} = \frac{1}{\rho}, \qquad \rho > 0 \tag{7.11}$$

Also, since multiplication by r corresponds to differentiation in the Laplace transform domain, we obtain

$$\mathscr{H}_0\{e^{-ar};\rho\} = \mathscr{L}\{rJ_0(\rho r); r \to a\}$$
$$= -\frac{d}{da}[(\rho^2 + a^2)^{-1/2}]$$

or

$$\mathscr{H}_0\{e^{-ar};\rho\} = \frac{a}{(\rho^2 + a^2)^{3/2}}, \qquad a > 0 \tag{7.12}$$

Example 7.3: Find the Hankel transform of order zero of $e^{-a^2r^2}$.

Solution: From the defining integral, we have

$$\mathscr{H}_0\{e^{-a^2r^2};\rho\} = \int_0^\infty r e^{-a^2r^2} J_0(\rho r)\, dr$$

By expressing the Bessel function as an infinite series and interchanging the order of integration and summation, we find*

$$\mathscr{H}_0\{e^{-a^2r^2};\rho\} = \sum_{n=0}^\infty \frac{(-1)^n(\rho/2)^{2n}}{(n!)^2} \int_0^\infty r^{2n+1} e^{-a^2r^2}\, dr$$
$$= \frac{1}{2a^2} \sum_{n=0}^\infty \frac{(-1)^n(\rho^2/4a^2)^n}{(n!)^2} \int_0^\infty t^n e^{-t}\, dt$$

*Also, note that $\mathscr{H}_0\{e^{-a^2r^2};\rho\} = \frac{1}{2}\mathscr{L}\{J_0(\rho\sqrt{t}); t \to a\}$, which can be evaluated through use of Exam. 4.15 in Sec. 4.4.

the last step of which follows a change of variables. This last integral is simply $n!$, and thus

$$\mathcal{H}_0\{e^{-a^2r^2};\rho\} = \frac{1}{2a^2}\sum_{n=0}^{\infty}\frac{(-1)^n(\rho^2/4a^2)^n}{n!}$$

or

$$\mathcal{H}_0\{e^{-a^2r^2};\rho\} = \frac{1}{2a^2}e^{-\rho^2/4a^2}, \qquad a > 0$$

Observe that the special case $a^2 = 1/2$ in Exam. 7.3 leads to

$$\mathcal{H}_0\{e^{-r^2/2};\rho\} = e^{-\rho^2/2} \tag{7.13}$$

which shows that $e^{-r^2/2}$ is *self-reciprocal*. It is interesting that this same function is self-reciprocal under the Fourier transform [see Eq. (2.52) in Sec. 2.4].

7.2.1 *Operational Properties*

Because the Hankel transform and inverse Hankel transform are exactly the same in functional form, it follows that each operational property of the transform is likewise a property of the inverse transform. For example, as a consequence of the linearity property of integrals, we deduce that

$$\mathcal{H}_\nu\{C_1f(r) + C_2g(r);\rho\} = C_1F(\rho) + C_2G(\rho) \tag{7.14a}$$

$$\mathcal{H}_\nu^{-1}\{C_1F(\rho) + C_2G(\rho);r\} = C_1f(r) + C_2g(r) \tag{7.14b}$$

where C_1 and C_2 are arbitrary constants and $F(\rho)$ and $G(\rho)$ are the Hankel transforms, respectively, of $f(r)$ and $g(r)$.

If we make the simple change of variable $x = ar$ in the defining integral

$$\mathcal{H}_\nu\{f(ar);\rho\} = \int_0^\infty rf(ar)J_\nu(\rho r)\,dr, \qquad a > 0$$

we obtain

$$\mathcal{H}_\nu\{f(ar);\rho\} = \frac{1}{a^2}\int_0^\infty xf(x)J_\nu(\rho x/a)\,dx$$

from which we deduce the *scaling property*

$$\mathcal{H}_\nu\{f(ar);\rho\} = (1/a^2)F(\rho/a), \qquad a > 0 \tag{7.15}$$

where $F(\rho)$ is the Hankel transform of $f(r)$.

It is not possible to derive a simple shift formula for the Hankel transform because the addition formula for the Bessel function, even for integer values of ν, takes a very complicated form. That is, in contrast with the simple addition formula

$$e^{x+y} = e^x e^y$$

for the exponential function, which is the basis of the shift formulas for the Fourier and Laplace transforms, we have the *Neumann–Lommel addition formula*

$$J_n(x + y) = \sum_{m=-\infty}^{\infty} J_m(x)J_{n-m}(y)$$

for the Bessel functions. Although it is possible to derive a shift formula for the Hankel transform based on this addition formula, it becomes unwieldly and not very useful.

The lack of a simple addition formula for the Bessel functions also precludes the existence of a simple convolution theorem for the Hankel transform. However, a simple relation of Parseval type can be easily derived for this transform. If $F(\rho)$ and $G(\rho)$ are Hankel transforms of $f(r)$ and $g(r)$, respectively, then

$$\begin{aligned}\int_0^\infty \rho F(\rho)G(\rho)\, d\rho &= \int_0^\infty \rho F(\rho) \int_0^\infty rg(r)J_\nu(\rho r)\, dr\, d\rho \\ &= \int_0^\infty rg(r) \int_0^\infty \rho F(\rho)J_\nu(\rho r)\, d\rho\, dr\end{aligned}$$

where we have interchanged the order of integration. It now follows that

$$\int_0^\infty \rho F(\rho)G(\rho)\, d\rho = \int_0^\infty rf(r)g(r)\, dr \tag{7.16}$$

Example 7.4: Evaluate the integral

$$I = \int_0^\infty \frac{1}{x} J_{\nu+1}(ax)J_{\nu+1}(bx)\, dx, \qquad 0 < a < b,\ \nu > -1/2$$

Solution: The given integral has the form

$$I = \int_0^\infty xF(x)G(x)\, dx$$

where

$$F(x) = (1/x)J_{\nu+1}(ax)$$

and

$$G(x) = (1/x)\mathrm{J}_{\nu+1}(bx)$$

We recognize from Ex. 7.1 that $F(x)$ and $G(x)$ are the Hankel transforms, respectively, of

$$f(r) = (r^\nu/a^{\nu+1})h(a - r), \qquad a > 0$$

and

$$g(r) = (r^\nu/b^{\nu+1})h(b - r), \qquad b > 0$$

Hence, based on the Parseval relation (7.16), we have

$$I = \frac{1}{(ab)^{\nu+1}} \int_0^a r^{2\nu+1}\, dr$$

which leads to

$$\int_0^\infty \frac{1}{x} J_{\nu+1}(ax)J_{\nu+1}(bx)\, dx = \frac{1}{2(\nu + 1)}\left(\frac{a}{b}\right)^{\nu+1}, \qquad 0 < a < b,\, \nu > -1/2$$

The Bessel functions satisfy the three-term recurrence formula [see Property (J9) in Sec. 1.4.1]

$$J_{\nu-1}(z) - (2\nu/z)J_\nu(z) + J_{\nu+1}(z) = 0 \tag{7.17}$$

Based on this relation, we obtain

$$\mathscr{H}_\nu\left\{\frac{1}{r}f(r);\rho\right\} = \frac{\rho}{2\nu}\int_0^\infty rf(r)\left[J_{\nu+1}(\rho r) + J_{\nu-1}(\rho r)\right] dr$$

or

$$\mathscr{H}_\nu\left\{\frac{1}{r}f(r);\rho\right\} = \frac{\rho}{2\nu}\left[\mathscr{H}_{\nu+1}\{f(r);\rho\} + \mathscr{H}_{\nu-1}\{f(r);\rho\}\right], \; \nu > -1/2 \tag{7.18}$$

In solving boundary-value problems by means of the Hankel transform it is necessary to develop formulas connecting the Hankel transform of functions to the Hankel transform of their derivatives. Most important among these formulas are the special cases

$$\mathscr{H}_1\{f'(r);\rho\} = -\rho\mathscr{H}_0\{f(r);\rho\} \tag{7.19}$$

and

$$\mathscr{H}_0\left\{\frac{1}{r}\frac{d}{dr}[rf(r)];\rho\right\} = \rho\mathscr{H}_1\{f(r);\rho\} \tag{7.20}$$

To derive (7.19), we start with an integration by parts to get

$$\begin{aligned}\mathscr{H}_1\{f'(r);\rho\} &= \int_0^\infty rf'(r)J_1(\rho r)\, dr \\ &= rf(r)J_1(\rho r)\Big|_0^\infty - \int_0^\infty f(r)\frac{d}{dr}[rJ_1(\rho r)]\, dr \end{aligned} \tag{7.21}$$

Recalling the asymptotic relations [see (J13) and (J14) in Sec. 1.4.1]

$$J_1(z) \sim z/2, \qquad z \to 0^+ \tag{7.22a}$$

$$J_1(z) \sim \sqrt{2/\pi z}\cos(z - 3\pi/4), \qquad z \to \infty \tag{7.22b}$$

we see that the first term on the right in (7.21) vanishes provided

$$\lim_{r\to\infty} \sqrt{r}f(r) = 0$$

$$\lim_{r\to 0} r^2 f(r) = 0.$$

The first limit condition is satisfied by any function whose Hankel transform exists, and the second condition imposes on $f(r)$ a certain behavior near $r = 0$. In addition, we note that [see Eq. (1.36) in Sec. 1.4.1]

$$\frac{d}{dr}[rJ_1(\rho r)] = \rho r J_0(\rho r)$$

so that (7.21) reduces to

$$\mathcal{H}_1\{f'(r);\rho\} = -\rho\int_0^\infty rf(r)J_0(\rho r)\,dr$$

which in turn leads to (7.19). Equation (7.20) follows in a similar fashion but is left to the reader to verify (see Prob. 25 in Exer. 7.2).

Both (7.19) and (7.20) have generalizations to Hankel transforms of order ν. For instance, Eq. (7.19) is a specialization of either

$$\mathcal{H}_\nu\{f'(r);\rho\} = -\frac{\rho}{2\nu}\Big[(\nu + 1)\mathcal{H}_{\nu-1}\{f(r);\rho\} - (\nu - 1)\mathcal{H}_{\nu+1}\{f(r);\rho\}\Big] \tag{7.23}$$

or

$$\mathcal{H}_\nu\{r^{\nu-1}\frac{d}{dr}[r^{1-\nu}f(r)];\rho\} = -\rho\mathcal{H}_{\nu-1}\{f(r);\rho\} \tag{7.24}$$

while (7.20) is a specialization of

$$\mathcal{H}_\nu\left\{\frac{1}{r^{\nu+1}}\frac{d}{dr}[r^{\nu+1}f(r)];\rho\right\} = \rho\mathcal{H}_{\nu+1}\{f(r);\rho\} \tag{7.25}$$

The verification of these results is also left to the exercises (see Probs. 26–28 in Exer. 7.2).

Hankel transforms provide significant simplifications in solving partial differential equations that lead to Bessel's equation

$$r^2y''(r) + ry'(r) + (r^2 - \nu^2)y(r) = 0 \tag{7.26}$$

To understand why this is so, let us first define the function

$$g_0(r) = y''(r) + \frac{1}{r}y'(r) = \frac{1}{r}\frac{d}{dr}[ry'(r)] \tag{7.27}$$

and apply the Hankel transform of order zero. This action leads to

$$\begin{aligned}\mathscr{H}_0\{g_0(r);\rho\} &= \rho\mathscr{H}_1\{y'(r);\rho\}\\ &= -\rho^2\mathscr{H}_0\{y(r);\rho\}\end{aligned} \tag{7.28}$$

where we have used (7.20) and (7.19), respectively. More generally, if we define

$$\begin{aligned}g_\nu(r) &= \frac{1}{r}\frac{d}{dr}[ry'(r)] - \frac{\nu^2}{r^2}y(r)\\ &= \frac{1}{r^{\nu+1}}\frac{d}{dr}\left[r^{2\nu+1}\frac{d}{dr}\left(\frac{1}{r^\nu}y(r)\right)\right]\end{aligned} \tag{7.29}$$

and apply the Hankel transform of order ν, we obtain the similar result

$$\mathscr{H}_\nu\{g_\nu(r);\rho\} = -\rho^2\mathscr{H}_\nu\{y(r);\rho\} \tag{7.30}$$

by use of (7.25) and (7.24), respectively. Hence, we see that the Hankel transforms of $g_0(r)$ or $g_\nu(r)$, which involve derivatives of $y(r)$, are related directly to Hankel transforms of $y(r)$.

EXERCISES 7.2

In Probs. 1–15, verify the given Hankel transform relation. When possible, use known integral transforms results from previous chapters.

1. $\mathscr{H}_\nu\{r^\nu e^{-ar};\rho\} = \dfrac{2a(2\rho)^\nu\,\Gamma(\nu + 3/2)}{\sqrt{\pi}(\rho^2 + a^2)^{\nu+3/2}}, \qquad a > 0$

Hint: Use Ex. 7.1.

2. $\mathscr{H}_\nu\left\{\dfrac{1}{r}e^{-ar};\rho\right\} = \dfrac{1}{\sqrt{\rho^2 + a^2}}\left(\dfrac{\rho}{a + \sqrt{\rho^2 + a^2}}\right), \qquad a > 0$

3. $\mathscr{H}_\nu\{r^\nu e^{-ar^2};\rho\} = \dfrac{\rho^\nu}{(2a)^{\nu+1}}e^{-\rho^2/4a}, \qquad a > 0$

4. $\mathscr{H}_\nu\{r^{\nu+2}e^{-ar^2};\rho\} = \dfrac{\rho^\nu}{a(2a)^{\nu+1}}\left(1 + \rho - \dfrac{\rho^2}{4a}\right)e^{-\rho^2/4a}, \qquad a > 0$

Hint: Differentiate both sides of Prob. 3 with respect to a.

5. $\mathscr{H}_0\{(a^2 - \rho^2)^{\mu-1}h(a - r);\rho\} = 2^{\mu-1}\Gamma(\mu)\left(\dfrac{a}{\rho}\right)^\mu J_\mu(a\rho), \qquad a > 0$

6. $\mathscr{H}_\nu\{r^\nu(a^2 - \rho^2)^{\mu-\nu-1}h(a - r);\rho\} = 2^{\mu-\nu-1}\Gamma(\mu - \nu)a^\mu\rho^{\nu-\mu}J_\mu(a\rho)$,
$a > 0,\ \mu > \nu \geq 0$

7. $\mathscr{H}_0\left\{\frac{1}{r} J_1(ar);\rho\right\} = \frac{1}{a} h(a - \rho), \qquad a > 0$

8. $\mathscr{H}_\nu\{r^{\nu-\mu}J_\mu(ar);\rho\} = \frac{\rho^\nu(a^2 - \rho^2)^{\mu-\nu-1}}{2^{\mu-\nu-1}\Gamma(\mu - \nu)a^\mu} h(a - \rho), \qquad a > 0, \mu > \nu \geq 0$

Hint: See Prob. 6.

9. $\mathscr{H}_0\left\{\frac{h(a - r)}{\sqrt{a^2 - r^2}};\rho\right\} = \sqrt{\frac{a\pi}{2\rho}} J_{1/2}(a\rho), \qquad a > 0$

Hint: See Prob. 8.

10. $\mathscr{H}_\nu\left\{\frac{r^\nu h(a - r)}{\sqrt{a^2 - r^2}};\rho\right\} = \sqrt{\frac{\pi}{2\rho}} a^{\nu+1/2} J_{\nu+1/2}(a\rho), \qquad a > 0$

Hint: See Prob. 8.

11. $\mathscr{H}_0\left\{\frac{1}{r}\sin ar;\rho\right\} = \frac{h(a - \rho)}{\sqrt{a^2 - \rho^2}}, \qquad a > 0$

12. $\mathscr{H}_1\left\{\frac{1}{r}\sin ar;\rho\right\} = \frac{ah(\rho - a)}{\rho\sqrt{\rho^2 - a^2}}, \qquad a > 0$

13. $\mathscr{H}_1\left\{\frac{1}{r} e^{-ar^2};\rho\right\} = \frac{1}{\rho}(1 - e^{-\rho^2/4a}), \qquad a > 0$

14. $\mathscr{H}_1\left\{\frac{1}{r}\cos(br^2);\rho\right\} = \frac{1}{\rho}\left[1 - \cos\left(\frac{\rho^2}{4b}\right)\right], \qquad b > 0$

15. $\mathscr{H}_1\left\{\frac{1}{r}\sin(br^2);\rho\right\} = \frac{1}{\rho}\sin\left(\frac{\rho^2}{4b}\right), \qquad b > 0$

16. Integrate both sides of the Hankel transform relation in Prob. 7 with respect to a from 0 to b and deduce that

$$\mathscr{H}_0\left\{\frac{1}{r^2}[1 - J_0(br)];\rho\right\} = h(b - \rho)\log\left(\frac{b}{\rho}\right), \qquad b > 0$$

Hint: $\int_0^b J_1(ar)\, da = \frac{1}{r}[1 - J_0(br)]$

17. Show that

(a) $\mathscr{H}_\nu\{r^{s-1};\rho\} = \mathscr{M}\{J_\nu(\rho r);r\to(s + 1)\}$

(b) Evaluate the Mellin transform in (a) and thus show that

$$\mathscr{H}_\nu\{r^{s-1};\rho\} = \frac{2^s\Gamma(s/2 + \nu/2 + 1/2)}{\rho^{s+1}\Gamma(\nu/2 - s/2 + 1/2)}, \qquad -1 - \nu < s < 1 + \nu$$

18. Starting with the integral formula

$$\int_0^\infty \cos(ax)J_0(bx)\,dx = 1/\sqrt{b^2 - a^2}, \qquad 0 < a < b$$

integrate both sides with respect to a to deduce that

$$\mathcal{H}_0\left\{\frac{1}{r^2}\sin r;\rho\right\} = \sin^{-1}\left(\frac{1}{\rho}\right), \qquad \rho > 1$$

19. Show that

$$\mathcal{H}_0\left\{\frac{1}{r}e^{-a^2r^2/4};\rho\right\} = \sqrt{\frac{\pi}{a}}e^{-\rho^2/2a}I_0\left(\frac{\rho^2}{2a}\right), \qquad a > 0$$

20. Show that

$$\mathcal{H}_0\{e^{-ar^2}J_0(br);\rho\} = \frac{a}{2}\exp\left(\frac{\rho^2 - b^2}{4a}\right)I_0\left(\frac{b\rho}{2a}\right), \qquad a > 0, b > 0$$

21. Use the integral representation

$$\frac{1}{\sqrt{r^2 + a^2}} = \frac{1}{\sqrt{\pi}}\int_0^\infty e^{-(r^2 + a^2)x}x^{-1/2}\,dx$$

to deduce that

$$\mathcal{H}_0\left\{\frac{1}{\sqrt{r^2 + a^2}};\rho\right\} = \frac{1}{\rho}e^{-a\rho}, \qquad a > 0$$

22. Use the result of Eq. (7.10) and other appropriate properties to show that

(a) $$\mathcal{H}_0\{re^{-ar};\rho\} = \frac{2a^2 - \rho^2}{(\rho^2 + a^2)^{5/2}}, \qquad a > 0$$

(b) $$\mathcal{H}_0\{e^{-ar};\rho\} = \frac{a}{(\rho^2 + a^2)^{3/2}}, \qquad a > 0$$

23. Use the result of Prob. 2 to deduce that

$$\mathcal{H}_\nu\{e^{-ar};\rho\} = \frac{a + \nu\sqrt{\rho^2 + a^2}}{(\rho^2 + a^2)^{3/2}}\left(\frac{\rho}{a + \sqrt{\rho^2 + a^2}}\right)^\nu, \qquad a > 0$$

24. Show that $r^\nu e^{-r^2/2}$ is a *self-reciprocal function* with respect to the Hankel transform of order ν.

25. Verify Eq. (7.20).

26. Verify Eq. (7.23).

27. Verify Eq. (7.24).

28. Verify Eq. (7.25).

7.3 Applications

One of the principal uses of the Hankel transform is in the solution of initial boundary-value problems involving cylindrical coordinates. We will briefly illustrate some of the classical examples of such problems where the Hankel transform is an effective tool.

7.3.1 Potential Problems

Let us first consider the axisymmetric Dirichlet problem for a half-space which is mathematically characterized by

$$u_{rr} + (1/r)\,u_r + u_{zz} = 0, \qquad 0 < r < \infty,\ z > 0$$

$$\text{B.C.:} \quad \begin{cases} u(r,0) = f(r), & 0 < r < \infty \\ u(r,z) \to 0 \text{ as } \sqrt{r^2 + z^2} \to \infty, & z > 0 \end{cases} \tag{7.31}$$

If we apply the Hankel transform of order zero to the variable r in (7.31), we obtain the transformed problem

$$U_{zz} - \rho^2 U = 0, \qquad z > 0$$

$$\text{B.C.:} \quad \begin{cases} U(\rho,0) = F(\rho) \\ U(\rho,z) \to 0 \text{ as } z \to \infty \end{cases} \tag{7.32}$$

where

$$\mathcal{H}_0\{u(r,z); r \to \rho\} = U(\rho,z) \tag{7.33}$$

$$\mathcal{H}_0\{f(r);\rho\} = F(\rho) \tag{7.34}$$

and we are using the result of Eq. (7.28). Clearly, the solution of (7.32) is

$$U(\rho,z) = F(\rho)e^{-\rho z} \tag{7.35}$$

and inverting this result by means of the Hankel inversion formula, we have

$$u(r,z) = \mathcal{H}_0^{-1}\{F(\rho)e^{-\rho z}; \rho \to r\}$$

or

$$u(r,z) = \int_0^\infty \rho F(\rho) e^{-\rho z} J_0(\rho r)\, d\rho \tag{7.36}$$

Example 7.5: Solve the problem described by (7.31) for the special case where

$$u(r,0) = f(r) = \frac{1}{\sqrt{r^2 + a^2}}, \qquad a > 0$$

Solution: Based on Eq. (7.10), we see that

$$\mathcal{H}_0\{f(r);\rho\} = F(\rho) = (1/\rho)\, e^{-a\rho}$$

Hence, substituting this result into (7.36) leads to the Laplace transform integral

$$\begin{aligned} u(r,z) &= \int_0^\infty e^{-(z+a)\rho} J_0(\rho r)\, d\rho \\ &= \mathcal{L}\{J_0(\rho r);\rho \to (z+a)\} \end{aligned}$$

from which we conclude

$$u(r,z) = \frac{1}{\sqrt{r^2 + (z+a)^2}}$$

Next, we consider the problem of finding a function $u(r,z)$ which is harmonic in the half-space $z > 0$, and which satisfies *mixed* boundary conditions (i.e., conditions where u is prescribed over part of the domain and the normal derivative of u prescribed over the remaining part of the domain). A specific example of this kind of problem is described by

$$u_{rr} + (1/r)u_r + u_{zz} = 0, \qquad 0 < r < \infty,\ z > 0$$

$$\text{B.C.:} \quad \begin{cases} u(r,0) = u_0, & 0 < r < 1 \\ u_z(r,0) = 0, & 1 < r < \infty \\ u(r,z) \to 0 \text{ as } \sqrt{r^2 + z^2} \to \infty, & z > 0 \end{cases} \tag{7.37}$$

where u_0 is a constant. Such a problem might describe the electrostatic potential of an electric field due to a uniformly charged flat circular disc of unit radius.

Because of the mixed boundary conditions, it is preferable to write down the general solution of the transformed problem and invert it before imposing the boundary conditions. Thus, taking the Hankel transform of order zero with respect to r, we obtain the transformed equation

$$U_{zz} - \rho^2 U = 0, \qquad z > 0, \tag{7.38}$$

where $U(\rho,z)$ is the Hankel transform of $u(r,z)$. The bounded solution of this equation is

$$U(\rho,z) = A(\rho)e^{-\rho z} \tag{7.39}$$

where $A(\rho)$ is an arbitrary function of ρ.

The inversion of (7.39) leads to the integral equation

$$u(r,z) = \int_0^\infty \rho A(\rho) e^{-\rho z} J_0(\rho r)\, d\rho \tag{7.40}$$

for the determination of the function $A(\rho)$. Now imposing on this formal solution the mixed boundary conditions in (7.37), we get the pair of equations

$$\int_0^\infty \rho A(\rho) J_0(\rho r)\, d\rho = u_0, \qquad 0 < r < 1 \tag{7.41a}$$

$$\int_0^\infty \rho^2 A(\rho) J_0(\rho r)\, d\rho = 0, \qquad 1 < r < \infty \tag{7.41b}$$

Equations of this variety are known as *dual integral equations,* but the general theory concerning them goes beyond the intended scope of this text.* However, to solve this particular pair of equations we simply start with the observations (see Probs. 13 and 14 in Exer. 1.4)

$$\int_0^\infty J_0(\rho r) \frac{\sin \rho}{\rho}\, d\rho = \frac{\pi}{2}, \qquad 0 < r < 1 \tag{7.42a}$$

$$\int_0^\infty J_0(\rho r) \sin \rho\, d\rho = 0, \qquad 1 < r < \infty \tag{7.42b}$$

and hence, by comparison with Eqs. (7.41a,b), we deduce that

$$A(\rho) = \frac{2u_0}{\pi \rho^2} \sin \rho \tag{7.43}$$

Thus, the solution of (7.37) is finally given by

$$u(r,z) = \frac{2u_0}{\pi} \int_0^\infty e^{-\rho z} J_0(\rho r) \frac{\sin \rho}{\rho}\, d\rho \tag{7.44}$$

7.3.2 *Vibration Problems*

Consider the transverse displacement $u(r,t)$ of a large thin membrane which is deformed symmetrically under the action of an external normally applied pressure $p(r,t)$. If T is the tension in the membrane, c is a physical constant proportional to $\sqrt{T}$ and having dimension of velocity, $f(r)$ and $g(r)$ denote the initial displacement and velocity, respectively, of the membrane, then the subsequent displacements of the membrane are described by solutions of the nonhomogeneous boundary-value problem

$$u_{rr} + (1/r)u_r = c^{-2}u_{tt} - (1/T)p(r,t), \qquad 0 < r < \infty,\ t > 0$$

B.C.: $u(r,t)$ finite as $r \to \infty$ (7.45)

I.C.: $u(r,0) = f(r), \qquad u_t(r,0) = g(r),\ 0 < r < \infty$

---

*For a discussion of such equations, see I. N. Sneddon, *Mixed Boundary-Value Problems in Potential Theory,* Amsterdam: North Holland, 1966.

Taking the Hankel transform of order zero with respect to r, we are led to the initial value problem

$$U_{tt} + c^2\rho^2 U = \frac{c}{\rho T}P(\rho,t), \qquad t > 0 \tag{7.46}$$

$$\text{I.C.:} \qquad U(\rho,t) = F(\rho), \qquad U_t(\rho,t) = G(\rho)$$

Using standard solution techniques, we find that (7.46) has the solution

$$U(\rho,t) = F(\rho)\cos c\rho t + \frac{1}{c\rho}G(\rho)\sin c\rho t + \frac{c}{\rho T}\int_0^t P(\rho,\tau)\sin[c\rho(t - \tau)]d\tau \tag{7.47}$$

The subsequent inversion of (7.47) then gives us the formal result for the displacements

$$u(r,t) = \int_0^\infty \rho U(\rho,t)J_0(\rho r)\,d\rho \tag{7.48}$$

EXERCISES 7.3

1. Find a formal solution of the heat-conduction problem

$$u_{rr} + \frac{1}{r}u_r = a^{-2}u_t, \qquad 0 < r < \infty,\ t > 0$$

B.C.: $u(r,t)$ finite as $r \to \infty$

I.C.: $u(r,0) = f(r)$

2. Solve the problem described by (7.31) for the special case where

$$u(r,0) = f(r) = 1/(r^2 + a^2)^{3/2}, \qquad a > 0$$

3. Show that the boundary-value problem

$$u_{rr} + (1/r)\,u_r + u_{zz} = -f(r), \qquad 0 < r < \infty,\ z > 0$$

$$\text{B.C.:} \quad \begin{cases} u(r,0), = 0 \\ u(r,z) \to 0 \text{ as } \sqrt{r^2 + z^2} \to \infty, & z > 0 \end{cases}$$

(a) has the formal solution

$$u(r,z) = \int_0^\infty F(\rho)\left(\frac{1 - e^{-\rho z}}{\rho}\right)J_0(\rho r)\,d\rho$$

(b) In the special case

$$f(r) = a/(r^2 + a^2)^{3/2}, \qquad a > 0$$

show that (a) reduces to

$$u(r,z) = \frac{1}{2}\left[\log\frac{w(r,0) - 1}{w(r,0) + 1} - \log\frac{w(r,z) - 1}{w(r,z) + 1}\right]$$

where

$$w(r,z) = \sqrt{\frac{r^2}{(z + a)^2} + 1}$$

4. Show that the steady-state temperature distribution problem

$$u_{rr} + (1/r)\, u_r + u_{zz} = 0, \qquad 0 < r < \infty,\ z > 0$$

$$\text{B.C.:} \quad \begin{cases} -\kappa u_z(r,0) = (Q/a^2)h(a - r), & a > 0 \\ u(r,z) \to 0 \text{ as } \sqrt{r^2 + z^2} \to \infty, & z > 0 \end{cases}$$

has the formal solution

$$u(r,z) = \frac{Q}{\kappa a}\int_0^a e^{-z\rho}\, J_1(a\rho)J_1(r\rho)\, d\rho$$

5. Show that the potential problem for

$$v(r,z) = Q/\sqrt{r^2 + z^2} + u(r,z)$$

where $u(r,z)$ satisfies

$$u_{rr} + (1/r)\, u_r + u_{zz} = 0, \qquad 0 < r < \infty,\ -a < z < a$$

$$\text{B.C.:} \quad u(r, \pm a) = -Q/\sqrt{r^2 + a^2}$$

has the formal solution

$$v(r,z) = \frac{Q}{\sqrt{r^2 + z^2}} - Q\int_0^\infty \frac{\cosh z\rho}{\cosh a\rho}\, e^{-a\rho}\, J_0(\rho r)\, d\rho$$

6. Let $u(r,z)$ denote the steady state temperature in a slab bounded by $0 < r < \infty$, $0 < z < 1$. If the face $z = 0$ is kept at $u = 0$ and the face $z = 1$ is insulated except that heat is supplied through a circular region, such that

$$u_z(r,1) = h(c - r)$$

find the subsequent temperature distribution in the slab.

7. The small transverse oscillations of a heavy chain of uniform line density σ, suspended vertically from one end, are determined by the equation

$$\sigma\frac{\partial^2 y}{\partial t^2} = g\sigma\frac{\partial}{\partial x}\left(x\frac{\partial y}{\partial x}\right) + p(x,t), \qquad 0 < x < \infty,\ t > 0$$

where the origin of the x coordinate is taken at the position of equilibrium of the lower end and the x axis is taken along the equilibrium position of the chain, pointing upward. The function $p(x,t)$ is the intensity of the external transverse force. If the initial conditions are given by

I.C.: $$y(x,0) = 0, \qquad \frac{\partial y}{\partial t}(x,0) = 0$$

show that for subsequent times

$$y(x,t) = \mathscr{H}_0\left\{\frac{1}{\rho}\int_0^t P(\rho,\tau)\sin[\rho(t-\tau)]\,d\tau;\rho \to 2\sqrt{\frac{x}{g}}\right\}$$

where

$$P(\rho,t) = \mathscr{H}_0\{p(gx^2/4,t);x \to \rho\}$$

Note: The constant g is simply the gravitational constant.

7.4 *Table of Hankel Transforms*

The following is a short table of Hankel transforms of order zero and their inverses.

Table 7.1 Table of Hankel Transforms

	$f(r) = \int_0^\infty \rho F(\rho)J_0(\rho r)\,d\rho$	$F(\rho) = \int_0^\infty r f(r)J_0(\rho r)\,dr$
1	$h(a-r), \quad a<0$	$(a/\rho)J_1(a\rho)$
2	$r^{s-1}, \quad \lvert s\rvert > 1$	$\dfrac{2^s\Gamma[(s+1)/2]}{\rho^{s+1}\Gamma[(1-s)/2]}$
3	$1/r$	$1/\rho$
4	$(1/r)e^{-ar}, \quad a<0$	$1/\sqrt{\rho^2+a^2}$
5	$e^{-ar}, \quad a>0$	$a/(\rho^2+a^2)^{3/2}$
6	$e^{-a^2r^2}, \quad a>0$	$(1/2a^2)\,e^{-\rho^2/4a^2}$
7	$(1/r)\sin ar, \quad a>0$	$h(a-\rho)/\sqrt{a^2-\rho^2}$
8	$(1/r^2)\sin ar, \quad a>0$	$\sin^{-1}(a/\rho)$

8

Finite Transforms

8.1 *Introduction*

The integral transforms considered thus far are applicable to problems involving either semiinfinite or infinite domains. However, in applying the method of integral transforms to problems formulated on finite domains it is necessary to introduce *finite* intervals on the transform integral. We then find that it is possible to derive their inverses from the theory of Fourier series. Transforms of this nature are called *finite transforms* and sometimes afford a more convenient method of solution than the classical methods which often require much ingenuity in assuming at the outset a correct solution form.

In this chapter we will introduce *finite Fourier transforms* and the *finite Hankel transform,* the latter being a special case of the more general *Sturm–Liouville transform*.

8.2 *Finite Fourier Transforms*

Let us begin by considering the simplest cases of finite transforms, which are known as finite sine or cosine transforms. The general theory of these transforms is based on the theory of Fourier series, with which we assume the reader is familiar.

According to the theory of Fourier series, a function f which is piecewise continuous and has a piecewise continuous derivative f' on the interval

$0 \le x \le \pi$ has the *Fourier sine series*

$$f(x) = \sum_{n=1}^{\infty} b_n \sin nx, \qquad 0 < x < \pi \tag{8.1}$$

where

$$b_n = \frac{2}{\pi}\int_0^{\pi} f(x)\sin nx\,dx, \qquad n = 1,2,3,\ldots \tag{8.2}$$

This series converges pointwise to $f(x)$ at points where f is continuous and to the average value $\frac{1}{2}[f(x^+) + f(x-)]$ at other points. A similar statement holds true for the *Fourier cosine series*

$$f(x) = \frac{1}{2}a_0 + \sum_{n=1}^{\infty} a_n \cos nx, \qquad 0 < x < \pi \tag{8.3}$$

where

$$a_n = \frac{2}{\pi}\int_0^{\pi} f(x)\cos nx\,dx, \qquad n = 0,1,2,\ldots \tag{8.4}$$

8.2.1 *Finite Sine Transform*

We define the *finite sine transform* of $f(x)$ by

$$S_\pi\{f(x);n\} = F_s(n) = \int_0^{\pi} f(x)\sin nx\,dx, \qquad n = 1,2,3,\ldots \tag{8.5}$$

which is distinct from previous transforms in that the transform function $F_s(n)$ is actually a sequence of numbers rather than a continuous function. By comparing this expression with (8.2), we see that

$$b_n = (2/\pi)\,F_s(n), \qquad n = 1,2,3,\ldots \tag{8.6}$$

and thus deduce from Eq. (8.1) that

$$S_\pi^{-1}\{F_s(n);x\} = f(x) = \frac{2}{\pi}\sum_{n=1}^{\infty} F_s(n)\sin nx, \qquad 0 < x < \pi \tag{8.7}$$

In this case Eq. (8.7) represents the *inversion formula* for the inverse finite sine transform, which is generally attributed to Doetsch.*

The finite sine transform is appropriate for solving differential equations containing only even-order derivatives. (Recall the discussion in Sec. (3.1.) Fortunately, many of the DEs of interest fall into this category.

*See G. Doetsch, "Integration von differentialgleichungen vermittels der endlichen Fourier-transformation," *Math. Ann.*, **112,** p. 52 (1935).

There are numerous examples of functions $f(x)$ whose finite sine transform can be found through elementary integration techniques applied to (8.5). Some of these are considered in the exercises. Our interest here is mostly concerned with the operational properties of this transform.

If f and f' are continuous on $0 \le x \le \pi$ and f'' is piecewise continuous on the same interval, then using integration by parts, we find

$$\begin{aligned} S_\pi\{f''(x);n\} &= \int_0^\pi f''(x)\sin nx\,dx \\ &= f'(x)\sin nx\Big|_0^\pi - n\int_0^\pi f'(x)\cos nx\,dx \end{aligned}$$

The first term on the right is identically zero and a further integration by parts leads to

$$S_\pi\{f''(x);n\} = -nf(x)\cos nx\Big|_0^\pi - n^2\int_0^\pi f(x)\sin nx\,dx$$

or, upon simplifying,

$$S_\pi\{f''(x);n\} = -n^2F_s(n) + n[f(0) - (-1)^n f(\pi)] \tag{8.8}$$

Hence, we see that the finite sine transform of $f''(x)$ depends upon the transform of $f(x)$ and upon the values of $f(x)$ at the boundary points $x = 0$ and $x = \pi$.

Although we have defined the finite sine transform for the particular interval $0 \le x \le \pi$, it is easy to generalize to other interval lengths. For instance, the generalization of (8.5) and (8.7) to an interval of length p, rather than π, merely involves a scale transformation with x replaced by $\pi x/p$. This leads to the more general sine transform

$$S_p\{f(x);n\} = F_s(n) = \int_0^p f(x)\sin\frac{n\pi x}{p}\,dx, \qquad n = 1,2,3,\ldots \tag{8.9}$$

with corresponding inversion formula

$$S_p^{-1}\{F_s(n);x\} = f(x) = \frac{2}{p}\sum_{n=1}^{\infty} F_s(n)\sin\frac{n\pi x}{p}, \qquad 0 < x < p \tag{8.10}$$

In the same manner, we find that (8.8) becomes

$$S_p\{f''(x);n\} = -\frac{n^2\pi^2}{p^2}F_s(n) + \frac{n\pi}{p}[f(0) - (-1)^n f(p)] \tag{8.11}$$

Still other generalizations are possible, but we will not discuss them.*

---

*See Chap. 11 in R. V. Churchill, *Operational Mathematics*, 3rd ed., New York: McGraw-Hill, 1972.

8.2.2 *Finite Cosine Transform*

In a similar fashion as we did for the finite sine transform, we define the *finite cosine transform* of $f(x)$ by

$$C_\pi\{f(x);n\} = F_c(n) = \int_0^\pi f(x)\cos nx\,dx, \qquad n = 0,1,2,\ldots \tag{8.12}$$

the *inversion formula* of which is

$$\begin{aligned} C_\pi^{-1}\{F_c(n);x\} &= f(x) \\ &= \frac{1}{\pi}F_c(0) + \frac{2}{\pi}\sum_{n=1}^{\infty} F_c(n)\cos nx, \qquad 0 < x < \pi \end{aligned} \tag{8.13}$$

The operational property concerning the second derivative of $f(x)$ takes the form

$$C_\pi\{f''(x);n\} = -n^2F_c(n) + (-1)^n f'(\pi) - f'(0) \tag{8.14}$$

the verification of which is left to the exercises (see Prob. 14 in Exer. 8.2).

The generalizations of these results to an interval of length p are

$$C_p\{f(x);n\} = F_c(n) = \int_0^p f(x)\cos\frac{n\pi x}{p}\,dx, \qquad n = 0,1,2,\ldots \tag{8.15}$$

$$\begin{aligned} C_p^{-1}\{F_c(n);x\} &= f(x) \\ &= \frac{1}{p}F_c(0) + \frac{2}{p}\sum_{n=1}^{\infty} F_c(n)\cos\frac{n\pi x}{p}, \qquad 0 < x < p \end{aligned} \tag{8.16}$$

and

$$C_p\{f''(x);n\} = -\frac{n^2\pi^2}{p^2}F_c(n) + (-1)^n f'(p) - f'(0) \tag{8.17}$$

8.2.3 *Applications*

The steady-state temperature distribution $u(x,y)$ in a long square bar with one face held at constant temperature T_0 and the other faces held at zero temperature is governed by the boundary-value problem

$$u_{xx} + u_{yy} = 0, \qquad 0 < x < \pi,\ 0 < y < \pi$$

$$\text{B.C.:}\quad \begin{cases} u(0,y) = 0, & u(\pi,y) = 0 \\ u(x,0) = 0, & u(x,\pi) = T_0 \end{cases} \tag{8.18}$$

If we apply the finite sine transform with respect to the variable x, we arrive at

$$U_{yy} - n^2U = 0, \qquad 0 < y < \pi$$

$$\text{B.C.:} \qquad U(n,0) = 0, \qquad U(n,\pi) = \begin{cases} 0, & n \text{ even} \\ 2T_0/n, & n \text{ odd} \end{cases} \tag{8.19}$$

where*

$$U(n,y) = \int_0^{\pi} u(x,y)\sin nx\, dx \tag{8.20}$$

The solution of (8.19) is readily found to be

$$U(n,y) = \frac{2T_0}{n}\frac{\sinh ny}{\sinh n\pi}, \qquad n = 1,3,5,\ldots \tag{8.21}$$

and $U(n,y) = 0$, $n = 2,4,6,\ldots$. The inversion of this result leads to

$$u(x,y) = \frac{4T_0}{\pi}\sum_{\substack{n=1 \\ (\text{odd})}}^{\infty} \frac{\sinh ny}{n \sinh n\pi} \sin nx \tag{8.22}$$

which is our solution of (8.18)

For a second example on the use of finite transforms, let us consider the problem of heat conduction in a solid bounded by the parallel planes $x = 0$ and $x = 1$. If the faces of the solid are thermally insulated and the initial temperature in the solid is $f(x)$, the subsequent temperatures $u(x,t)$ are solutions of

$$u_{xx} = a^{-2}u_t, \qquad 0 < x < 1,\ t > 0$$

$$\text{B.C.:} \qquad u_x(0,t) = 0, \qquad u_x(1,t) = 0,\ t > 0 \tag{8.23}$$

$$\text{I.C.:} \qquad u(x,0) = f(x), \qquad 0 < x < 1$$

where a^2 is the thermal diffusivity of the material of the solid.

Because the boundary conditions in (8.23) involve the derivative of u, we must use the finite cosine transform this time. Also, the length of the x-interval is unity instead of π, so we use the general form of the transform given by (8.15) with $p = 1$. In this case the transformed problem is

$$U_t + a^2n^2U = 0, \qquad t > 0 \tag{8.24}$$

$$\text{I.C.:} \qquad U(n,0) = F(n)$$

where

$$U(n,t) = \int_0^1 u(x,t)\cos n\pi x\, dx \tag{8.25}$$

---

*For simplicity of notation, we will drop the s subscript on the transform function.

and

$$F(n) = \int_0^1 f(x)\cos n\pi x \, dx \tag{8.26}$$

The solution of (8.24) is

$$U(n,t) = F(n)e^{-a^2n^2t} \tag{8.27}$$

the inversion of which yields

$$u(x,t) = F(0) + 2\sum_{n=1}^{\infty} F(n)\, e^{-a^2n^2t}\cos n\pi x \tag{8.28}$$

However, using (8.26) we can express this solution in the more convenient form

$$u(x,t) = \int_0^1 f(\xi)d\xi + 2\sum_{n=1}^{\infty}\left[\int_0^1 f(\xi)\cos n\pi\xi \, d\xi\right] e^{-a^2n^2t}\cos n\pi x \tag{8.29}$$

EXERCISES 8.2

In Probs. 1–10, evaluate the finite sine transform and finite cosine transform of the given function.

1. $f(x) = 1, \quad 0 < x < \pi$

2. $f(x) = x, \quad 0 < x < \pi$

3. $f(x) = (x/2)(\pi - x), \quad 0 < x < \pi$

4. $f(x) = e^x, \quad 0 < x < 4$

5. $f(x) = x^3, \quad 0 < x < 1$

6. $f(x) = x(p - x), \quad 0 < x < p$

7. $f(x) = \sin kx, \quad 0 < x < \pi, \; k \neq \pm 1, \pm 2, \pm 3, \ldots$

8. $f(x) = \cos kx, \quad 0 < x < \pi, \; k \neq \pm 1, \pm 2, \pm 3, \ldots$

9. $f(x) = x(1 - x^2), \quad 0 < x < 1$

10. $f(x) = (x/6)(x^2 - 3x + 2), \quad 0 < x < 1$

11. If $f(\pi - x) = f(x)$, show that $F_s(n) = 0$ when n is even.

12. Verify the identities:

(a) $S_\pi\{f(x)\cos kx;n\} = \frac{1}{2}[F_s(n + k) + F_s(n - k)]$

(b) $C_\pi\{f(x)\cos kx;n\} = \frac{1}{2}[F_c(n - k) + F_c(n + k)]$

(c) $S_\pi\{f(x)\sin kx;n\} = \frac{1}{2}[F_c(n - k) - F_c(n + k)]$

(d) $C_\pi\{f(x)\sin kx;n\} = \frac{1}{2}[F_s(n + k) - F_s(n - k)]$

13. Show that

(a) $S_\pi\{f(\pi - x);n\} = (-1)^{n-1}F_s(n)$

(b) $C_\pi\{f(\pi - x);n\} = (-1)^{n}F_c(n)$

14. Show that

(a) $C_\pi\{f''(x);n\} = -n^2F_c(n) + (-1)^n f'(\pi) - f'(0)$

(b) $C_p\{f''(x);n\} = -\dfrac{n^2\pi^2}{p^2}F_c(n) + (-1)^n f'(p) - f'(0)$

15. Show that

(a) $S_\pi\{f'(x);n\} = -n\,C_\pi\{f(x);n\}$

(b) $C_\pi\{f'(x);n\} = n\,S_\pi\{f(x);n\} - f(0) + (-1)^n f(\pi)$

In Probs. 16–20, use the finite sine transform or finite cosine transform to solve the given boundary-value problem

16. $u_{xx} = a^{-2}u_t, \quad 0 < x < 1, t > 0$

B.C.: $u(0,t) = 0, \quad u(1,t) = 0$

I.C.: $u(x,0) = 3\sin \pi x - 5\sin 4\pi x$

17. $u_{xx} = a^{-2}u_t, \quad 0 < x < 10, t > 0$

B.C.: $u(0,t) = 10, \quad u(10,t) = 30$

I.C.: $u(x,0) = 0$

18. $u_{xx} = a^{-2}u_t, \quad 0 < x < p, t > 0$

B.C.: $u_x(0,t) = 0, \quad u_x(p,t) = 0$

I.C.: $u(x,0) = T_0 \sin^2(\pi x/p)$

19. $u_{tt} = c^2u_{xx} - 2ku_t, \quad 0 < x < 1, t > 0\ (0 < k < \pi c)$

B.C.: $u(0,t) = 0, \quad u(1,t) = 0$

I.C.: $u(x,0) = \sin \pi x, \quad u_t(x,0) = 0$

20. $u_{xx} + u_{yy} = 0, \quad 0 < x < \pi, 0 < y < 1$

B.C.: $\begin{cases} u_x(0,y) = 0, & u_x(\pi,y) = 0 \\ u(x,0) = T_0\cos x, & u(x,1) = T_0\cos^2 x \end{cases}$

21. Find a formal solution of the boundary-value problem

$u_t = u_{xx} + g(x,t), 0 < x < \pi$

B.C.: $u(0,t) = 0, \quad u(\pi,t) = 0$

I.C.: $u(x,0) = f(x)$

22. A uniform string of length p is stretched tightly between two fixed points at $x = 0$ and $x = p$. It is displaced a small distance ε at a point $x = b$, $0 < b < p$, and released from rest at time $t = 0$. Starting with the equation of motion

$$u_{xx} = c^{-2}u_{tt}$$

show that subsequent displacements are described by

$$u(x,t) = \frac{2\varepsilon p^2}{\pi^2 b(p-b)} \sum_{n=1}^{\infty} \frac{1}{n^2} \sin\frac{n\pi b}{p} \sin\frac{n\pi x}{p} \cos\frac{n\pi ct}{p}$$

8.3 *Sturm–Liouville Transforms*

A *Sturm–Liouville problem* is a boundary-value problem of the general form

$$\begin{gathered} \frac{d}{dx}[p(x)y'] + [q(x) + \lambda r(x)]y = 0, \qquad a < x < b \\ h_1 y(a) + y'(a) = 0, \qquad h_2 y(b) + y'(b) = 0 \end{gathered} \tag{8.30}$$

where $p(x) > 0$ and $r(x) > 0$ in $a < x < b$, and $p'(x)$, $q(x)$, and $r(x)$ are all continuous functions in the interval $a \leq x \leq b$. Here h_1 and h_2 are given constants. (Other types of boundary conditions are also possible.*) It is known that, under appropriate conditions, a collection of nontrivial solutions $\{\phi_n(x)\}$, $n = 1,2,3,\ldots$, exists corresponding to a set of values $\{\lambda_n\}$, $n = 1,2,3,\ldots$, for the parameter λ. We refer to the set $\{\lambda_n\}$ as the *eigenvalues* of the problem and the corresponding nontrivial solutions $\{\phi_n(x)\}$ as the *eigenfunctions*.

Sturm–Liouville problems provide a direct method of determining a linear integral transformation that is appropriate for a given linear boundary-value problem, with the finite Fourier sine and cosine transforms as special cases. In such situations the integral transform assumes the general form

$$F(n) = \int_a^b r(x)\phi_n(x)f(x)\,dx, \qquad n = 1,2,3,\ldots \tag{8.31}$$

where the function $r(x)$ is called a weighting function. We call $F(n)$ the *Sturm–Liouville (S–L) transform* of the function $f(x)$, and $K(x,n) = r(x)\phi_n(x)$ is the kernel of the transform.†

For example, the finite Fourier sine transform (Sec. 8.2) is a S–L transform associated with the Sturm–Liouville problem

$$\begin{gathered} y'' + \lambda y = 0, \qquad 0 < x < \pi \\ y(0) = 0, \qquad y(\pi) = 0 \end{gathered} \tag{8.32}$$

---

*For a general discussion of Sturm–Liouville problems, see Chap. 1 in L. C. Andrews, *Elementary Partial Differential Equations with Boundary Value Problems*, Orlando: Academic Press, 1986.

†Sometimes the kernel of the transform is defined as simply the eigenfunction $\phi_n(x)$.

Here we identify $p(x) = 1$, $q(x) = 0$, and $r(x) = 1$. The eigenvalues and eigenfunctions of (32) are

$$\lambda_n = n^2; \qquad \phi_n(x) = \sin nx, \qquad n = 1,2,3,\ldots \tag{8.33}$$

Similarly, the finite Fourier cosine transform is closely associated with the Sturm–Liouville problem

$$\begin{aligned} y'' + \lambda y &= 0, \qquad 0 < x < \pi \\ y'(0) &= 0, \qquad y'(\pi) = 0 \end{aligned} \tag{8.34}$$

which has eigenvalues and eigenfunctions given by

$$\lambda_n = n^2; \qquad \phi_n(x) = \cos nx, \qquad n = 0,1,2,\ldots \tag{8.35}$$

The most important property of the eigenfunctions of a Sturm–Liouville problem is the *orthogonality relation*

$$\int_a^b r(x)\phi_m(x)\phi_n(x)\,dx = 0, \qquad m \neq n \tag{8.36}$$

which is useful in the development of an inversion formula for the S–L transform (8.31). To obtain this inversion formula, let us assume that the function $f(x)$ in (8.31) can be represented by the *generalized Fourier series*

$$f(x) = \sum_{n=1}^{\infty} c_n\phi_n(x), \qquad a < x < b \tag{8.37}$$

for some set of constants $\{c_n\}$, $n = 1,2,3,\ldots$. If we formally multiply both sides of (8.37) by $r(x)\phi_m(x)$, $m = 1,2,3,\ldots$, and integrate the result over the inverval $a \le x \le b$, we get

$$\int_a^b r(x)\phi_m(x)f(x)\,dx = \sum_{n=1}^{\infty} c_n \int_a^b r(x)\phi_m(x)\phi_n(x)\,dx$$

where we have interchanged the order of integration and summation. Based on the orthogonality property (8.36), it is clear that all terms of the above series vanish except for the one corresponding to $n = m$. Also, based on (8.31) we recognize the left-hand side as $F(m)$, and thus the expression reduces to

$$F(m) = c_m \int_a^b r(x)[\phi_m(x)]^2\,dx, \qquad m = 1,2,3,\ldots$$

Solving now for c_m (and changing the dummy index m to n), we find that

$$c_n = F(n)\,\|\phi_n(x)\|^{-2}, \qquad n = 1,2,3,\ldots \tag{8.38}$$

where

$$\|\phi_n(x)\|^2 = \int_a^b r(x)[\phi_n(x)]^2\,dx, \qquad n = 1,2,3,\ldots \tag{8.39}$$

Because the integrand in (8.39) is positive, it is clear that this integral never vanishes. Finally, the substitution of (8.38) into (8.37) yields the *inversion formula*

$$f(x) = \sum_{n=1}^{\infty} F(n)\,\|\phi_n(x)\|^{-2}\,\phi_n(x), \qquad a < x < b \tag{8.40}$$

It can be shown that when f, f', and f'' are all continuous functions over the closed interval $a \le x \le b$, the series (8.40) converges to $f(x)$ over the open interval $a < x < b$ (and possibly at the endpoints).

8.3.1 *Generalized Finite Fourier Transforms*

In addition to the finite Fourier transforms, the next most useful transforms are those in the category of *generalized* finite Fourier transforms. These transforms are associated with Sturm–Liouville problems usually having the same differential equation as in (8.32) and (8.34), but with different boundary conditions. For example, let us consider the Sturm–Liouville problem

$$\begin{gathered} y'' + \lambda y = 0, \qquad 0 < x < b \\ y(0) = 0, \qquad hy(b) + y'(b) = 0, \qquad h > 0 \end{gathered} \tag{8.41}$$

It can be shown that the only nontrivial solutions of (8.41) correspond to the case $\lambda > 0$. By setting $\lambda = k^2 > 0$, we find that the solution of the DE satisfying $y(0) = 0$ is

$$y = C \sin kx$$

where C is an arbitrary constant. The second boundary condition in (8.41) leads to

$$h \sin kb + k \cos kb = 0 \tag{8.42}$$

Hence, denoting the nth solution of this transcendental equation by k_n,* the eigenvalues and eigenfunctions of (8.41) are represented by

$$\lambda_n = k_n^2; \qquad \phi_n(x) = \sin k_n x, \qquad n = 1,2,3,\ldots \tag{8.43}$$

In this case we can define the *generalized finite Fourier transform*

$$T\{f(x);n\} = F(n) = \int_0^b f(x) \sin k_n x\,dx, \qquad n = 1,2,3,\ldots \tag{8.44}$$

---

*The actual values k_n, $n = 1,2,3,\ldots$, must be determined by a numerical procedure.

To obtain the corresponding inversion formula for (8.44), we first calculate

$$\|\phi_n(x)\|^2 = \int_0^b \sin^2 k_n x\, dx$$
$$= \frac{1}{2}\left(b - \frac{\sin 2k_n b}{2k_n}\right)$$
$$= \frac{1}{2}\left(b - \frac{\sin k_n b \cos k_n b}{k_n}\right)$$

but since $\sin k_n b = -(k_n/h)\cos k_n b$ [see Eq. (8.42)], we have

$$\|\phi_n(x)\|^2 = \frac{1}{2}\left(b + \frac{1}{h}\cos^2 k_n b\right), \qquad n = 1,2,3,\ldots \tag{8.45}$$

It now follows from (8.40) that the *inversion formula* is

$$T^{-1}\{F(n);x\} = f(x)$$
$$= 2\sum_{n=1}^{\infty}\left(\frac{F(n)}{b + (1/h)\cos^2 k_n b}\right)\sin k_n x, \qquad 0 < x < b \tag{8.46}$$

In solving differential equations with this transform, it is important to know the transforms of derivatives of a given function $f(x)$. In particular, it can be shown that (see Prob. 5 in Exer. 8.3)

$$T\{f''(x);n\} = -k_n^2 F(n) + [hf(b) + f'(b)]\sin k_n b + k_n f(0) \tag{8.47}$$

where $F(n)$ is the transform of $f(x)$.

Other generalized finite Fourier transforms are introduced in the exercises (see Probs. 6–8 in Exer. 8.3).

8.3.2 *Applications*

To illustrate the use of the generalized finite Fourier transform (8.44) in solving partial differential equations, let us consider the heat conduction problem

$$\begin{aligned} &u_{xx} = a^{-2}u_t, && 0 < x < 1,\ t > 0 \\ \text{B.C.:}\quad &u(0,t) = 0, && u(1,t) + u_x(1,t) = 0,\ t > 0 \\ \text{I.C.:}\quad &u(x,0) = u_0, && 0 < x < 1 \end{aligned} \tag{8.48}$$

where u_0 is a known constant.

Introducing the transform

$$T\{u(x,t);x \to n\} = U(n,t) \tag{8.49}$$

it follows that

$$T\{u_{xx}(x,t);x \to n\} = -k_n^2\, U(n,t) \tag{8.50}$$

$$T\{u_t(x,t);x \to n\} = U_t(n,t) \tag{8.51}$$

and

$$T\{u_0;x \to n\} = u_0 \int_0^1 \sin k_n x\, dx = \frac{u_0}{k_n}(1 - \cos k_n) \tag{8.52}$$

Hence, the transformed problem is

$$U_t + a^2k_n^2\, U = 0,\ t > 0 \tag{8.53}$$

I.C.: $U(n,0) = (u_0/k_n)(1 - \cos k_n)$

with solution

$$U(n,t) = (u_0/k_n)(1 - \cos k_n)e^{-a^2k_n^2t} \tag{8.54}$$

Using (8.46) to invert (8.54), we finally deduce that

$$u(x,t) = 2u_0 \sum_{n=1}^{\infty} \frac{(1 - \cos k_n)\sin k_n x}{k_n(1 + \cos^2 k_n)} e^{-a^2k_n^2t} \tag{8.55}$$

EXERCISES 8.3

In Probs. 1–4, determine the generalized finite Fourier transform defined by (8.44) of the given function.

1. $f(x) = x$

2. $f(x) = 1 - x$

3. $f(x) = h(a - x)$, $0 < a < b$

4. $f(x) = x^2$

5. Show that

$$T\{f''(x);n\} = -k_n^2\, F(n) + [hf(b) + f'(b)]\sin k_n b + k_n f(0)$$

where $F(n)$ is the transform of $f(x)$ and each k_n satisfies (8.42).

6. Develop a generalized finite Fourier transform and inversion formula associated with the Sturm–Liouville problem

$$y'' + \lambda y = 0, \quad 0 < x < b$$
$$y'(0) = 0, \quad hy(b) + y'(b) = 0$$

7. Develop a generalized finite Fourier transform and inversion formula associated with the Sturm–Liouville problem

$$y'' + \lambda y = 0, \quad 0 < x < 1$$
$$hy(0) - y'(0) = 0, \quad y'(1) = 0$$

8. Develop a generalized finite Fourier transform and inversion formula associated with the Sturm–Liouville problem

$$\frac{d}{dx}(e^x y') + \lambda e^x y = 0, \qquad 0 < x < 1$$

$$y(0) = 0, \qquad y(1) = 0$$

9. Use the generalized finite Fourier transform (8.44) to solve the heat conduction problem (u_1 and u_2 are known constants)

$$u_{xx} = a^{-2}u_t, \qquad 0 < x < 1,\ t > 0$$

B.C.: $\quad u(0,t) = u_1, \qquad u(1,t) + u_x(1,t) = u_2$

I.C.: $\quad u(x,0) = u_1$

10. Use the generalized finite Fourier transform defined in Prob. 6 to solve the heat conduction problem (u_0 is a known constant)

$$u_{xx} = a^{-2}u_t, \qquad 0 < x < 1,\ t > 0$$

B.C.: $\quad u_x(0,t) = 0, \qquad 2u(1,t) + u_x(1,t) = u_0$

I.C.: $\quad u(x,0) = 0$

8.4 *Finite Hankel Transform*

A particular Sturm–Liouville transform that has proven useful in the solution of certain boundary-value problems formulated in cylindrical coordinates is the finite Hankel transform. The Sturm–Liouville problem that motivates the introduction of this transform is

$$\frac{d}{dx}(xy') - \frac{\nu^2}{x}y + \lambda xy = 0, \qquad 0 < x < b,\ |y(0)| < \infty,\ y(b) = 0 \tag{8.56}$$

where the DE is *Bessel's equation*. The eigenvalues and eigenfunctions of (8.56) are given, respectively, by

$$\lambda_n = k_n^2; \qquad \phi_n(x) = J_\nu(k_n x), \qquad n = 1,2,3,\ldots \tag{8.57}$$

where $k_1, k_2, k_3, \ldots, k_n, \ldots,$ are chosen to satisfy

$$J_\nu(k_n b) = 0, \qquad n = 1,2,3,\ldots \tag{8.58}$$

Recognizing from (8.56) that the weighting function is $r(x) = x$, we define the *finite Hankel transform of order* ν of the function $f(x)$ by

$$H_\nu\{f(x);n\} = F(n) = \int_0^b xf(x)J_\nu(k_n x)\,dx, \qquad n = 1,2,3,\ldots \tag{8.59}$$

Based on the known relationship*

$$\|J_\nu(k_n x)\|^2 = \int_0^b x[J_\nu(k_n x)]^2\,dx$$
$$= \tfrac{1}{2}\,b^2[J_{\nu+1}(k_n b)]^2 \tag{8.60}$$

it follows that the *inversion formula* for the finite Hankel transform assumes the form

$$H_\nu^{-1}\{F(n);x\} = f(x) = \frac{2}{b^2}\sum_{n=1}^{\infty}\frac{F(n)\,J_\nu\,(k_n x)}{[J_{\nu+1}(k_n b)]^2}, \qquad 0 < x < b \tag{8.61}$$

It is known that if f and f' are piecewise continuous functions on the interval $0 \le x \le b$, the Bessel series (8.61) converges pointwise to $f(x)$ at points x where f is continuous and to $\frac{1}{2}[f(x^+) + f(x^-)]$ at points x where f has finite discontinuities.

8.4.1 *Basic Properties*

There are several properties of the finite Hankel transform that are mere consequences of the properties of Bessel functions. First, let us consider the finite Hankel transform of $f'(x)$. Using integration by parts, we find

$$H_\nu\{f'(x);n\} = \int_0^b xf'(x)J_\nu(k_n x)\,dx$$
$$= xf(x)J_\nu(k_n x)\Big|_0^b - \int_0^b f(x)\frac{d}{dx}[xJ_\nu(k_n x)]\,dx \tag{8.62}$$

The first term on the right in (8.62) for $\nu \ge 0$ is zero at both endpoints [recall Eq. (8.58)], and by using the identity (see Prob. 8 in Exer. 8.4)

$$\frac{d}{dx}[xJ_\nu(kx)] = \left(\frac{\nu+1}{2\nu}\right)kxJ_{\nu-1}(kx) - \left(\frac{\nu-1}{2\nu}\right)kxJ_{\nu+1}(kx)$$

we see that (8.62) reduces to

$$H_\nu\{f'(x);n\} = \left(\frac{\nu+1}{2\nu}\right)k_n H_{\nu-1}\{f(x);n\}$$
$$- \left(\frac{\nu-1}{2\nu}\right)k_n H_{\nu+1}\{f(x);n\}, \qquad \nu > 0 \tag{8.63}$$

Clearly, the case $\nu = 0$ has to be handled separately, but it leads to (see Prob. 9 in Exer. 8.4)

*See Chap. 6 in L. C. Andrews, *Special Functions for Engineers and Applied Mathematicians,* New York: Macmillan, 1985.

$$H_0\{f'(x);n\} = H_0\left\{\frac{1}{x}f(x);n\right\} - k_nH_1\{f(x);n\} \tag{8.64}$$

In a similar manner we can show that (see Probs. 10 and 11 in Exer. 8.4)

$$H_\nu\left\{\frac{1}{x}f(x);n\right\} = \frac{k_n}{2\nu}\left[H_{\nu-1}\{f(x);n\} + H_{\nu+1}\{f(x);n\}\right], \quad \nu > 0 \tag{8.65}$$

and

$$H_\nu\{x^{\nu-1}f'(x);n\} = -k_nH_{\nu-1}\{x^{\nu-1}f(x);n\}, \quad \nu > 0 \tag{8.66}$$

Also, by defining the function

$$g(x) = f''(x) + \frac{1}{x}f'(x) = \frac{1}{x}\frac{d}{dx}[xf'(x)] \tag{8.67}$$

we find that two integrations by parts leads to

$$\begin{aligned} H_0\{g(x);n\} &= \int_0^b \frac{d}{dx}[xf'(x)]J_0(k_nx)\,dx \\ &= xf'(x)J_0(k_nx)\Big|_0^b - k_nxf(x)J_0'(k_nx)\Big|_0^b \\ &\quad + \int_0^b xf(x)\left[k_n^2J_0''(k_nx) + \frac{k_n}{x}J_0'(k_nx)\right]dx \end{aligned} \tag{8.68}$$

The first term on the right above vanishes at both endpoints provided that $f'(x)$ is bounded at $x = 0$. Also, using the identity $J_0'(x) = -J_1(x)$ and the fact that (see Prob. 13 in Exer. 8.4)

$$k_n^2J_0''(k_nx) + \frac{k_n}{x}J_0'(k_nx) = -k_n^2J_0(k_nx) \tag{8.69}$$

the remaining expressions on the right in (8.68) simplify to

$$H_0\{g(x);n\} = k_nbf(b)J_1(k_nb) - k_n^2\int_0^b xf(x)J_0(k_nx)\,dx$$

which we can write as

$$H_0\{f''(x) + (1/x)f'(x);n\} = k_nbf(b)J_1(k_nb) - k_n^2F(n) \tag{8.70}$$

More generally, it can likewise be shown that (see Prob. 12 in Exer. 8.4)

$$\begin{aligned} &H_\nu\{f''(x) + (1/x)f'(x) - (\nu^2/x^2)f(x);n\} \\ &\qquad = k_nbf(b)J_{\nu+1}(k_nb) - k_n^2F(n) \end{aligned} \tag{8.71}$$

Other versions of the finite Hankel transform are also possible, but we will not discuss them. The interested reader may consult Chap. 8 in I. N. Sneddon, *The Use of Integral Transforms*, New York: McGraw-Hill, 1972.

8.4.2 *Applications*

Consider the heating of a long circular cylinder of radius b, whose initial temperature throughout is zero. If the temperature on the lateral surface is described by the function $f(t)$, the subsequent temperature $u(r,t)$ throughout the cylinder is governed by

$$\begin{aligned} & u_{rr} + (1/r)u_r = a^{-2}u_t, \qquad 0 < r < b,\ t > 0 \\ \text{B.C.:}\quad & u(b,t) = f(t), \qquad t > 0 \\ \text{I.C.:}\quad & u(r,0) = 0, \qquad 0 < r < b \end{aligned} \tag{8.72}$$

Taking the finite Hankel transform of order zero, we obtain

$$\begin{aligned} & U_t + a^2k_n^2U = a^2k_nbf(t)J_1(k_nb), \qquad t > 0 \\ \text{I.C.:}\quad & U(n,0) = 0 \end{aligned} \tag{8.73}$$

where

$$U(n,t) = \int_0^b ru(r,t)J_0(k_nr)\,dr \tag{8.74}$$

and we have used (8.70). The solution of (8.73) is readily found to be

$$U(n,t) = a^2k_nbJ_1(k_nb)\int_0^t f(\tau)e^{-a^2k_n^2(t-\tau)}\,d\tau \tag{8.75}$$

the inversion of which yields the formal solution

$$u(r,t) = \frac{2a^2}{b}\sum_{n=1}^{\infty}\frac{k_n}{J_1(k_nb)}\left[\int_0^t f(\tau)e^{-a^2k_n^2(t-\tau)}\,d\tau\right]J_0(k_nr) \tag{8.76}$$

EXERCISES 8.4

In Probs. 1–5, find the finite Hankel transform of order zero of the given function.

1. $f(x) = 1, \quad 0 < x < 1$

2. $f(x) = \begin{cases} x, & 0 < x < 1 \\ 0, & 1 < x < 2 \end{cases}$

3. $f(x) = 1 - x^2, \quad 0 < x < 1$

4. $f(x) = x^4, \quad 0 < x < 1$

5. $f(x) = \log x, \quad 0 < x < 1$

6. Show that $(0 < x < b)$

$$H_\nu\{x^\nu;n\} = (b^{\nu+1}/k_n)J_{\nu+1}(k_n b)$$

7. Show that $(0 < x < 1)$

$$H_\nu\{x^{\nu+1};n\} = \frac{2(\nu + 1)}{k_n^2} J_{\nu+1}(k_n)$$

8. Verify the identity

$$\frac{d}{dx}[xJ_\nu(kx)] = \left(\frac{\nu + 1}{2\nu}\right) kxJ_{\nu-1}(kx) - \left(\frac{\nu - 1}{2\nu}\right) kxJ_{\nu+1}(kx), \qquad \nu > 0$$

In Probs. 9–12, verify the given property of the finite Hankel transform.

9. $H_0\{f'(x);n\} = H_0\{(1/x)f(x);n\} = k_n H_1\{f(x);n\}$

10. $H_\nu\{(1/x)f(x);n\} = (k_n/2\nu)[H_{\nu-1}\{f(x);n\} + H_{\nu+1}\{f(x);n\}], \qquad \nu > 0$

11. $H_\nu\{x^{\nu-1}f'(x);n\} = -k_n H_{\nu-1}\{x^{\nu-1}f(x);n\}, \qquad \nu > 0$

12. $H_\nu\{f''(x) + (1/x)f'(x) - (\nu^2/x^2)f(x);n\} = k_n b f(b) J_{\nu+1}(k_n b) - k_n^2 F(n)$

13. Given that $y = J_0(x)$ is a solution of Bessel's equation

$$y'' + (1/x)y' + y = 0$$

show that $J_0(kr)$ satisfies

$$J_0''(kr) + (1/kr)J_0'(kr) + J_0(kr) = 0$$

14. The displacements of a thin circular membrane of unit radius are approximately governed by the boundary-value problem

$$u_{rr} + (1/r)u_r = c^{-2}u_{tt}, \qquad 0 < r < 1,\ t > 0$$

B.C.: $u(1,t) = 0$

I.C.: $u(r,0) = \begin{cases} 1, & 0 < r < 1/2 \\ 0, & 1/2 < r < 1 \end{cases} \quad u_t(r,0) = 0$

Show that the displacements are given by

$$u(r,t) = \sum_{n=1}^{\infty} \frac{J_1(k_n/2)}{k_n[J_1(k_n)]^2} J_0(k_n r)\cos k_n ct$$

where $J_0(k_n) = 0$, $n = 1,2,3,\ldots$

15. Given the boundary-value problem

$$u_{rr} + (1/r)u_r = c^{-2}u_{tt}, \qquad 0 < r < 1,\ t > 0$$

B.C.: $u(1,t) = 0$

I.C.: $u(r,0) = 0,\ u_t(r,0) = 1$

show that

$$u(r,t) = \frac{2}{c}\sum_{n=1}^{\infty} \frac{\sin(k_n ct)}{k_n^2 J_1(k_n)} J_0(k_n r)$$

where $J_0(k_n) = 0$, $n = 1,2,3,\ldots$.

16. Find a formal solution of

$$u_{rr} + (1/r)u_r = a^{-2}u_t, \qquad 0 < r < b,\ t > 0$$

B.C.: $u(b,t) = 0$

I.C.: $u(r,0) = f(r)$

17. The axisymmetric motion of a viscous fluid in concentric circles about the axis of rotation of an infinitely long cylinder is governed by the boundary-value problem

$$u_{rr} + (1/r)u_r - (1/r^2)u = (1/\eta)u_t, \qquad 0 < r < b,\ t > 0$$

B.C.: $u(b,t) = u_0$

I.C.: $u(r,0) = 0$

where η is the coefficient of kinematic viscosity of the fluid.

(a) By taking the finite Hankel transform of order one, obtain the transformed problem

$$U_t + \eta k_n^2 U = -\eta u_0 b k_n J_0(k_n b), \qquad t > 0$$

I.C.: $U(n,0) = 0$

(b) Solve the problem in (a) and show that its inversion leads to

$$u(r,t) = -\frac{2u_0}{b}\sum_{n=1}^{\infty} \frac{J_1(k_n r)}{k_n J_0(k_n b)}(1 - e^{-\eta k_n^2 t})$$

(c) Using the identity

$$r = -2\sum_{n=1}^{\infty} \frac{J_1(k_n r)}{k_n J_0(k_n b)}, \qquad 0 < r < b$$

show that the long-time solution in (b) simplifies to $u(r,t) \simeq u_0 r/b$, $t \to \infty$. This shows that eventually all of the fluid will rotate as a rigid body.

18. The equation governing the horizontal deflection $y(x,t)$ of a heavy chain of uniform line density σ, fixed at $x = 1$, and acted on by an external transverse force of intensity $p(x,t)$ is

$$\sigma\frac{\partial^2 y}{\partial t^2} = g\,\sigma\frac{\partial}{\partial x}\left(x\frac{\partial y}{\partial x}\right) + p(x,t), \qquad 0 < x < 1,\ t > 0$$

(a) If the chain is initially at rest under gravity, show that the horizontal deflection at any subsequent time is given by

$$y(x,t) = \frac{4}{\sigma\sqrt{g}} \sum_{n=1}^{\infty} \frac{J_0(k_n\sqrt{x})}{[J_1(k_n)]^2} \int_0^t q(t - \tau)\sin(\tfrac{1}{2}k_n\sqrt{g}\tau)\, d\tau$$

where

$$q(t) = \frac{1}{2}\int_0^1 p(x,t)J_0(k_n\sqrt{x})\, dx$$

Hint: Under the substitution $x = z^2$, show that the governing equation takes the form

$$\sigma \frac{\partial^2 y}{\partial t^2} = \frac{1}{4} g\, \sigma \left(\frac{\partial^2 y}{\partial z^2} + \frac{1}{z}\frac{\partial y}{\partial z}\right) + p(z^2,t)$$

(b) If the chain is released from rest in the position $y = \varepsilon(1 - x)$, $0 < x < 1$, and swings freely, show that the horizontal deflections at subsequent times are described by

$$y(x,t) = 8\varepsilon \sum_{n=1}^{\infty} \frac{J_0(k_n\sqrt{x})}{k_n^3 J_1(k_n)} \cos(\tfrac{1}{2}k\sqrt{g}\, t)$$

9

Discrete Transforms

9.1 *Introduction*

In many engineering applications the function (signal) under consideration is a continuous function of time that needs to be processed by a digital computer. The only way this can be accomplished is to sample the continuous function at discrete intervals of time. The sampled signal $x^*(t)$ is then processed as an approximation to the true signal $x(t)$.

The relation between a continuous function $x(t)$ and its sample values $x(kT)$, $k = 0, \pm 1, \pm 2, \ldots$, where T is a fixed interval of time, is one of prime importance in digital processing techniques. If the Fourier transform of $x(t)$ is nonzero only over a finite range of the transform variable, it turns out that the continuous function $x(t)$ can always be recovered (theoretically) from knowledge of only its sample values $x(kT)$, provided that the sampling rate is fast enough. This remarkable result is known as the *sampling theorem* and plays a central role in digital processing techniques. Functions whose transform is zero everywhere except for a finite interval are known as *band-limited* waveforms in signal analysis. Such functions do not actually exist in the real world, but theoretical considerations of band-limited waveforms is fundamental to the digital field. If the function under consideration is closely approximated by a band-limited waveform, then the sampled version of the function gives a reasonably accurate description of that function, provided the sample values are taken at a rate that is at least twice the highest frequency that is significant in the continuous waveform. This restriction is important in that it eliminates the problem of "aliasing" (high frequency components impersonating low frequencies).

9.2 *Discrete Fourier Transform*

Suppose we are able to sample the continuous function $x(t)$ at the discrete times $t = kT$, $k = 0, \pm 1, \pm 2, \ldots$ (see Fig. 9.1).* The sampled function $x^*(t)$ then consists solely of the sample values $\{x(kT)\}$, obtained through our sampling procedure. If we idealize the situation by assuming the sampling is done instantaneously, it is convenient to represent the sampled function by

$$x^*(t) = \sum_{k=-\infty}^{\infty} x(t)\delta(t - kT) = \sum_{k=-\infty}^{\infty} x(kT)\delta(t - kT) \tag{9.1}$$

where δ is the impulse function (see Sec. 1.5.2). The sampled function $x^*(t)$ is really a train of impulses in this sense, but is treated as if it were a continuous function of t through use of the properties of the impulse function.

In reality, we cannot obtain an infinite number of samples as suggested by Eq. (9.1). Hence, in practice we must settle for N samples over a total time duration NT, and in this case, Eq. (9.1) is replaced by the finite sum

$$x^*(t) = \sum_{k=0}^{N-1} x(kT)\delta(t - kT) \tag{9.2}$$

9.2.1 *Discrete Fourier Transform Pairs*

In engineering applications involving time and frequency domain analysis it is customary to define Fourier transforms pairs by the expressions

$$X(\omega) = \int_{-\infty}^{\infty} x(t)e^{-i\omega t}\,dt \tag{9.3}$$

and

$$x(t) = \frac{1}{2\pi}\int_{-\infty}^{\infty} X(\omega)e^{i\omega t}\,d\omega \tag{9.4}$$

---

*We are using the term "continuous" here merely to distinguish the function $x(t)$ from its sampled version $x^*(t)$. Our only real assumption regarding $x(t)$ is that it has a Fourier transform.

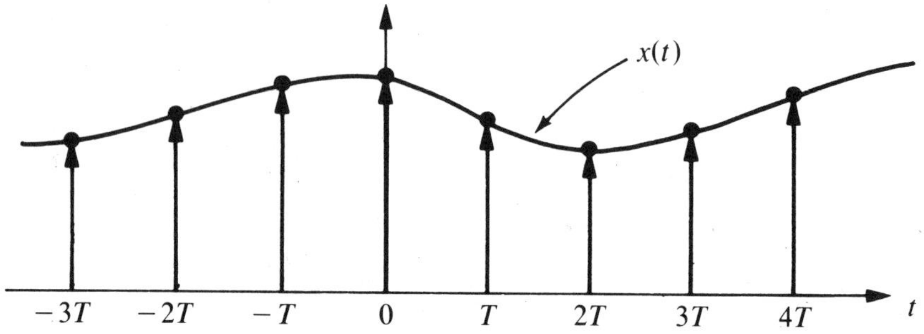

Figure 9.1 Sampled function

or, in terms of frequency $f = \omega/2\pi$,

$$X(f) = \int_{-\infty}^{\infty} x(t)e^{-2\pi ift}\, dt \tag{9.5}$$

and

$$x(t) = \int_{-\infty}^{\infty} X(f)e^{2\pi ift}\, df \tag{9.6}$$

Equations (9.5) and (9.6) as transform pairs have a certain appeal because the constant $1/2\pi$ has been absorbed. In our discussion to follow, we will henceforth convert to these definitions of Fourier transforms in order to be consistent with the majority of the literature on discrete transforms.

Let us begin by taking the Fourier transform of the sampled function $x^*(t)$ given by (9.2); thus, we obtain

$$\begin{aligned} X^*(f) &= \sum_{k=0}^{N-1} \int_{-\infty}^{\infty} x(kT)\delta(t - kT)e^{-2\pi ift}\, dt \\ &= \sum_{k=0}^{N-1} x(kT)e^{-2\pi ifkT} \end{aligned} \tag{9.7}$$

At this point $X^*(f)$ is a continuous function of frequency f, and represents our approximation to the true frequency function $X(f)$ associated with the continuous function $x(t)$. If one is interested in a few frequency samples of $X^*(f)$ taken at various values of f, then (9.7) is convenient to use. Owing to the periodic nature of the complex exponential, however, we find that $X^*(f)$ is a periodic function with period $1/T$, the reciprocal of the sampling interval. Hence we can obtain frequency information only up to this value and then the function repeats itself. If we desire to calculate $X^*(f)$ at N points, it is usually best to choose these N points evenly spaced over one period. Thus, we pick the discrete frequencies

$f = j/NT, j = 0,1,\ldots,N-1$, which cover one period. At this point it is convenient to introduce the simplified notation

$$x(k) \equiv x(kT), \qquad X(j) \equiv X^*(j/NT)$$

so that (9.7) becomes

$$X(j) = \sum_{k=0}^{N-1} x(k)e^{-2\pi ijk/N}, \qquad j = 0,1,\ldots,N-1 \tag{9.8}$$

which is called a *discrete Fourier transform* (DFT).

In order to derive the discrete inverse Fourier transform, we first make the observation that $(m = 0,1,\ldots,N-1)$

$$\sum_{j=0}^{N-1} e^{-2\pi i(k-m)j/N} = \begin{cases} 0, & k \neq m \\ N, & k = m \end{cases} \tag{9.9}$$

the verification of which is left to the reader (see Prob. 6 in Exer. 9.2). Now multiplying (9.8) by $e^{2\pi imj/N}$ and summing from $j = 0$ to $N-1$, we find

$$\begin{aligned} \sum_{j=0}^{N-1} X(j)e^{2\pi imj/N} &= \sum_{k=0}^{N-1} x(k) \sum_{j=0}^{N-1} e^{-2\pi i(k-m)j/N} \\ &= Nx(m) \end{aligned} \tag{9.10}$$

where we have made use of (9.9). Changing the free index m to k in (9.10), we have derived the *inverse transform* relation

$$x(k) = \frac{1}{N}\sum_{j=0}^{N-1} X(j)e^{2\pi ijk/N}, \qquad k = 0,1,\ldots,N-1 \tag{9.11}$$

Equations (9.8) and (9.11) constitute what are called DFT pairs. In some areas of the literature, however, the factor $1/N$ that appears in (9.11) is found in (9.8) instead. Thus, once again, we caution the reader to carefully check the definitions when using properties of these transforms found in other reference sources.

Example 9.1: Find the DFT of the four-point sequence $\{x(k)\} = \{1,1,0,0\}$, and then find the inverse DFT of the result.

Solution: The DFT in this case is given by

$$X(j) = \sum_{k=0}^{3} x(k)e^{-i\pi jk/2}, \qquad j = 0,1,2,3,$$

which leads to

$$X(0) = \sum_{k=0}^{3} x(k) = 2$$

$$X(1) = \sum_{k=0}^{3} x(k)e^{-i\pi k/2} = 1 - i$$

$$X(2) = \sum_{k=0}^{3} x(k)e^{-i\pi k} = 0$$

$$X(3) \sum_{k=0}^{3} x(k)e^{-3i\pi k/2} = 1 + i$$

Using these four values to calculate the inverse DFT, we have

$$x(k) = \frac{1}{4}\sum_{k=0}^{3} X(j)e^{i\pi jk/2}, \qquad k = 0,1,2,3$$

from which we recover the original sample values

$$x(0) = 1, \qquad x(1) = 1, \qquad x(2) = 0, \qquad x(3) = 0$$

As already pointed out, one of the major distinctions in the DFT as compared with the continuous Fourier transform is that both $\{x(k)\}$ and $\{X(j)\}$ form *periodic sequences* with period N; that is,

$$x(k + N) = x(k), \qquad \text{all } k \tag{9.12}$$

and

$$X(j + N) = X(j), \qquad \text{all } j \tag{9.13}$$

the verification of which we leave to the exercises (see Prob. 8 in Exer. 9.2). To geometrically display this periodicity property, the sample values are often represented as equally spaced marks around a circle as depicted in Fig. 9.2.

Other than the periodicity property, the *operational properties* of the DFT correlate very closely with the corresponding operational properties

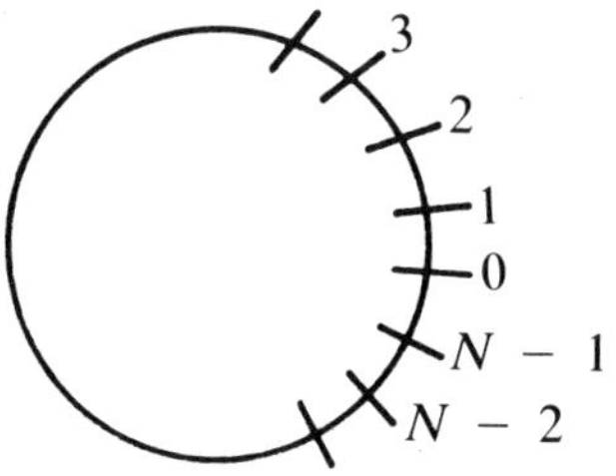

Figure 9.2 Sample values arranged on a circle

of the continuous Fourier transform. For example, the DFT has the *shift properties*

$$\sum_{k=0}^{N-1} [x(k)e^{-2\pi ijk/N}]\, e^{-2\pi imk/N} = X(j + m) \tag{9.14}$$

and

$$\sum_{k=0}^{N-1} x(k + m)e^{-2\pi ijk/N} = X(j)e^{2\pi ijm/N} \tag{9.15}$$

The *convolution theorem* takes the form

$$\frac{1}{N}\sum_{j=0}^{N-1} X(j)Y(j)e^{2\pi ijk/N} = (x * y)(k) = \sum_{m=0}^{N-1} x(m)y(k - m) \tag{9.16}$$

and *Parseval's relation* is given by

$$\frac{1}{N}\sum_{j=0}^{N-1} |X(j)|^2 = \sum_{k=0}^{N-1} |x(k)|^2 \tag{9.17}$$

The proofs of these properties, and some additional ones, are left to the exercises.

Example 9.2: Compute the convolution of the two four-point sequences

$$\{x(k)\} = \{1,2,3,4\} \text{ and } \{y(k)\} = \{5,6,7,8\}$$

Solution: The convolution is defined by

$$(x * y)(k) = \sum_{m=0}^{N-1} x(m)y(k - m)$$

To evaluate this convolution in a simple fashion, we display the two sequences around two concentric circles as shown in Fig. 9.3. The

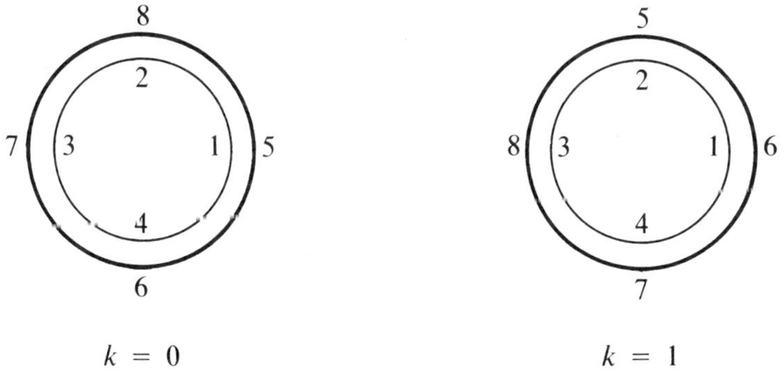

Figure 9.3 Convolution

value of the convolution at $k = 0$ is then

$$(x * y)(0) = (1)(5) + (2)(8) + (3)(7) + (4)(6) = 66,$$

while at $k = 1$ the outer circle is rotated counterclockwise by one position so that

$$(x * y)(1) = (1)(6) + (2)(5) + (3)(8) + (4)(7) = 68$$

Continuing in this fashion, we obtain the remaining values

$$(x * y)(2) = (1)(7) + (2)(6) + (3)(5) + (4)(8) = 66$$

and

$$(x * y)(3) = (1)(8) + (2)(7) + (3)(6) + (4)(5) = 60$$

9.2.2 *Fast Fourier Transform*

The DFT and its inverse have been defined, respectively, by

$$X(J) = \sum_{K=0}^{N-1} x(K)W^{JK}, \qquad J = 0,1,\ldots,N-1 \tag{9.18}$$

and

$$x(K) = \frac{1}{N}\sum_{J=0}^{N-1} X(J)W^{-JK}, \qquad K = 0,1,\ldots,N-1 \tag{9.19}$$

where

$$W = e^{-2\pi i/N} \tag{9.20}$$

Because of subscripting, etc., in the material to follow, it is more convenient in this setting to use capital indices J and K rather than lower case as before. To actually perform the indicated operations in a direct computation of the DFT would necessitate N operations for each sample output, where an operation is defined as one complex multiplication and addition. The complete DFT of a signal of length N would then require N^2 operations. For applications involving large N, the required computational time can often be prohibitive, even on a high-speed computer. For real-time analysis the computational time may be prohibitive even for moderate size N.

In 1965, Cooley and Tukey published an algorithm that, under certain conditions, significantly reduces the number of computations required to compute a DFT.* This algorithm, which has become known as the *fast Fourier transform* (FFT), is one of the most important contributions

*J. W. Cooley and J. W. Tukey, "An algorithm for the machine calculation of complex Fourier series," *Math. Comp.*, **19**, pp. 297–301, April 1965.

in this century to the field of numerical computations. Whereas N^2 operations are required for computing a DFT, the FFT only requires $N \log_2 N$ operations, when N is a power of 2. Today there are several variations of this algorithm, but they are all based on the same principle.

In order to derive the original form of the algorithm let us assume that $N = 2^m$, where m is a positive integer. Using base 2, the indices J and K can be represented by

$$J = 2^{m-1}J_{m-1} + \dots + 2J_1 + J_0 = (J_{m-1}\dots J_1J_0) \tag{9.21}$$

and

$$K = 2^{m-1}K_{m-1} + \dots + 2K_1 + K_0 = (K_{m-1}\dots K_1K_0) \tag{9.22}$$

where each J_j and K_j is either zero or one. For example, the number 121 in base 2 is represented by

$$121 = 2^6\cdot1 + 2^5\cdot1 + 2^4\cdot1 + 2^3\cdot1 + 2^2\cdot0 + 2\cdot0 + 1 = (1111001)$$

Thus if we introduce the notations

$$x(K) = x(K_{m-1}\dots K_0)$$

and

$$X(J) = X(J_{m-1}\dots J_0)$$

then Eq. (9.18) takes the form

$$X(J_{m-1}\dots J_0) = \sum_{K_0}\dots\sum_{K_{m-1}} x(K_{m-1}\dots K_0)W^{J(2^{m-1}K_{m-1}+\dots+K_0)} \tag{9.23}$$

Certain simplifications can now take place by making use of the following theorem and its obvious generalizations.

Theorem 9.1. For $N = 2^m$, the function $W = e^{-2\pi i/N}$ satisfies the identity

$$W^{J2^{m-1}K_{m-1}} = W^{J_0 2^{m-1}K_{m-1}}$$

Proof: Consider

$$W^{J2^{m-1}K_{m-1}} = \exp\left[\frac{-2\pi i(2^{m-1}J_{m-1} + \dots + J_0)2^{m-1}K_{m-1}}{2^m}\right]$$
$$= \exp[-\pi i(2^{m-1}J_{m-1} + \dots + J_0)K_{m-1}]$$

From properties of the complex exponential, it follows that

$$\exp(-\pi i 2^{m-1}J_{m-1}K_{m-1}) = 1$$
$$\exp(-\pi i 2^{m-2}J_{m-2}K_{m-1}) = 1$$
$$\dots$$
$$\exp(-\pi i 2J_1K_{m-1}) = 1$$

and thus we deduce our result

$$W^{J2^{m-1}K_{m-1}} = \exp(-\pi i J_0 K_{m-1}) = W^{J_0 2^{m-1}K_{m-1}} \qquad \blacksquare$$

The result of Theorem 9.1 is readily generalized to

$$W^{J2^{m-p}K_{m-p}} = W^{(2^{p-1}J_{p-1}+\ldots+J_0)2^{m-p}K_{m-p}}, \quad p = 1,\ldots,m \tag{9.24}$$

although we will not present the proof. Hence, starting with the innermost summation in (9.23), we obtain successively

$$\begin{aligned}
A_1(J_0K_{m-2}\ldots K_0) &= \sum_{K_{m-1}} x(K_{m-1}\ldots K_0)W^{J_0 2^{m-1}K_{m-1}} \\
A_2(J_0J_1K_{m-3}\ldots K_0) &= \sum_{K_{m-2}} A_1(J_0K_{m-2}\ldots K_0)W^{(2J_1+J_0)2^{m-2}K_{m-2}} \\
&\quad \cdots \\
A_p(J_0\ldots J_{p-1}K_{m-p-1}\ldots K_0) &= \sum_{K_{m-p}} A_{p-1}(J_0\ldots J_{p-2}K_{m-p}\ldots K_0) \\
&\qquad \cdot W^{(2^{p-1}J_{p-1}+\ldots+J_0)2^{m-p}K_{m-p}} \\
A_m(J_0\ldots J_{m-1}) &= \sum_{K_0} A_{m-1}(J_0\ldots J_{m-2}K_0)W^{(2^{m-1}J_{m-1}+\ldots+J_0)K_0}
\end{aligned} \tag{9.25}$$

The last result $A_m(J_0\ldots J_{m-1})$ is our desired output, but in bit-reversed order; that is,

$$X(J_{m-1}\ldots J_0) = A_m(J_0\ldots J_{m-1}) \tag{9.26}$$

This bit-reversal is an inherent property of the algorithm. To obtain the output in proper order, the input could first be scrambled. Either way, the scrambling of input or output data is simply part of the entire FFT process.

Example 9.3: Use a four-point FFT to compute the Fourier transform of

$$\{x(K)\} = \{1,2,3,4\}$$

Solution: We first note that the process consists of three steps of calculations:

$$\begin{aligned}
A_1(J_0K_0) &= \sum_{K_1=0}^{1} x(K_1K_0)W^{2J_0K_1} \\
A_2(J_0J_1) &= \sum_{K_0=0}^{1} A_1(J_0K_0)W^{2J_1K_0+J_0K_0} \\
X(J_1J_0) &= A_2(J_0J_1)
\end{aligned}$$

where $W = e^{-i\pi/2} = i$.

Step 1:

$$A_1(00) = \sum_{K_1=0}^{1} x(K_1 0) = x(00) + x(10) = 1 + 3 = 4$$

$$A_1(01) = \sum_{K_1=0}^{1} x(K_1 1) = x(01) + x(11) = 2 + 4 = 6$$

$$A_1(10) = \sum_{K_1=0}^{1} x(K_1 0)W^{2K_1} = x(00) - x(10) = 1 - 3 = -2$$

$$A_1(11) = \sum_{K_1=0}^{1} x(K_1 1)W^{2K_1} = x(01) - x(11) = 2 - 4 = -2$$

Step 2:

$$A_2(00) = \sum_{K_0=0}^{1} A_1(0K_0) = A_1(00) + A_1(01) = 4 + 6 = 10$$

$$A_2(01) = \sum_{K_0=0}^{1} A_1(0K_0)W^{2K_0} = A_1(00) - A_1(01) = 4 - 6 = -2$$

$$A_2(10) = \sum_{K_0=0}^{1} A_1(1K_0)W^{K_0} = A_1(10) + A_1(11)i = -2 + 2i$$

$$A_2(11) = \sum_{K_0=0}^{1} A_1(1K_0)W^{3K_0} = A_1(10) - A_1(11)i = -2 - 2i$$

Step 3:

$$X(0) = X(00) = A_2(00) = 10$$

$$X(1) = X(01) = A_2(10) = -2 + 2i$$

$$X(2) = X(10) = A_2(01) = -2$$

$$X(3) = X(11) = A_2(11) = -2 - 2i$$

Hence, the desired transform is

$$\{X(J)\} = \{10, -2 + 2i, -2, -2 - 2i\}$$

EXERCISES 9.2

In Probs. 1–4, find the DFT of the given four-point sequence and then find the inverse DFT of the result.

1. $\{1,2,3,4\}$

2. $\{0,1,0,-1\}$

3. $\{1,2/3,1/3,0\}$

4. $\{1,2,1,2\}$

5. Calculate the convolution of the sequences
(a) from Probs. 1 and 2.
(b) from Probs. 3 and 4.

6. Prove that

$$\sum_{j=0}^{N-1} e^{-2\pi i(k-m)j/N} = \begin{cases} 0, & k \neq m \\ N, & k = m \end{cases}$$

Hint: Sum the series as a geometric series.

7. Prove the *linearity* properties

(a) $\sum_{k=0}^{N-1} [x(k) + y(k)]e^{-2\pi ijk/N} = X(j) + Y(j), \qquad j = 0,1,\ldots,N-1$

(b) $\frac{1}{N}\sum_{j=0}^{N-1} [X(j) + Y(j)]e^{2\pi ijk/N} = x(k) + y(k), \qquad k = 0,1,\ldots,N-1$

8. Prove the *periodicity* properties

(a) $x(k + N) = x(k), \qquad \text{all } k$

(b) $X(j + N) = X(j), \qquad \text{all } j$

9. Prove the *shift* properties

(a) $\sum_{k=0}^{N-1} [x(k)e^{-2\pi ijk/N}]e^{-2\pi imk/N} = X(j + m)$

(b) $\sum_{k=0}^{N-1} x(k + m)e^{-2\pi ijk/N} = X(j)e^{2\pi ijm/N}$

10. Prove the *convolution theorem*

(a) $\frac{1}{N}\sum_{j=0}^{N-1} X(j)Y(j)e^{2\pi ijk/N} = \sum_{m=0}^{N-1} x(m)y(k - m)$

(b) From (a), deduce *Parseval's relation*

$$\frac{1}{N}\sum_{j=0}^{N-1} |X(j)|^2 = \sum_{k=0}^{N-1} |x(k)|^2$$

11. Verify Parseval's relation for the sequences given in Exam. 9.1.

12. A sequence $\{x(k)\}$ is said to be *even* if $x(N - k) = x(-k) = x(k)$. We say the sequence is *odd* if $x(N - k) = x(-k) = -x(k)$. Prove the following properties concerning even and odd sequences involving DFT pairs:

(a) $\{X(j)\}$ is even if and only if $\{x(k)\}$ is even.

(b) $\{X(j)\}$ is odd if and only if $\{x(k)\}$ is odd.
(c) $\{X(j)\}$ is real and even if and only if $\{x(k)\}$ is real and even.
(d) $\{X(j)\}$ is real and odd if and only if $\{x(k)\}$ is pure imaginary and odd.
(e) $\{X(j)\}$ is pure imaginary and even if and only if $\{x(k)\}$ is pure imaginary and even.
(f) $\{X(j)\}$ is pure imaginary and odd if and only if $\{x(k)\}$ is real and odd.

In Probs. 13–16, use the FFT to evaluate the Fourier transform of the given four-point sequence.

13. $\{1,1,0,0\}$

14. $\{0,1,0,-1\}$

15. $\{1,2/3,1/3,0\}$

16. $\{1,2,1,2\}$

9.3 *The Z Transform*

Communication systems using pulse modulation techniques, where several different messages may be interlaced on a time-sharing basis, rely on sample values of the signals taken at regular spaced intervals of time. In this case the sample values constitute the full available information about the signals. Another area of application based on sample values includes control systems in which feedback is applied on a basis of sample values of some quantity which is to be controlled. The Z transform is utilized heavily in these areas of application, although it has also proven useful in other applications as well.

The Z transform is an operation that converts a discrete signal into a complex frequency domain representation. In this regard it is the discrete analog of the Laplace transform, and thus will have many properties in common with the Laplace transform.

Since the Laplace transform is ordinarily associated with causal functions, i.e., functions which are identically zero for $t < 0$, we will consider only that class of functions in the present discussion. Let $x(t)$ be a continuous function on $t \geq 0$ and of exponential order c_0. The related sampled function $x^*(t)$ then has the representation

$$x^*(t) = \sum_{n=0}^{\infty} x(nT)\delta(t - nT) \tag{9.27}$$

where T denotes the time interval between samples. In developing properties of the Z transform, we normally assume that the sampled function consists of an infinite number of samples as indicated in Eq. (9.27). Of course, in practice there may be only a finite number of these sample values that are actually available for processing. The Laplace transform of the above sampled function formally leads to

$$\mathscr{L}\{x^*(t);p\} = \sum_{n=0}^{\infty} \int_0^{\infty} e^{-pt}\, x(nT)\delta(t - nT)\, dt$$

$$= \sum_{n=0}^{\infty} x(nT)e^{-pnT} \tag{9.28}$$

It is now notationally convenient to make the substitution

$$z = e^{pT} \tag{9.29}$$

which transforms the axis of convergence $\mathrm{Re}(p) = \sigma$ of the Laplace transform into a circle in the z plane. Hence, in the z plane we have the new function

$$X(z) = \mathscr{L}\{x^*(t);p\} = X^*(p)$$

and (9.28) becomes

$$X(z) = \sum_{n=0}^{\infty} x(nT)z^{-n}, \qquad |z| > e^{\sigma T} \tag{9.30}$$

We say that $X(z)$ is the *Z transform* of $x(nT)$, but which we will also call the Z transform of the continuous function $x(t)$. Using the principle of analytic continuation, we can deduce that $X(z)$ is an analytic function of z outside the circle $|z| = e^{\sigma T}$. The series on the right in (9.30) is the Laurent series expansion of $X(z)$ about the origin. Because the spacing T between samples has no effect on developing properties and on the use of the Z transform, it is conventional to set $T = 1$. Also, corresponding to the notation $\mathscr{L}\{f(t);p\} = F(p)$ that is used in the Laplace transform, we introduce the similar notation

$$Z\{x(t);z\} = X(z) = \sum_{n=0}^{\infty} x(n)z^{-n} \tag{9.31}$$

in which we have set $T = 1$.

Remark: Although our discussion of Laplace transforms is usually based on the possibility of discontinuities in the time function, we normally assume in the use of the Z transform that the function $x(t)$ is continuous. However, if $x(t)$ has discontinuities between the sampling points, the sampled function $x^*(t)$ will be insensitive to these points. Also, if a discontinuity in $x(t)$ occurs at kT, we can simply replace $x(kT)$ by $x(kT^+)$.

9.3.1 *Evaluating Z Transforms*

If the function to be transformed consists of only its sample values $\{x(n)\}$, then we can immediately write down its Z transform. For instance, if $\{x(n)\} = \{2,1,0,5\}$, then

$$X(z) = 2 + z^{-1} + 5z^{-3}$$

When we start with a continuous function $x(t)$, the Z transform normally leads to an infinite series that in many cases can be summed exactly. Let us consider some examples.

Example 9.4: Given that $x(t) = 1$, find $Z\{1;z\}$.

Solution: We first note that the sample values of $x(t)$ are given by

$$x(n) = 1, \qquad n = 0,1,2,\ldots$$

Hence,

$$Z\{1;z\} = \sum_{n=0}^{\infty} z^{-n} = \frac{1}{1 - z^{-1}}, \qquad |z| > 1$$

where we have summed the series as a geometric series. We also write this result as

$$Z\{1;z\} = z/(z - 1), \qquad |z| > 1$$

Example 9.5: Find $Z\{a^t;z\}$, $a \neq 0$.

Solution: Here $x(n) = a^n$, $n = 0,1,2,\ldots$, so that

$$Z\{a^t;z\} = \sum_{n=0}^{\infty} a^n z^{-n} = \sum_{n=0}^{\infty} (a/z)^n$$

or

$$Z\{a^t;z\} = z/(z - a), \qquad |z| > a$$

Example 9.6: Find $Z\{t;z\}$.

Solution: Since $x(n) = n$, $n = 0,1,2,\ldots$, we get

$$Z\{t;z\} = \sum_{n=0}^{\infty} nz^{-n}$$

$$= z \sum_{n=0}^{\infty} nz^{-(n+1)}$$

$$= -z \frac{d}{dz} \sum_{n=0}^{\infty} z^{-n}$$

from which we deduce (using the result of Example 9.4)

$$Z\{t;z\} = z/(z - 1)^2, \qquad |z| > 1$$

Example 9.7: Find $Z\{1/\Gamma(t + 1);z\}$

Solution: Here $x(n) = 1/\Gamma(n + 1) = 1/n!$, and thus

$$Z\{1/\Gamma(t + 1);z\} = \sum_{n=0}^{\infty} \frac{z^{-n}}{n!} = e^{1/z}, \qquad \text{all } z$$

Other Z transforms can be derived in a similar manner. A table of Z transforms involving elementary functions can be found at the end of this chapter.

Based upon the definition of the Z transform, it is clear that whenever two functions $x(t)$ and $y(t)$ have the same values at $t = n$, they will have the same Z transform. Hence the Z transform is not one-to-one. For example, if $y(t) = x(t) + \sin \pi t$, then $x(t)$ and $y(t)$ have the same Z transform. There is, however, a one-to-one relationship between the samples $\{x(n)\}$ and the transform function $X(z)$. Only when the conditions of the sampling theorem are satisfied will we be able to uniquely determine the time function $x(t)$ from its sample values $\{x(n)\}$.

9.3.2 *Properties of the Z Transform*

Many of the operational properties of the Z transform are simple discrete analogs of the properties of the Laplace transform. For example, directly from definition we have the *linearity property*

$$Z\{C_1x(t) + C_2y(t);z\} = C_1X(z) + C_2Y(z) \tag{9.32}$$

where C_1 and C_2 are arbitrary constants.

The *first shift theorem* takes the form

$$Z\{x(t + 1);z\} = zX(z) - zx(0) \tag{9.33}$$

To derive this property, we note that

$$\begin{aligned} Z\{x(t + 1);z\} &= \sum_{n=0}^{\infty} x(n + 1)z^{-n} \\ &= z \sum_{n=0}^{\infty} x(n + 1)z^{-(n+1)} \\ &= z \sum_{n=1}^{\infty} x(n)z^{-n} \end{aligned}$$

the last step of which follows the change of index $n \to (n - 1)$. By adding the $n = 0$ term to the sum and subtracting it back out, we obtain (9.33). The *second shift property* is given by

$$Z\{x(t - a)h(t - a);z\} = z^{-a} X(z) \tag{9.34}$$

the verification of which is left to the exercises (see Prob. 11 in Exer. 9.3).

If $Z\{x(t);z\} = X(z)$, then

$$\begin{aligned} Z\{tx(t);z\} &= \sum_{n=0}^{\infty} nx(n)z^{-n} \\ &= -z\frac{d}{dz}\sum_{n=0}^{\infty} x(n)z^{-n} \end{aligned}$$

from which we deduce

$$Z\{tx(t);z\} = -z\, X'(z) \tag{9.35}$$

Also, it is easily shown that (see Prob. 12 in Exer. 9.3)

$$Z\{e^{-at}x(t);z\} = X(e^{a}z) \tag{9.36}$$

Additional properties are taken up in the exercises.

9.3.3 *Inverse Z Transforms*

There are several ways in which inverse Z transforms can be evaluated. First, if it is known that

$$X(z) = \sum_{n=0}^{\infty} c_n z^{-n} \tag{9.37}$$

then we immediately deduce*

$$x(n) = Z^{-1}\{X(z);n\} = c_n, \qquad n = 0,1,2,\ldots \tag{9.38}$$

In some cases we try to represent $X(z)$ in terms of functions whose inverse transforms are known. Partial fraction techniques are particularly useful in such cases.

Example 9.8: Find the inverse Z transform of

$$X(z) = \frac{z + 3}{z - 2}$$

*Because the uniqueness of the Z transform and its inverse extends only to the sample values $\{x(n)\}$ and not to $x(t)$, we will consider $\{x(n)\}$ as the inverse Z transform of $X(z)$ rather than $x(t)$.

Solution: By writing

$$X(z) = \frac{z}{z-2} + \frac{3}{z-2}$$

it follows from Exam. 9.5 and the second shift property (9.34) that

$$\begin{aligned} x(n) &= Z^{-1}\{X(z);n\} \\ &= 2^n + 3 \cdot 2^{n-1}h(n-1) \\ &= \begin{cases} 1, & n = 0 \\ 5 \cdot 2^{n-1}, & n = 1,2,3,\ldots \end{cases} \end{aligned}$$

Example 9.9: Find the inverse Z transform of

$$X(Z) = \frac{z^3 - 9z^2 + 5z - 1}{4z^3 - 8z^2 + 5z - 1}$$

Solution: Since the numerator is of the same degree as the denominator, we first divide and then apply a partial fraction expansion to obtain

$$\begin{aligned} X(z) &= \frac{1}{4} + \frac{-7z^2 + 15z/4 - 3/4}{4z^3 - 8z^2 + 5z - 1} \\ &= \frac{1}{4} - \frac{4}{z-1} + \frac{9/4}{z-1/2} + \frac{5/16}{(z-1/2)^2} \end{aligned}$$

Therefore, with the aid of entries 13, 15, and 5 in the table and the second shift property, we deduce that

$$\begin{aligned} x(n) &= \frac{1}{4}[h(n) - h(n-1)] - 4h(n-1) + \frac{9}{4}\left(\frac{1}{2}\right)^{n-1} h(n-1) \\ &\quad + \frac{5}{16}(n-1)\left(\frac{1}{2}\right)^{n-2} h(n-1) \\ &= \frac{1}{4}h(n) + \left[\frac{(13+5n)}{4}\left(\frac{1}{2}\right)^n - \frac{17}{4}\right]h(n-1), \qquad n=0,1,2,\ldots \end{aligned}$$

An alternate way of finding the inverse transform in Exam. 9.9 is to use long division to obtain

$$\frac{z^3 - 9z^2 + 5z - 1}{4z^3 - 8z^2 + 5z - 1} = \frac{1}{4} - \frac{7}{4}z^{-1} - \frac{41}{16}z^{-2} + \ldots$$

from which we conclude $x(0) = 1/4$, $x(1) = -7/4$, $x(2) = -41/16,\ldots$. This method, of course, provides no general formula for $x(n)$ but it does eliminate the need to factor the polynomial in the denominator.

The *convolution theorem* of the Z transform is (see Prob. 21 in Exer. 9.3)

$$Z^{-1}\{X(z)Y(z);n\} = (x*y)(n) \tag{9.39}$$

where

$$(x*y)(n) = \sum_{k=0}^{n} x(k)y(n-k) \tag{9.40}$$

Example 9.10: Use the convolution theorem to find the inverse Z transform of the product

$$\frac{z^2}{(z-2)(z-3)}$$

Solution: By setting

$$X(z) = \frac{z}{z-2}, \qquad Y(z) = \frac{z}{z-3}$$

we can immediately obtain the individual inverse transform relations

$$Z^{-1}\{X(z);n\} = x(n) = 2^n$$
$$Z^{-1}\{Y(z);n\} = y(n) = 3^n$$

Thus, from the convolution theorem (9.39), we have

$$\begin{aligned} Z^{-1}\{X(z)Y(z);n\} &= \sum_{k=0}^{n} 2^k \cdot 3^{n-k} \\ &= 3^n \sum_{k=0}^{n} \left(\frac{2}{3}\right)^k \end{aligned}$$

We recognize this last finite series as a geometric series, from which we deduce the result

$$Z^{-1}\left\{\frac{z^2}{(z-2)(z-3)};n\right\} = 3^{n+1}\left[1-\left(\frac{2}{3}\right)^{n+1}\right], \qquad n = 0,1,2,\ldots$$

The inverse Z transform can be evaluated in many cases by using a complex inversion formula similar to that used in evaluating inverse Laplace transforms. To derive this formula, we return to Eq. (9.37) written as

$$X(z) = c_0 + c_1z^{-1} + c_2z^{-2} + \cdots + c_nz^{-n} + \ldots$$

and then multiply both sides by z^{n-1} to get

$$X(z)z^{n-1} = c_0z^{n-1} + c_1z^{n-2} + \cdots + c_nz^{-1} + \ldots$$

We then integrate both sides of this expression around a closed contour

$|z| = R$ such that $X(z)$ is analytic *on and outside* the closed contour. On the right all terms will vanish except the one involving z^{-1}, and we are left with the inversion formula

$$c_n = \frac{1}{2\pi i} \oint_{|z|=R} X(z)z^{n-1}\, dz, \qquad n = 0,1,2,... \tag{9.41}$$

Thus, if $X(z)z^{n-1}$ has poles at $z = a_k$, $k = 1,2,...,N$, within the circle $|z| = R$, we deduce that

$$x(n) = c_n = \sum_{k=1}^{N} \text{Res}\{X(z)z^{n-1};a_k\}, \qquad n = 0,1,2,... \tag{9.42}$$

Example 9.11: Using (9.42), obtain the inverse Z transform of

$$X(z) = \frac{z+3}{z-2}$$

Solution: We first note that

$$X(z)z^{n-1} = \frac{(z+3)z^{n-1}}{z-2}$$

has simple poles at $z = 0$ and $z = 2$ when $n = 0$, and only the simple pole $z = 2$ when $n \geq 1$. Thus, as in Exam. 9.8, we find

$$x(0) = \text{Res}\{0\} + \text{Res}\{2\} = -\frac{3}{2} + \frac{5}{2} = 1$$

$$x(n) = \text{Res}\{2\} = 5 \cdot 2^{n-1}, \qquad n = 1,2,3,...$$

EXERCISES 9.3

In Probs. 1–10, evaluate the Z transform of the given function.

1. e^{ct}

2. t^2

3. $\cos bt$

4. $\sin bt$

5. $h(t-1)$

6. $h(t) - h(t-1)$

7. $h(t-1) - h(t-2)$

8. $(1/t)h(t-1)$

9. $h(t-k), \quad k = 1,2,3,...$

10. $a^{t-1}\, h(t-1)$

11. Show that

$$Z\{x(t-a)h(t-a);z\} = z^{-a}X(z), \qquad a > 0$$

12. Show that

$$Z\{e^{-at}x(t);z\} = X(e^{a}z)$$

13. Use Eq. (9.35) to find the Z transform of te^{ct}.

14. Use the result of Problem 12 to find the Z transform of $e^{ct}\cos bt$.

In Probs. 15–20, find the inverse Z transform of the given function without the use of residues. If necessary, use the table of Z transforms on p. 334.

15. $\dfrac{z}{z^2 + 1}$

16. $\dfrac{z^2 + 1}{z^2 - 1}$

17. $\dfrac{z^2}{(z + 1)(z + 3)}$

18. $\dfrac{z}{(z + 1)(z + 3)}$

19. $\dfrac{z^2 + 1}{(z - 1)^3}$

20. $\dfrac{z^3}{(z^2 + 1)(z - 2)}$

21. Verify the convolution theorem

$$Z^{-1}\{X(z)Y(z);n\} = \sum_{k=0}^{n} x(k)y(n - k)$$

where $X(z)$ and $Y(z)$ are, respectively, the Z transforms of $x(n)$ and $y(n)$.

22. Verify the initial and final value theorems

(a) $\lim\limits_{|z|\to\infty} X(z) = x(0)$

(b) $\lim\limits_{z\to\infty} (z - 1)X(z) = \lim\limits_{t\to\infty} x(t) = x(\infty)$

In Probs. 23–30, use Eq. (9.42) to evaluate the inverse Z transform of the given function.

23. $\dfrac{z}{z^2 + 1}$

24. $\dfrac{z^2 + 1}{z^2 - 1}$

25. $\dfrac{z^2}{(z + 1)(z + 3)}$

26. $\dfrac{z}{(z + 1)(z + 3)}$

27. $\dfrac{z^2 + 1}{(z - 1)^3}$

28. $\dfrac{z^3}{(z^2 + 1)(z - 2)}$

29. $\dfrac{z^2 - 1}{z^2 - 2z}$

30. $\dfrac{z \sin b}{z^2 - 2z \cos b + 1}$

9.4 *Difference Equations*

Difference equations arise in a variety of applications. In particular, they are closely related to differential equations and their theory basically parallels that of differential equations.

To get some idea of how differences and derivatives are related, let us start by considering the definition of derivative,

$$y'(t) = \lim_{T\to 0} \frac{y(t+T) - y(t)}{T}$$

Rather than passing to the limit, suppose we now think of T as fixed, i.e., $T = 1$. We then define

$$\Delta y(t) = y(t+1) - y(t) \tag{9.43}$$

which is called the *first-order difference*. Using (9.43) as a definition, we can formally define a *second-order difference* by

$$\begin{aligned}\Delta^2 y(t) &= \Delta[\Delta y(t)]\\ &= \Delta[y(t+1) - y(t)]\\ &= [y(t+2) - y(t+1)] - [y(t+1) - y(t)]\end{aligned}$$

or

$$\Delta^2 y(t) = y(t+2) - 2y(t+1) + y(t) \tag{9.44}$$

Continuing in this fashion, we can construct $\Delta^3 y(t)$, $\Delta^4 y(t)$, and so forth.

Although we can interpret t as a continuous variable in (9.43) and (9.44), it is generally regarded as a discrete variable n in most applications. Also, for notational convenience it is customary to adopt the notation $y(n) = y_n$, $y(n+1) = y_{n+1}, \ldots$. Such notation is suggestive of the recurrence formulas that occur in many applications, e.g., the power series method of solving ordinary differential equations. Adopting this notation, we will henceforth define Z transforms by

$$Z\{y_n; z\} = \sum_{n=0}^{\infty} y_n z^{-n} \tag{9.45}$$

In solving difference equations by the method of Z transforms, we will find the results of the following theorem helpful. [Note that theorem 9.2 is simply special cases of (9.33).]

Theorem 9.2. If $Z\{y_n; z\} = Y(z)$, then

(a) $Z\{y_{n+1}; z\} = z[Y(z) - y_0]$

(b) $Z\{y_{n+2}; z\} = z^2[Y(z) - y_0] - zy_1$

Proof: From definition,

$$Z\{y_{n+1};z\} = \sum_{n=0}^{\infty} y_{n+1}z^{-n}$$

$$\downarrow$$

$$n \to n - 1$$

$$= z \sum_{n=1}^{\infty} y_n z^{-n}$$

$$= z\left[\sum_{n=0}^{\infty} y_n z^{-n} - y_0\right]$$

$$= z[Y(z) - y_0]$$

Similarly,

$$Z\{y_{n+2};z\} = \sum_{n=0}^{\infty} y_{n+2}z^{-n}$$

$$\downarrow$$

$$n \to n - 2$$

$$= z^2 \sum_{n=2}^{\infty} y_n z^{-n}$$

$$= z^2\left[\sum_{n=0}^{\infty} y_n z^{-n} - y_0 - z^{-1}y_1\right]$$

from which (b) follows. ■

To compare one of the subtle distinctions between differential equations and difference equations, let us start with the first-order linear differential equation

$$y' - y = 0, \qquad y(0) = 3 \tag{9.46}$$

whose solution is

$$y(t) = 3e^t \tag{9.47}$$

Replacing y' in (9.46) with the difference (9.43), we get the corresponding difference equation

$$\Delta y - y_n = 0, \qquad y_0 = 3 \tag{9.48}$$

or

$$y_{n+1} - 2y_n = 0, \qquad y_0 = 3 \tag{9.49}$$

Applying the Z transform to (9.49) leads to

$$z[Y(z) - 3] - 2Y(z) = 0$$

from which we obtain*

$$Y(z) = 3z/(z - 2)$$

The inversion of this function leads to the solution

$$y_n = 3 \cdot 2^n, \qquad n = 0, 1, 2,... \tag{9.50}$$

One of the things that this simple example is illustrating is that the "natural base" for exponential functions in difference calculus is 2 rather than e as in ordinary calculus. There are several other correspondences of this nature that become evident by pursuing how the difference operator Δ works on various types of functions. However, here we will not develop such correspondences since our primary interest is in solving difference equations.†

Example 9.12: Use the Z transform to solve the difference equation

$$y_{n+2} + 3y_{n+1} + 2y_n = 0; \qquad y_0 = 1, \quad y_1 = 2$$

Solution: Application of the Z transform yields

$$z^2[Y(z) - 1] - 2z + 3z[Y(z) - 1] + 2Y(z) = 0$$

or

$$(z^2 + 3z + 2)Y(z) = z^2 + 5z$$

from which we find

$$Y(z) = \frac{z^2 + 5z}{(z + 1)(z + 2)} = \frac{4z}{z + 1} - \frac{3z}{z + 2}$$

Inverting this last result leads to the desired solution

$$y_n = 4(-1)^n - 3(-2)^n = (-1)^n[4 - 3 \cdot 2^n], \qquad n = 0,1,2,...$$

EXERCISES 9.4

In Probs. 1–10, use the Z transform to solve the given difference equation.

1. $y_{n+2} - 5y_{n+1} + 6y_n = 0; \quad y_0 = 0, \; y_1 = 1$

2. $y_{n+2} + 2y_{n+1} - 3y_n = 0; \quad y_0 = 0, \; y_1 = 1$

---

*On the other hand, the Z transform applied directly to (9.47) yields $Y(z) = 3z/(z - e)$.

†For more discussion of finite differences, see K. S. Miller, *An Introduction to the Calculus of Finite Differences and Difference Equations*, New York: Dover, 1960.

3. $3y_{n+2} - 5y_{n+1} + 2y_n = 0; \quad y_0 = 1, \ y_1 = 0$

4. $y_{n+2} - 4y_{n+1} + 4y_n = 0; \quad y_0 = 1, \ y_1 = 4$

5. $y_{n+2} - 2y_{n+1} + 2y_n = 0; \quad y_0 = 0, \ y_1 = 1$

6. $y_{n+2} - 5y_{n+1} + 6y_n = 4^n; \quad y_0 = 0, \ y_1 = 1$

7. $y_{n+2} - 7y_{n+1} + 10y_n = 16n; \quad y_0 = 6, \ y_1 = 2$

8. $y_{n+2} - 5y_{n+1} + 6y_n = 2n + 1; \quad y_0 = 0, y_1 = 1$

9. $y_{n+2} + 4y_{n+1} - 5y_n = 24n - 8; \quad y_0 = 3, \ y_1 = -5$

10. $y_{n+2} - 6y_{n+1} + 5y_n = 2^n; \quad y_0 = 0, \ y_1 = 0$

11. Use the Z transform to solve the variable-coefficient difference equation

$$(n + 1)y_{n+1} - y_n = 0; \qquad y_0 = 1$$

12. Show that

$$\Delta^3 y(t) = y(t + 3) - 3y(t + 2) + 3y(t + 1) - y(t)$$

13. Prove that

$$Z\{y_{n+3};z\} = z^3[Y(z) - y_0] - z^2 y_1 - zy_2$$

In Probs. 14 and 15, use the result of Prob. 13 to solve the given difference equation.

14. $y_{n+3} - 2y_{n+2} - y_{n+1} + 2y_n = 0; \quad y_0 = 0, \ y_1 = 1, \ y_2 = 1$

15. $y_{n+3} - 2y_{n+2} - y_{n+1} + 2y_n = n^2 + 2^n; \quad y_0 = 0, \ y_1 = 1, \ y_2 = 1$

16. The sequence of numbers 0,1,1,2,3,5,8,13,21,..., where each number after the first is the sum of the two preceeding numbers, is called a *Fibonocci sequence*. Find a formula for y_n such that $y_0 = 0$, $y_1 = 1$, $y_2 = 1$, $y_3 = 2$, ...

9.5 *Table of Z Transforms*

The following is a short table of Z transforms for reference purposes.

Table 9.1. *Z* Transforms

$$Z\{x(t);z\} = \sum_{n=0}^{\infty} x(n)z^{-n} = X(z)$$

No.	$x(t), \quad t \geq 0$	$x(n), \quad n = 0,1,\ldots$	$X(z)$
1	1	$h(n)$	$\dfrac{z}{z-1}$
2	t	n	$\dfrac{z}{(z-1)^2}$
3	t^2	n^2	$\dfrac{z(z+1)}{(z-1)^3}$
4	$a^t, \quad a > 0$	a^n	$\dfrac{z}{z-a}$
5	$ta^t, \quad a > 0$	na^n	$\dfrac{az}{(z-a)^2}$
6	$\cos bt$	$\cos bn$	$\dfrac{z(z-\cos b)}{z^2 - 2z\cos b + 1}$
7	$\sin bt$	$\sin bn$	$\dfrac{z\sin b}{z^2 - 2z\cos b + 1}$
8	$\cosh bt$	$\cosh bn$	$\dfrac{z(z-\cosh b)}{z^2 - 2z\cosh b + 1}$
9	$\sinh bt$	$\sinh bn$	$\dfrac{z\sinh b}{z^2 - 2z\cosh b + 1}$
10	$e^{ct}\cos bt$	$e^{cn}\cos bn$	$\dfrac{z(z-e^c\cos b)}{z^2 - 2ze^c\cos b + e^{2c}}$
11	$e^{ct}\sin bt$	$e^{cn}\sin bn$	$\dfrac{ze^c\sin b}{z^2 - 2ze^c\cos b + e^{2c}}$
12	$h(t-1)$	$h(n-1)$	$\dfrac{1}{z-1}$
13	$h(t) - h(t-1)$	$h(n) - h(n-1)$	1
14	$h(t-k), \quad k = 1,2,\ldots$	$h(n-k)$	$\dfrac{1}{z^{k-1}(z-1)}$
15	$a^{t-1}h(t-1), \quad a > 0$	$a^{n-1}h(n-1)$	$\dfrac{1}{z-a}$
16	$\dfrac{1}{t}h(t-1)$	$\dfrac{1}{n}h(n-1)$	$\log\dfrac{z}{z-1}$
17	$\dfrac{1}{\Gamma(t+1)}$	$\dfrac{1}{n!}$	$e^{1/z}$

Bibliography

Listed below are some of the standard references on integral transforms. Each of these references supplies numerous additional references, including many of the related research papers.

R. N. Bracewell, *The Fourier Transform and Its Applications,* New York: McGraw-Hill, 1978.

E. O. Brigham, *The Fast Fourier Transform,* New Jersey: Prentice-Hall, 1974.

H. S. Carslaw and J. C. Jaeger, *Operational Methods in Applied Mathematics,* Oxford: Oxford University Press, 1941.

R. V. Churchill, *Operational Mathematics,* New York: McGraw-Hill, 1972.

B. Davies, *Integral Transforms and Their Applications,* New York: Springer-Verlag, 1985.

D. F. Elliott and K. Rao, *Fast Transforms: Algorithms, Analyses, Applications,* New York: Academic Press, 1982.

W. R. LePage, *Complex Variables and the Laplace Transform for Engineers,* New York: Dover, 1980.

J. W. Miles, *Integral Transforms in Applied Mathematics,* London: Cambridge University Press, 1973.

A. Papoulis, *The Fourier Integral and Its Applications,* New York: McGraw-Hill, 1963.

I. N. Sneddon, *Fourier Transforms,* New York: McGraw-Hill, 1951.

I. N. Sneddon, *The Use of Integral Transforms,* New York: McGraw-Hill, 1972.

C. J. Tranter, *Integral Transforms in Mathematical Physics,* London: Methuen, New York: John Wiley & Sons, 1951.

E. C. Titchmarsh, *Introduction to the Theory of Fourier Integrals,* 2nd ed., Oxford: Oxford University Press, 1948.

H. J. Weaver, *Applications of Discrete and Continuous Fourier Analysis,* New York: John Wiley & Sons, 1983.

Appendix A Review of Complex Variables

In this appendix we present a very brief review of some of the most basic concepts and theorems from complex variables that have direct bearing on material in this text. There is no attempt for completeness, and proofs of the theorems will not be provided. The reader who desires a more thorough coverage of these topics can consult any introductory text on complex variables.

A *complex variable* is one that can be represented by

$$z = x + iy \tag{A.1}$$

where x and y are real variables and $i = \sqrt{-1}$. The variable x is called the *real part* of z, denoted by $x = \text{Re}(z)$, and y is called the *imaginary part* of z, also written as $y = \text{Im}(z)$.

If f is a function depending on the complex variable z, we say that f is a complex function. Such functions can always be represented in the form

$$f(z) = u(x, y) + iv(x, y) \tag{A.2}$$

where both u and v are real functions.

If the derivative $f'(z)$ exists at all points z of a region D in the complex plane, then $f(z)$ is said to be *analytic* in D. Necessary and sufficient conditions for $f(z)$ to be an analytic function are the *Cauchy–Riemann equations*

$$\frac{\partial u}{\partial x} = \frac{\partial v}{\partial y}, \qquad \frac{\partial u}{\partial y} = -\frac{\partial v}{\partial x} \tag{A.3}$$

where these partial derivatives are assumed to be continuous. A point where $f(z)$ is not analytic is called a *singular point*.

Integration of a complex function is actually *line integration* along some rectifiable curve (or path, or contour), in the complex plane. The most fundamental integral theorem concerning analytic functions bears the name of Cauchy.

Theorem A.1 (*Cauchy's integral theorem*). If $f(z)$ is analytic on and inside a closed path C in the complex plane, then

$$\oint_C f(z)dz = 0$$

If there is some neighborhood of a singular point $z = a$ of a function $f(z)$ throughout which $f(z)$ is analytic, except at the point itself, then $z = a$ is called an *isolated singularity*. Every function $f(z)$ has a *residue* at each of its isolated singular points, the value of which may be zero. In general, the value of the residue is the value of the integral

$$\frac{1}{2\pi i}\oint_C f(z)\,dz$$

around any closed contour containing the isolated singularity. Multiple-valued functions have singularities that are called *branch points*. For example, the function $f(z) = z^{1/2}$ has a branch point at $z = 0$.

One kind of isolated singular point is called a *pole*. If $f(z)$ is not finite at some point $z = a$, but the product $(z - a)^m f(z)$ is analytic for some integer m, we say that $f(z)$ has a *pole of order m* at $z = a$. A pole of order one is also called a *simple pole*. The residue for a function having a pole of order m at $z = a$ is given by

$$\operatorname{Res}\{f(z); a\} = \frac{1}{(m-1)!}\lim_{z\to a}\frac{d^{m-1}}{dz^{m-1}}[(z-a)^m f(z)] \tag{A.4}$$

If $f(z) = p(z)/q(z)$ has a simple pole at $z = a$, then also

$$\operatorname{Res}\{f(z); a\} = \frac{p(a)}{q'(a)} \tag{A.5}$$

The evaluation of an integral around a closed contour containing a finite number of poles of a given function $f(z)$ relies on the following residue theorem.

Theorem A.2 (*Residue theorem*). If $f(z)$ is a single-valued function which is analytic on and inside a closed path C, except for finitely-many isolated singular points $a_1, a_2, \ldots, a_N$ inside C, then

$$\oint_C f(z)dz = 2\pi i \sum_{k=1}^{N} \text{Res}\{f(z); a_k\}$$

The evaluation of real definite integrals can often be accomplished by using the residue theorem together with a suitable function $f(z)$ and a suitable contour C, the choice of which in some cases may require a certain amount of ingenuity. In the evaluation of such integrals it is frequently necessary to rely on some of the following theorems.

Theorem A.3. If, on a circular arc C with radius R and center at $z = 0$, we have $zf(z) \to 0$ uniformly as $R \to \infty$, then

$$\lim_{R\to\infty} \int_{C_R} f(z)dz = 0$$

Theorem A.4. If, on a circular arc C_ρ with radius ρ and center at $z = a$, we have $(z - a)f(z) \to 0$ uniformly as $\rho \to 0$, then

$$\lim_{\rho\to 0} \int_{C_\rho} f(z)dz = 0.$$

Theorem A.5. If, on a circular arc C_ρ with radius ρ and center at $z = a$, and intercepting at an angle α at $z = a$, $f(z)$ has a simple pole at $z = a$, then

$$\lim_{\rho\to 0} \int_{C_\rho} f(z)\, dz = \alpha i \,\text{Res}\{f(z); a\}$$

where $\alpha > 0$ if integration is counterclockwise and $\alpha < 0$ otherwise.

Theorem A.6. If, on a circular arc C_R with radius R and center at $z = 0$, we have $f(z) \to 0$ uniformly as $R \to \infty$, then

(a) $\displaystyle\lim_{R\to\infty} \int_{C_R} e^{imz} f(z)\, dz = 0 \qquad (m > 0)$

provided that C_R lies in the first and/or second quadrants.

(b) $\displaystyle\lim_{R\to\infty} \int_{C_R} e^{mz} f(z)\, dz = 0 \qquad (m > 0)$

provided that C_R lies in the second and/or third quadrants.

Appendix B
Table of Fourier Transforms

Listed below are short tables of Fourier transforms, cosine transforms, and sine transforms.

Table B.1. Fourier Transforms

No.	$f(t) = \frac{1}{\sqrt{2\pi}} \int_{-\infty}^{\infty} e^{-ist} F(s) ds$	$F(s) = \frac{1}{\sqrt{2\pi}} \int_{-\infty}^{\infty} e^{ist} f(t) dt$
1	1	$\sqrt{2\pi}\, \delta(s)$
2	$\frac{1}{t}$	$\sqrt{\frac{\pi}{2}}\, i \operatorname{sgn}(s)$
3	$\frac{1}{t^2 + a^2}, \quad a > 0$	$\sqrt{\frac{\pi}{2}} \frac{1}{a} e^{-a\|s\|}$
4	$\frac{t}{t^2 + a^2}, \quad a > 0$	$-\sqrt{\frac{\pi}{2}} \frac{is}{2a} e^{-a\|s\|}$
5	$\frac{t^2 - a^2}{t(t^2 + a^2)}, \quad a > 0$	$i\sqrt{\frac{2}{\pi}} (2e^{-a\|s\|} - 1) \operatorname{sgn}(s)$
6	e^{iat}	$\sqrt{2\pi}\, \delta(s + a)$
7	$e^{-a\|t\|}, \quad a > 0$	$\sqrt{\frac{2}{\pi}} \frac{a}{s^2 + a^2}$
8	$te^{-a\|t\|}, \quad a > 0$	$\sqrt{\frac{2}{\pi}} \frac{2ais}{(s^2 + a^2)^2}$

Table B.1. *(continued)*

No.	$f(t) = \frac{1}{\sqrt{2\pi}} \int_{-\infty}^{\infty} e^{-ist} F(s) ds$	$F(s) = \frac{1}{\sqrt{2\pi}} \int_{-\infty}^{\infty} e^{ist} f(t) dt$
9	$\lvert t\rvert e^{-a\lvert t\rvert}, \quad a > 0$	$\sqrt{\frac{2}{\pi}} \frac{a^2 - s^2}{(s^2 + a^2)^2}$
10	$e^{-a^2t^2}, \quad a > 0$	$\frac{1}{\sqrt{2}a} e^{-s^2/4a^2}$
11	$\cos(t^2/2)$	$\frac{1}{\sqrt{2}} [\cos(s^2/2) + \sin(s^2/2)]$
12	$\sin(t^2/2)$	$\frac{1}{\sqrt{2}} [\cos(s^2/2) - \sin(s^2/2)]$
13	$e^{-a\lvert t\rvert/\sqrt{2}}[\cos(at/\sqrt{2}) + \sin(a\lvert t\rvert/\sqrt{2})],$ $a > 0$	$\frac{2a^3}{\sqrt{\pi}} \frac{1}{s^4 + a^4}$
14	$e^{-\lvert t\rvert} \frac{\sin t}{t}$	$\frac{1}{\sqrt{2\pi}} \arctan(2/s^2)$
15	$\operatorname{sgn}(t)$	$\sqrt{\frac{2}{\pi}} \frac{i}{s}$
16	$t \operatorname{sgn}(t)$	$-\sqrt{\frac{2}{\pi}} \frac{1}{s^2}$
17	$h(1 - \lvert t\rvert)$	$\sqrt{\frac{2}{\pi}} \frac{\sin s}{s}$
18	$(1 - \lvert t\rvert)h(1 - \lvert t\rvert)$	$\frac{1}{\sqrt{2\pi}} \left(\frac{\sin s/2}{s/2}\right)^2$
19	$h(t)$	$\sqrt{\frac{\pi}{2}} \left[\delta(s) + \frac{i}{\pi s}\right]$
20	$\delta(t - a)$	$\frac{1}{\sqrt{2\pi}} e^{ias}$
21	$\lvert t\rvert^{-\alpha}, \quad 0 < \alpha < 1$	$\sqrt{\frac{2}{\pi}} \frac{\Gamma(1 - \alpha)}{\lvert s\rvert^{1-\alpha}} \sin\frac{\pi\alpha}{2}$
22	$\lvert t\rvert^{-\alpha} \operatorname{sgn}(t), \quad 0 < \alpha < 1$	$i\sqrt{\frac{2}{\pi}} \frac{\Gamma(1 - \alpha)}{\lvert s\rvert^{1-\alpha}} \cos\frac{\pi\alpha}{2}$
23	$P_n(t)h(1 - \lvert t\rvert)$	$\frac{i^n}{\sqrt{s}} J_{n+1/2}(s)$

Table B.2. Cosine Transforms

No.	$f(t) = \sqrt{\frac{2}{\pi}} \int_0^\infty F_C(s) \cos st\, ds$	$F_C(s) = \sqrt{\frac{2}{\pi}} \int_0^\infty f(t) \cos st\, dt$
1	1	$\sqrt{2\pi}\,\delta(s)$
2	$\frac{1}{\sqrt{t}}$	$\frac{1}{\sqrt{s}}$
3	$\frac{1}{t^2 + a^2}, \quad a > 0$	$\sqrt{\frac{\pi}{2}} \frac{1}{a} e^{-as}$
4	$e^{-at}, \quad a > 0$	$\sqrt{\frac{2}{\pi}} \frac{a}{s^2 + a^2}$
5	$te^{-at}, \quad a > 0$	$\sqrt{\frac{2}{\pi}} \frac{a^2 - s^2}{(s^2 + a^2)^2}$
6	$e^{-a^2t^2}, \quad a > 0$	$\frac{1}{\sqrt{2}a} e^{-s^2/4a^2}$
7	$e^{-at} \cos at, \quad a > 0$	$\sqrt{\frac{2}{\pi}} \frac{as^2 + 2a^3}{s^4 + 4a^4}$
8	$e^{-at} \sin at, \quad a > 0$	$\sqrt{\frac{2}{\pi}} \frac{2a^3 - as^2}{s^4 + 4a^4}$
9	$\cos(t^2/2)$	$\frac{1}{\sqrt{2}} [\cos(s^2/2) + \sin(s^2/2)]$
10	$\sin(t^2/2)$	$\frac{1}{\sqrt{2}} [\cos(s^2/2) - \sin(s^2/2)]$
11	$t^{p-1}, \quad 0 < p < 1$	$\sqrt{\frac{2}{\pi}} \frac{\Gamma(p)}{s^p} \cos(\pi p/2)$
12	$(a^2 - t^2)^{p-1/2} h(a - t),$ $a > 0$	$2^{p-1/2}\Gamma(p + 1/2)(a/s)^p J_p(as),$ $p > -1/2$

Table B.3. Sine Transforms

No.	$f(t) = \sqrt{\frac{2}{\pi}} \int_0^\infty F_S(s) \sin st\, ds$	$F_S(s) = \sqrt{\frac{2}{\pi}} \int_0^\infty f(t) \sin st\, dt$
1	$\frac{1}{t}$	$\sqrt{\frac{\pi}{2}} \operatorname{sgn}(s)$
2	$\frac{1}{\sqrt{t}}$	$\frac{1}{\sqrt{s}}$
3	$\frac{t}{t^2 + a^2}, \quad a > 0$	$\sqrt{\frac{\pi}{2}} e^{-as}$

Table B.3. *(continued)*

No.	$f(t) = \sqrt{\frac{2}{\pi}} \int_0^\infty F_S(s) \sin st\, ds$	$F_S(s) = \sqrt{\frac{2}{\pi}} \int_0^\infty f(t) \sin st\, dt$
4	$\frac{t}{(t^2 + a^2)^2}, \quad a > 0$	$\frac{1}{\sqrt{2\pi}} \frac{s}{a} e^{-as}$
5	$\frac{1}{t(t^2 + a^2)}, \quad a > 0$	$\sqrt{\frac{\pi}{2}} \frac{1}{a^2}(1 - e^{-as})$
6	$e^{-at}, \quad a > 0$	$\sqrt{\frac{2}{\pi}} \frac{s}{s^2 + a^2}$
7	$te^{-at}, \quad a > 0$	$\sqrt{\frac{2}{\pi}} \frac{2as}{(s^2 + a^2)^2}$
8	$te^{-a^2t^2}, \quad a > 0$	$\frac{s}{2\sqrt{2}a^3} e^{-s^2/4a^2}$
9	$\frac{1}{t} e^{-at}$	$\sqrt{\frac{2}{\pi}} \arctan(s/a)$
10	$e^{-at} \cos at, \quad a > 0$	$\sqrt{\frac{2}{\pi}} \frac{s^3}{s^4 + 4a^4}$
11	$e^{-at} \sin at, \quad a > 0$	$\sqrt{\frac{2}{\pi}} \frac{2a^2 s}{s^4 + 4a^4}$
12	$t^{p-1/2}, \quad 0 < p < 1$	$\sqrt{\frac{\pi}{2}} \frac{\Gamma(p)}{s^p} \sin(\pi p/2)$
13	$t(a^2 - t^2)^{p-3/2} h(a - t), \quad a > 0$	$2^{p-3/2} a^p s^{1-p} \Gamma(p - 1/2) J_p(as), \quad p > 1/2$

Appendix C
Table of Laplace Transforms

Listed below is a short table of Laplace transforms and their inverses.

Table C.1. Laplace Transforms

No.	$F(p) = \int_0^\infty e^{-pt} f(t)dt$	$f(t) = \frac{1}{2\pi i}\int_{c-i\infty}^{c+i\infty} e^{pt}F(p)dp$
1	$\frac{1}{p}$	1
2	$\frac{1}{p^2}$	t
3	$\frac{1}{p^n} \quad (n = 1, 2, 3, \ldots)$	$\frac{t^{n-1}}{(n-1)!}$
4	$\frac{1}{\sqrt{p}}$	$\frac{1}{\sqrt{\pi t}}$
5	$\frac{1}{p^{3/2}}$	$2\sqrt{t/\pi}$
6	$\frac{1}{p^x}, \quad x > 0$	$\frac{t^{x-1}}{\Gamma(x)}$
7	$\frac{1}{p-a}$	e^{at}
8	$\frac{1}{(p-a)^2}$	te^{at}

Table C.1. *(continued)*

No.	$F(p) = \int_0^\infty e^{-pt} f(t)dt$	$f(t) = \frac{1}{2\pi i}\int_{c-i\infty}^{c+i\infty} e^{pt}F(p)dp$
9	$\frac{1}{(p-a)^n} \quad (n = 1, 2, 3, \ldots)$	$\frac{t^{n-1}e^{at}}{(n-1)!}$
10	$\frac{1}{(p-a)^x}, \quad x > 0$	$\frac{t^{x-1}e^{at}}{\Gamma(x)}$
11	$\frac{1}{(p-a)(p-b)}, \quad a \neq b$	$\frac{e^{at} - e^{bt}}{a - b}$
12	$\frac{p}{(p-a)(p-b)}, \quad a \neq b$	$\frac{ae^{at} - be^{bt}}{a - b}$
13	$\frac{1}{p^2 + a^2}$	$\frac{1}{a}\sin at$
14	$\frac{p}{p^2 + a^2}$	$\cos at$
15	$\frac{1}{p^2 - a^2}$	$\frac{1}{a}\sinh at$
16	$\frac{p}{p^2 - a^2}$	$\cosh at$
17	$\frac{1}{(p-a)^2 + b^2}$	$\frac{1}{b}e^{at}\sin bt$
18	$\frac{p-a}{(p-a)^2 + b^2}$	$e^{at}\cos bt$
19	$\frac{1}{p(p^2 + a^2)}$	$\frac{1}{a^2}(1 - \cos at)$
20	$\frac{1}{p^2(p^2 + a^2)}$	$\frac{1}{a^3}(at - \sin at)$
21	$\frac{1}{(p^2 + a^2)^2}$	$\frac{1}{2a^3}(\sin at - at\cos at)$
22	$\frac{p}{(p^2 + a^2)^2}$	$\frac{t}{2a}\sin at$
23	$\frac{p^2}{(p^2 + a^2)^2}$	$\frac{1}{2a}(\sin at + at\cos at)$
24	$\frac{p}{(p^2 + a^2)(p^2 + b^2)}, \quad a^2 \neq b^2$	$\frac{1}{b^2 - a^2}(\cos at - \cos bt)$
25	$\frac{1}{p^4 + 4a^4}$	$\frac{1}{4a^3}(\sin at\cosh at - \cos at\sinh at)$

Table C.1. *(continued)*

No.	$F(p) = \int_0^\infty e^{-pt} f(t)dt$	$f(t) = \frac{1}{2\pi i}\int_{c-i\infty}^{c+i\infty} e^{pt}F(p)dp$
26	$\frac{p}{p^4 + 4a^4}$	$\frac{1}{2a^2}\sin at \sinh at$
27	$\frac{1}{p^4 - a^4}$	$\frac{1}{2a^3}(\sinh at - \sin at)$
28	$\frac{p}{p^4 - a^4}$	$\frac{1}{2a^2}(\cosh at - \cos at)$
29	$\frac{1}{\sqrt{p^2 + a^2}}$	$J_0(at)$
30	$\frac{1}{(p^2 + a^2)^\nu}, \quad a > 0, \quad \nu > 0$	$\frac{\sqrt{\pi}}{\Gamma(\nu)}\left(\frac{t}{2a}\right)^{\nu - 1/2} J_{\nu-1/2}(at)$
31	$\frac{1}{\sqrt{p}(p - a)}$	$\frac{1}{\sqrt{a}} e^{at} \operatorname{erf}(\sqrt{at})$
32	$\frac{1}{\sqrt{p}(\sqrt{p} + \sqrt{a})}$	$e^{at} \operatorname{erfc}(\sqrt{at})$
33	$e^{-ap}, \quad a > 0$	$\delta(t - a)$
34	$\frac{1}{p} e^{-ap}, \quad a > 0$	$h(t - a)$
35	$\frac{1}{p^2} e^{-ap}, \quad a > 0$	$(t - a)h(t - a)$
36	$\frac{1}{\sqrt{p}} e^{-a/p}, \quad a > 0$	$\frac{1}{\sqrt{\pi t}} \cos(2\sqrt{at})$
37	$\frac{1}{p\sqrt{p}} e^{-a/p}, \quad a > 0$	$\frac{1}{\sqrt{\pi a}} \sin(2\sqrt{at})$
38	$\frac{1}{p^\nu} e^{-a/p}, \quad a > 0, \quad \nu > 0$	$\left(\frac{t}{a}\right)^{(\nu-1)/2} J_{\nu-1}(2\sqrt{at})$
39	$e^{-a\sqrt{p}}, \quad a > 0$	$\frac{a}{2\sqrt{\pi t^3}} e^{-a^2/4t}$
40	$\frac{1}{\sqrt{p}} e^{-a\sqrt{p}}, \quad a \geq 0$	$\frac{1}{\sqrt{\pi t}} e^{-a^2/4t}$
41	$\frac{1}{p} e^{-a\sqrt{p}}, \quad a \geq 0$	$\operatorname{erfc}(a/2\sqrt{t})$
42	$\arctan(a/p)$	$\frac{\sin at}{t}$

Table C.1. *(continued)*

No.	$F(p) = \int_0^\infty e^{-pt} f(t)dt$	$f(t) = \frac{1}{2\pi i}\int_{c-i\infty}^{c+i\infty} e^{pt}F(p)dp$
43	$\frac{1}{p}\arctan(a/p)$	$\mathrm{Si}(at)$
44	$e^{a^2p^2}\,\mathrm{erfc}(as), \quad a > 0$	$\frac{1}{a\sqrt{\pi}}e^{-t^2/4a^2}$
45	$\frac{1}{p}e^{a^2p^2}\,\mathrm{erfc}(as), \quad a \geq 0$	$\mathrm{erf}(t/2a)$
46	$e^{ap}\,\mathrm{erfc}(\sqrt{ap}), \quad a > 0$	$\frac{\sqrt{a}}{\pi\sqrt{t}(t+a)}$
47	$\frac{1}{\sqrt{p}}e^{ap}\,\mathrm{erfc}(\sqrt{ap}), \quad a \geq 0$	$\frac{1}{\sqrt{\pi}(t+a)}$
48	$\mathrm{erf}(a/\sqrt{p})$	$\frac{1}{\pi t}\sin(2a\sqrt{t})$
49	$\frac{1}{\sqrt{p}}e^{a/p}\,\mathrm{erfc}(\sqrt{a/p}), \quad a \geq 0$	$\frac{1}{\sqrt{\pi t}}e^{-2\sqrt{at}}$

Index